Climate Change Impacts and Adaptation Strategies in Japan

Nobuo Mimura • Satoshi Takewaka
Editors

Climate Change Impacts and Adaptation Strategies in Japan

Integrated Research toward Climate Resilient Society

Springer

Editors
Nobuo Mimura
Global and Local Environment Co-creation Institute
Ibaraki University
Mito, Ibaraki, Japan

Satoshi Takewaka
Institute of Systems and Information Engineering
University of Tsukuba
Tsukuba, Ibaraki, Japan

ISBN 978-981-96-2438-6 ISBN 978-981-96-2436-2 (eBook)
https://doi.org/10.1007/978-981-96-2436-2

The original submitted manuscript has been translated into English. The translation was done using artificial intelligence. A subsequent revision was performed by the author(s) to further refine the work and to ensure that the translation is appropriate concerning content and scientific correctness. It may, however, read stylistically different from a conventional translation.
This work was supported by Environmental Restoration and Conservation Agency.

This Springer imprint is published by the registered company Springer Nature Singapore Pte Ltd.
The registered company address is: 152 Beach Road, #21-01/04 Gateway East, Singapore 189721, Singapore

Preface

Climate change's far-reaching impacts are now unmistakably evident across the globe. In response to the extreme heat, floods, and droughts that affect many regions every year, in July 2023, UN Secretary-General António Guterres warned that "the era of global boiling has arrived." Mitigation efforts are underway to achieve carbon neutrality; however, at current levels, the global average temperature is expected to surpass a 1.5 °C increase in the near future, making more severe weather extremes inevitable. Thus, implementing adaptation measures is not only about preparing for future risks but also about responding to the emergency we face today.

In Japan, the Climate Change Adaptation Act was enacted in 2018, establishing a framework for the government, local governments, the business sector, and local communities to collaborate on adaptation efforts. Japan's adaptation plan covers seven areas: agriculture, forestry, and fisheries; water environment and water resources; natural ecosystems; natural disasters; health; industrial and economic activities; and people's lives. An adaptive policy cycle has also been introduced, where a national impact assessment is conducted every 5 years, and the adaptation plans of both central and local governments are revised based on the updated impact assessment. These advancements in adaptation governance have presented new challenges for the research community, particularly in impact projection and adaptation studies. This book highlights how the Japanese research community is addressing these new challenges.

The impacts of climate change vary from country to country and even within different regions of a country. Therefore, it is essential to design impact projections and adaptation measures with higher spatial resolution that also reflect the context of the local community, such as geographical conditions, natural environment, industry, people's lives, social traditions, and culture. Adaptation measures may also encounter "limits to adaptation" or lead to the risk of "maladaptation." Further research is needed to provide answers to key questions about the scientific information necessary for effective climate change risk management.

- How are sectoral and cross-sectoral climate change impacts distributed nationally?
- How can impact projections be downscaled with higher spatial resolution?
- Which regions are particularly vulnerable to sectoral impacts and compound impacts?
- How do climate change and societal changes, such as population decline and aging in Japan, interact with each other?
- How can socioeconomic scenarios be envisaged for this?
- How can effective measures for local adaptation be designed?
- What are the costs of impacts and adaptation measures?
- To what extent can adaptation and mitigation efforts reduce impacts?
- How can synergies between adaptation measures, or between mitigation and adaptation, be exploited to limit trade-offs?
- How can climate change action contribute to climate-resilient and sustainable development?

To answer these broad questions, diverse, systematic, and integrated research involving a wide range of disciplines is essential. In this context, the Strategic Research S-18 project, Comprehensive Research on Projection of Climate Change Impacts and Evaluation of Adaptation (FY2020–2024), supported by the Ministry of the Environment, was launched and is now compiling its results. The S-18 project consists of five themes and nineteen sub-themes, involving over 170 researchers. What makes this project unique is the collaboration of researchers from different fields, working together to provide solutions to climate change from a broad perspective.

This book presents research achievements on impact projections and adaptation in Japan, focusing on the S-18 Project. The emphasis is on showcasing unique research rather than reporting general findings. The book introduces the development of a new research framework, impact models, and assessment results for various affected areas, as well as the effects of mitigation and adaptation measures in reducing impacts and the economic assessment of those impacts.

Given the future projections of climate change, there is a growing concern that we would face increasingly severe impacts. To minimize the effects of climate change, it is essential to closely integrate two climate change measures: mitigation, which aims to limit the progress of global warming to a level that humanity and the global natural system can adapt to, and adaptation, which addresses the negative impacts. We need to strengthen these efforts further and move toward building a climate-resilient and sustainable society. To achieve this, we need to develop ways to project the impacts of climate change in various regions systematically and understand how to implement countermeasures more effectively. This book introduces the approaches and outcomes of such research conducted in Japan. We hope this book provides valuable insights to people in many countries and regions around the world striving to address climate change risks, even though their environments and social characteristics may differ.

Mito, Ibaraki, Japan — Nobuo Mimura
Tsukuba, Ibaraki, Japan — Satoshi Takewaka
February 2025

Acknowledgments

We would like to express our deep gratitude to the Ministry of the Environment (MOE) and the Environmental Restoration and Conservation Agency (ERCA) for their continuous support over the past 5 years. Many of the research findings presented in this book were made possible through the support of the MOE/ERCA Environmental Research Promotion Fund S-18 project (JPMEERF12S11800). The S-18 project was carried out through five thematic teams: Theme 1 (JPMEERF12S11810), Theme 2 (JPMEERF12S11820), Theme 3 (JPMEER F12S11830), Theme 4 (JPMEERF12S11840), and Theme 5 (JPMEERF12S11850). Additionally, the publication of this book was made possible through this funding. We are sincerely grateful for this valuable opportunity and for the research funding that enabled the generation of many significant outcomes.

During the S-18 project, we engaged in joint research, collaborative symposiums, and exchanges with many research institutions and projects. In particular, we have developed a close cooperative relationship with the Center for Climate Change Adaptation (CCCA) at the National Institute for Environmental Studies (NIES). As a result, the three chapters in Part 3 of this book, which focus on natural ecosystems, were written by members of the CCCA at NIES. We would like to express our gratitude for their cooperation and contributions to this book.

Acknowledgments also go to the Academic Advisory Board, which supported the S-18 project with constructive advice and suggestions. Dr. Keisuke Hanaki (Professor Emeritus, The University of Tokyo and Toyo University), Mr. Takashi Hongo (Senior Fellow, Global Strategic Studies Institute, Mitsui & Co., Ltd.), Dr. Koji Ikeuchi (President, Foundation of River and Basin Integrated Communications; Professor Emeritus, The University of Tokyo), and Dr. Kaoru Nakata (Fellow, Japan Fishery Research and Education Agency) regularly assessed the research activities. Their insights greatly contributed to the development and maturation of the project.

Many people, including researchers and members of the S-18 Office at the Global and Local Environment Co-creation Institute (GLEC), Ibaraki University, contributed to the successful implementation of this project. The preparation of this book was made possible by the dedicated work of the authors, the Editorial Committee, the Editorial Advisory Board, and the contributions of Dr. Kohei Imamura, Ms. Yoko Makita, and Ms. Izumi Tani from GLEC. We express our sincere gratitude for their tremendous efforts and dedication.

Editorial Team

Editors

Nobuo Mimura, Ibaraki University, S-18 Project Leader, and S-18 Theme 1 Leader
Satoshi Takewaka, University of Tsukuba

Editorial Committee

Yuki Imai, Kyoto University
Shuichi Kure, Toyama Prefectural University
Yoshifumi Masago, National Institute for Environmental Studies
Gen Sakurai, National Agriculture and Food Research Organization
Hitomi Wakatsuki, National Agriculture and Food Research Organization
Feifan Xu, Nagoya University
Junya Yamasaki, Nagoya University
Jun Yoshida, Tohoku Gakuin University

Editorial Advisory Board

Toshihiro Hasegawa, National Agriculture and Food Research Organization, S-18 Theme 2 Leader
Hiromune Yokoki, Ibaraki University, S-18 Theme 3 Leader
Kiyo Kurisu, The University of Tokyo, S-18 Theme 4 Leader
Akira Hibiki, Tohoku University, S-18 Theme 5 Leader

Secretary

Sachiko Sakamoto, S-18 Office, Ibaraki University

Contents

Contributors

Hiroya Abe National Institute for Environmental Studies, Tsukuba, Japan

Naota Hanasaki National Institute for Environmental Studies, Tsukuba, Japan

Toshihiro Hasegawa National Agriculture and Food Research Organization (NARO), Tsukuba, Japan

Shoji Hashimoto Forestry and Forest Products Research Institute, Tsukuba, Japan

Seiichiro Hasui Ibaraki University, Mito, Japan

Katsuma Hayashi Kyoto University, Kyoto, Japan

Michiya Hayashi National Institute for Environmental Studies, Tsukuba, Japan

Seiji Hayashi National Institute for Environmental Studies, Tsukuba, Japan

Akira Hibiki Tohoku University, Sendai, Japan

Hironori Higashi National Institute for Environmental Studies, Tsukuba, Japan

Yasumasa Hirata Forestry and Forest Products Research Institute, Tsukuba, Japan

Atsuya Ikemoto Tohoku University, Sendai, Japan

Yoko Imai The University of Tokyo, Tokyo, Japan
National Institute of Environmental Studies, Tsukuba, Japan

Kohei Imamura Global and Local Environment Co-creation Institute, Ibaraki University, Mito, Japan

Tomomi Inoue National Institute for Environmental Studies, Tsukuba, Japan

Yasushi Ishigooka National Agriculture and Food Research Organization, Tsukuba, Japan

Fumiko Ishihama National Institute for Environmental Studies, Tsukuba, Japan

Noriko N. Ishizaki National Institute for Environmental Studies, Tsukuba, Japan

Ken Itakura Nagoya City University, Nagoya, Japan

Shigeho Kakehi Japan Fisheries Research and Education Agency, Shiogama, Miyagi, Japan

Hirokazu Kato Nagoya University, Nagoya, Japan

So Kazama Tohoku University, Sendai, Japan

Hideaki Kidokoro Japan Fisheries Research and Education Agency, Niigata, Japan

Ayato Kohzu National Institute for Environmental Studies, Tsukuba, Japan

Dai Koide National Institute for Environmental Studies, Tsukuba, Japan

Naoki Kumagai National Institute for Environmental Studies, Tsukuba, Japan

Kiyo Kurisu The University of Tokyo, Tokyo, Japan

Yuji Masutomi National Institute for Environmental Studies, Tsukuba, Japan

Keisuke Matsuhashi National Institute for Environmental Studies, Tsukuba, Japan

Nobuo Mimura Ibaraki University, Mito, Japan

Takuya Miyashita Disaster Prevention Research Institute, Kyoto University, Uji, Kyoto, Japan

Hideki Mori Forestry and Forest Products Research Institute, Tsukuba, Japan

Nobuhito Mori Disaster Prevention Research Institute, Kyoto University, Uji, Kyoto, Japan

Wataru Murakami Forestry and Forest Products Research Institute, Tsukuba, Japan

Akito Murayama The University of Tokyo, Tokyo, Japan

Yoshiaki Nakagawa Waseda University, Tokorozawa, Saitama, Japan

Katsuhiro Nakao Kansai Research Center, Forestry and Forest Products Research Institute, Kyoto, Japan

Kou Nakazono Central Region Agricultural Research Center, National Agriculture and Food Research Organization, Tsukuba, Japan

Hiroshi Nishiura Kyoto University, Kyoto, Japan

Tomohiro Nishizono Forestry and Forest Products Research Institute, Tsukuba, Japan

Katsuhito Nohara Rikkyo University, Niiza, Saitama, Japan

Kazutaka Oka National Institute for Environmental Studies, Tsukuba, Japan

Masashi Okada National Institute for Environmental Studies, Tsukuba, Japan

Goh Onitsuka Japan Fisheries Research and Education Agency, Hatsukaichi, Hiroshima, Japan

Gen Sakurai National Agriculture and Food Research Organization, Tsukuba, Japan

Hideo Shiogama National Institute for Environmental Studies, Tsukuba, Japan

Kosuke Shirai The University of Tokyo, Tokyo, Japan

Toshihiko Sugiura National Agriculture and Food Research Organization, Tsukuba, Japan

Asari Takada National Agriculture and Food Research Organization, Tsukuba, Japan

Kiyoshi Takahashi National Institute for Environmental Studies, Tsukuba, Japan

Makoto Tamura Global and Local Environment Co-creation Institute, Ibaraki University, Mito, Japan

Hiroki Tanikawa Nagoya University, Nagoya, Japan

Jumpei Toriyama Kyushu Research Center, Forestry and Forest Products Research Institute, Kumamoto, Japan

Haruka Tsunetaka Forestry and Forest Products Research Institute, Tsukuba, Japan

Shinsuke Uchida Nagoya City University, Nagoya, Japan

Kentaro Uchiyama Forestry and Forest Products Research Institute, Tsukuba, Japan

Tokuko Ujino-Ihara Forestry and Forest Products Research Institute, Tsukuba, Japan

Hitomi Wakatsuki National Agriculture and Food Research Organization, Tsukuba, Japan

Yasutaka Wakazuki Ibaraki University, Mito, Japan

Qinxue Wang National Institute for Environmental Studies, Tsukuba, Japan

Feifan Xu Nagoya University, Nagoya, Japan

Hiroya Yamano National Institute for Environmental Studies, Tsukuba, Japan

The University of Tokyo, Tokyo, Japan

Junya Yamasaki Nagoya University, Nagoya, Japan

Naho Yamashita The University of Tokyo, Tokyo, Japan

Hayata Yanagihara Tohoku University, Sendai, Japan

Hiromune Yokoki Ibaraki University, Hitachi, Japan

Masayuki Yokozawa Waseda University, Tokorozawa, Saitama, Japan

Jun Yoshida Tohoku Gakuin University, Sendai, Miyagi, Japan

Takeo Yoshida National Agriculture and Food Research Organization, Tsukuba, Japan

Sayaka Yoshikawa Nagasaki University, Nagasaki, Japan

Tetsuro Yoshikawa Osaka Metropolitan University, Osaka, Japan

Part I
Framework of Integrated Research and Future Scenarios

Chapter 1
Climate Change Impacts and Adaptation Strategies: Japan's Challenges to Integrated Research

Nobuo Mimura

Abstract This book introduces the research efforts of the Japanese academic community to address the adverse effects of climate change. This chapter begins with an overview of recent climate change impacts and adaptation initiatives in Japan. The Climate Change Adaptation Act, enacted in 2018, established an adaptive policy cycle, where a national impact assessment is conducted every five years. Based on these assessments, both central and local government adaptation plans are revised. These advancements in adaptation governance have created a series of new challenges for the research community. To tackle these complex issues, diverse, systematic, and interdisciplinary research is crucial. In this context, the Strategic Research S-18 project, Comprehensive Research on the Projection of Climate Change Impacts and Evaluation of Adaptation (FY2020–2024), supported by the Ministry of the Environment, was initiated and is now compiling its findings. The latter half of this chapter outlines the purposes and research framework of the S-18 Project.

Keywords Climate change impacts · Climate Change Adaptation Act · Local adaptation plan · Nation-wide impact projections · Effect of adaptation measures · S-18 research project · Integrated research

Abbreviations of the Organizations

IPCC	Intergovernmental Panel on Climate Change
IPSS	National Institute of Population and Social Security Research
JMA	Japan Meteorological Agency
JMASTEC	Japan Agency for Marine-Earth Science and Technology
MAFF	Ministry of Agriculture, Forestry and Fisheries
METI	Ministry of Economy, Trade and Industry
MILT	Ministry of Land, Infrastructure, Transport and Tourism

N. Mimura (✉)
Ibaraki University, Mito, Japan
e-mail: nobuo.mimura.iu@vc.ibaraki.ac.jp

N. Mimura, S. Takewaka (eds.), *Climate Change Impacts and Adaptation Strategies in Japan*, https://doi.org/10.1007/978-981-96-2436-2_1

MEXT	Ministry of Education, Culture, Sports, Science and Technology
MOE	Ministry of the Environment
NIES	National Institute for Environmental Studies
UNEP	United Nations Environment Programme

1.1 Introduction

Impacts of climate change are rapidly becoming apparent across the globe. Extreme high temperatures, wildfires, heavy rain and flooding, droughts, and other severe disasters are occurring around the world, even leading to the emergence of environmental refugees. The threat of climate change is clear to everyone, with the global average temperature having risen by 1.2 °C compared to the end of the nineteenth century. Japan is no exception to this change, experiencing unprecedented heatwaves and severe rainfall damage.

Japan has long coexisted with nature and enjoyed its blessings. It is also characteristic of Japan to have suffered from various disasters. A wide variety of natural disasters occur, including geoscientific disasters such as earthquakes, tsunamis, and volcanic eruptions, as well as heavy rain during the rainy season, typhoons, landslides, storm surges, and droughts. Therefore, Japan has historically developed disaster prevention and mitigation measures over time. However, since the 2010s, the intensity of meteorological disasters has increased, and damage has spread to unprecedentedly wide areas. It has expanded the impacts on agriculture, fisheries, and heatstroke deaths. This situation is largely due to the progress of climate change, prompting society as a whole to take a strategic response to global warming and climate change (Mimura and Hijioka 2022).

This book introduces the results of research being conducted by the Japanese research community to combat the adverse effects of climate change in collaboration with society. In this chapter, we first discuss the impacts of climate change and efforts to adapt to them in Japan. Then, we introduce the current state of the latest research, focusing on the S-18 Project, which forms the basis of this book.

1.2 Climate Change in Japan and Its Impacts

Let's explore the historical changes in weather in Japan. Figure 1.1 shows the annual changes in Japan's mean temperature published by the Japan Meteorological Agency (JMA) (JMA 2024). This figure shows the changes in the annual mean temperature from 1898 to 2023 at 15 observation points where the impact of urbanization is relatively small. According to this, the average temperature in Japan has risen by 1.35 °C over the past 100 years. The increase in 2023 was particularly rapid, recording the highest annual average temperature in the observation history. This record also shows natural fluctuations on the scale of several years to several

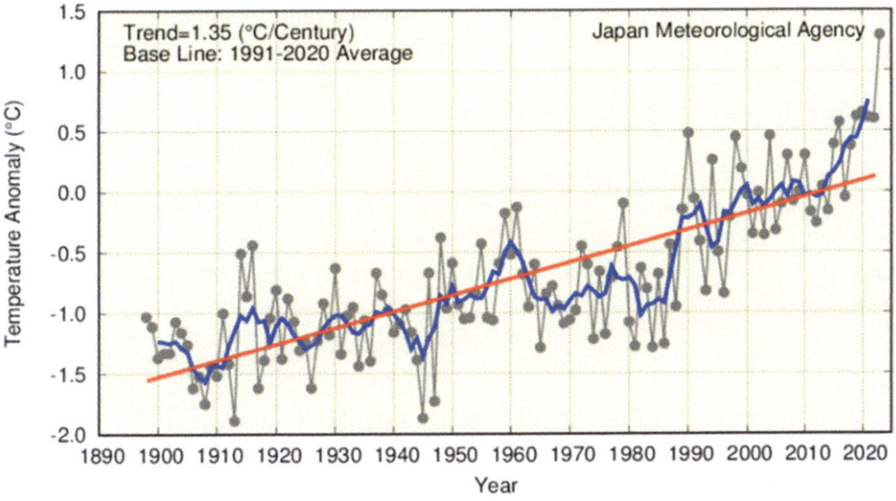

Fig. 1.1 Annual mean surface temperature from 1898 to 2023 in Japan. Anomalies are deviations from the baseline (i.e., the 1991–2020 average). Thin black line indicates the surface temperature anomaly for each year. Blue line shows the five-year running mean, and the red line represents the long-term linear trend. (Reproduced from JMA 2024)

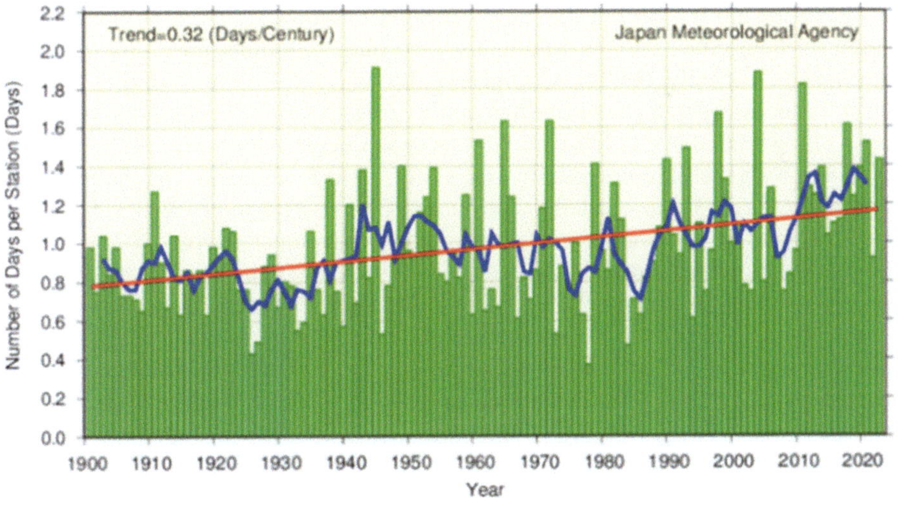

Fig. 1.2 Annual number of days with precipitation >100mm from 1901–2023. Green bars represent the annual number of days per station for each year, blue line shows the five-year running mean, and red line is the long-term liner trend (Reproduced from JMA 2024).

decades. Nevertheless, all of the top 10 high-temperatures in the past 120 years have occurred after 1990, indicating the significant impact of global warming.

Figure 1.2 shows the long-term changes in heavy rainfall (JMA 2024). While the number of days with a daily rainfall of 100 mm or more is increasing, the number of days with precipitation exceeding 1 mm is decreasing. This implies that the frequency of heavy rain and dry days are both increasing, suggesting a trend towards more extreme rainfall patterns.

These changes in weather conditions have a significant impact on ecosystems and the people's daily lives. Since 1953, the cherry blossom blooming date has been advancing by 1.0 day per decade, and the autumn leaf color change date for maple trees has been delayed by 2.8 days per decade (MEXT, JMA 2020). Furthermore, the five-year average from 2019 to 2022 shows that the number of people hospitalized due to heatstroke is 69,300 per year, with 1313 deaths. The impacts extend to broader areas, confirming the rapid manifestation of climate change.

1.3 Climate Change Policies and Initiatives in Japan

1.3.1 Mitigation and Adaptation

Global efforts to address climate change have made significant progress with the 2015 Paris Agreement. There are two main approaches to climate change response: mitigation and adaptation, both of which are explicitly mentioned as key pillars in the Paris Agreement.

Mitigation aims to remove the causes of global warming by reducing Greenhouse Gases (GHGs) emissions, as well as by absorbing and capturing CO_2 from the atmosphere. Globally, the 1.5 °C target has become a shared goal, and to achieve this, it is necessary to realize carbon neutrality (net zero GHGs emissions) by 2050. Japan also declared in 2020 its commitment to achieving carbon neutrality by 2050. However, with the current reduction targets of various countries, the global mean temperature in 2100 is projected to rise by 2.8 °C (with a range of 1.9–3.3 °C), making it difficult to achieve the goals of the Paris Agreement (UNEP 2022). As the average temperature is likely to exceed a 1.5 °C increase in the near future, the challenge lies in how to strengthen the efforts of emission reduction.

Adaptation is a strategy to prevent the adverse impacts of climate change or to utilize new climatic conditions. Since achieving carbon neutrality would take more than 20 years, the need for adaptation measures to suppress impacts and reduce damage is becoming increasingly important. The following will introduce the situation in Japan focusing on adaptation.

1.3.2 Initiatives for Climate Change Adaptation

Figure 1.3 shows the progress of climate change adaptation in Japan. The first "Adaptation Plan for the Impacts of Climate Change" of the government was formulated in 2015. Subsequently, in 2018, the Climate Change Adaptation Act was

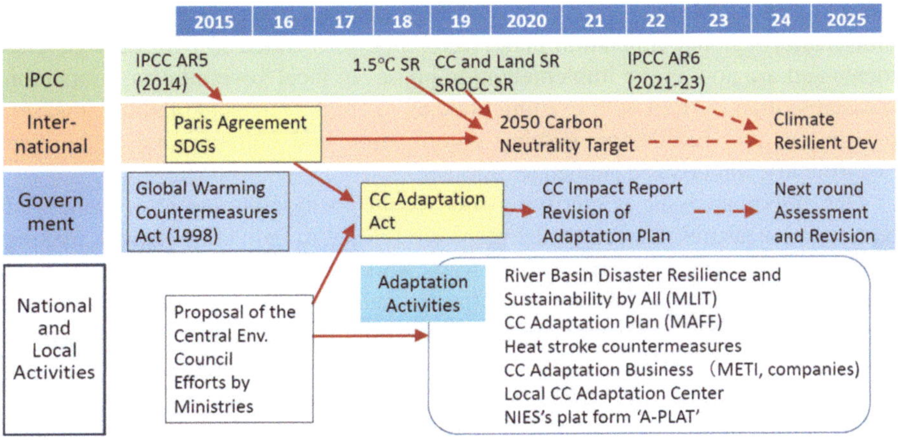

Fig. 1.3 Progress of Climate Change Adaptation in Japan. (Adapted from Mimura (2022) with permission from The Japan Science Support Foundation)

enacted as the world's first law solely aimed at climate change adaptation. This Climate Change Adaptation Act has the following contents.

1. This law defines climate change adaptation as an effort that the national government, local authorities, businesses, and all citizens need to undertake, and it clarifies the roles of each. The areas targeted by the climate change adaptation plan are agriculture, forestry and fisheries, water environment and resources, natural ecosystems, natural disasters and coastal areas, human health, industrial and economic activities, and the lives of the people and urban life.
2. To address the uncertainty of climate change projection, impact assessments are conducted approximately every five years, and the climate change adaptation plan will be revised accordingly.
3. Prefectures and municipalities are encouraged to formulate local climate change adaptation plans and establish local climate change adaptation centers (LCCAC). The LCCAC is an organization that provides information and technical advice on adaptation in the region. As of June 2024, it has been established in 63 out of 1788 local authorities.
4. Since scientific knowledge across a wide range of fields is necessary to predict the impacts of climate change and implement adaptation measures, the Climate Change Adaptation Center was established in the National Institute for Environmental Studies. Its role is to provide local governments and other stakeholders with scientific knowledge and information, as well as technical support.

Since the enactment of the Climate Change Adaptation Act, efforts toward adaptation have been accelerating nationwide. At the national level, in addition to the government's climate change adaptation plan, individual adaptation plans have been formulated by the Ministry of Land, Infrastructure, Transport and Tourism

(MLIT) and the Ministry of Agriculture, Forestry and Fisheries (MAFF). On the other hand, regional adaptation plans have been formulated in 310 local governments and are now in the implementation phase. In local governments, adaptation measures often overlap with existing policies in areas such as agriculture, disaster prevention, and healthcare, making coordination and collaboration with other departments' initiatives a challenge.

One important issue is clarifying the overall positioning and effectiveness of adaptation measures. Currently, the plans are limited to individual sectors, making it necessary to create a comprehensive framework that includes broader capacity to manage climate change risks and prioritizes target sectors. Furthermore, developing strategies to strengthen society's resilience in responding to climate change will be essential.

Next, we introduce the efforts in the field of natural disasters as an example of policy development for adaptation. Among recent weather-related disasters, Typhoon Hagibis in 2019 had a significant policy impact. Along its path from central to northeastern Japan, 142 river embankments were breached, resulting in extensive flood damage. This suggests that weather-related hazards are now exceeding the protective standards Japan has built over its history. In response, the Ministry of Land, Infrastructure, Transport, and Tourism (MLIT) introduced a disaster prevention and national land conservation plan in 2020 that accounts for climate change.

In the "Flood Control Measures Considering Climate Change," future rainfall changes were examined using ensemble climate prediction data from the Database for Policy Decision Making for Future Climate Change (d4PDF 2021), assuming that under the RCP2.6 scenario, the intensity of rainfall would increase by 1.1 times nationwide. Based on this, a policy of basin-wide flood control, which is called "River Basin Disaster Resilience and Sustainability by All", was introduced to prepare for flood risk through a combination of measures such as strengthening embankments, improving the water storage of river basins, and land use regulations (Fig. 1.4). The basin-wide flood control approach, therefore, involves enhancing infiltration and water storage functions using forests and rice paddies, as well as relocating residences from areas at high risk of flooding. For its implementation, it is necessary to develop specific project plans that reflect local circumstances and to build consensus among stakeholders. Promoting these projects is expected to take a considerable amount of time.

1.4 Research for Climate Change Impacts and Adaptation

1.4.1 Research Challenges at the New Stage

The progress of these adaptation measures is closely supported by research in the field of climate change science. Since the 1990s, research on predicting the climate change impacts has been conducted across various fields, gradually allowing a

Fig. 1.4 Overall structure of "River Basin Disaster Resilience and Sustainability by All". (Adapted from MLIT (2020)).

clearer understanding of its effects on Japan (MOE et al. 2018). A key characteristic of climate change impacts is that they vary by region in terms of the affected sectors and the intensity of those effects. Additionally, adaptation measures need to incorporate the local context such as the natural environment, history, and social characteristics of each region. As a result, there is an increasing demand for scientific knowledge with higher resolution that reflects the realities of each region. In other words, as climate change responses have entered a new stage, research is also expected to provide relevant perspectives and knowledge to improve the planning and implementation of adaptation measures.

The first challenge for research is the high-resolution impact prediction, as very high spatial resolution data is required for planning regional adaptation measures. Therefore, it is necessary to develop advanced downscaling techniques for both climate models and impact predictions. Additionally, it is increasingly important for impact studies to consider the interactions between climate change and societal changes such as population decline, aging, and land use change, for they alter the exposure and vulnerability of the society. Furthermore, research on compound impacts and cascading effects is also critical.

The second challenge relates to research on adaptation measures. Adaptation measures encompass a wide range of approaches, including spatial planning and land use, infrastructures, technical solutions, institutional responses, economic measures, social initiative, and research and development. It is essential to select the most appropriate adaptation measures for each sector and region. Moreover, evaluating the effectiveness of adaptation measures is crucial for the formulation of adaptation plans, which requires the development of assessment methodologies. Additionally, for adaptation measures to be sustainable, research is needed on the co-benefits and trade-offs between adaptation and mitigation actions. These efforts contribute to strengthening societal resilience, which is the ultimate goal of climate change adaptation.

The third is the challenge of how to connect the scientific community and society. This includes research on how to apply research findings and data to policies at the government, local authority, and corporate levels, as well as efforts to improve public understanding through outreach. While the relationship between science and society is important in many fields, it is particularly crucial for climate change responses, as it is closely linked to a wide range of societal issues. Therefore, communication and collaboration between science and society are of paramount importance.

1.4.2 Population Decline, Aging, and Climate Change Research

A particularly important issue in Japan is the declining population and aging. As shown in Fig. 1.5, the National Institute of Population and Social Security Research (IPSS) predicts that Japan's population will continue to decline (IPSS 2023). In its medium projection, Japan's total population of 126.15 million in 2020 is estimated to decline to 104.69 million by 2050, and to 87 million by 2070. Alongside this

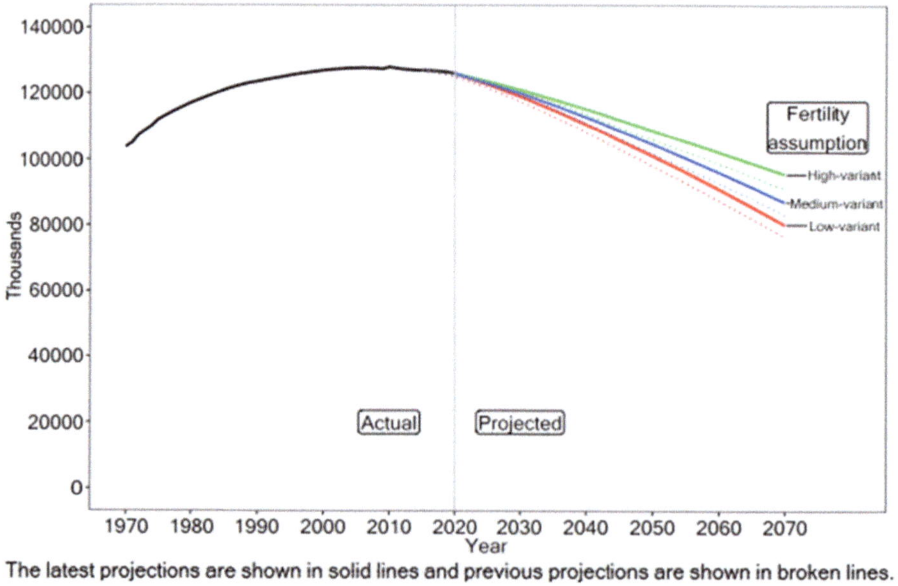

Fig. 1.5 Actual and projected population trends of Japan: Medium-, high-, and low-fertility (medium-mortality). (Reproduced from IPSS (2023))

decline, the aging of the population is also advancing, with seniors aged 65 and over expected to comprise 37% of the population by 2050. Due to these demographic changes, it has even been suggested that 45% of municipalities (774 municipalities) could become so-called "disappearing municipalities" unable to maintain their populations. These population changes will have an extremely significant impact on the future of Japanese society.

Climate change is progressing alongside these societal changes, exerting synergistic effects. For example, when considering both the increase in storm surges and sea level rise along with the population decline in coastal areas, it is predicted that the potential population affected by coastal flooding will peak around 2050, after which it will decrease (Shion et al. 2024). Given these results, we should reconsider the target timeline for its storm surge countermeasures, shifting from the end of the twenty-first century to the middle of it. Additionally, agriculture and fisheries, which are facing a decline in workforce and an aging population, face the significant challenge of securing successors for farmers and fishermen in order to maintain production. Such demographic changes impact every aspect of society. Therefore, in formulating effective climate change adaptation measures, it is necessary to consider the interactions with these social challenges. Adaptation strategies should be planned in a way that integrates them with the vision of local communities aimed at solving social issues and building a sustainable society. This approach, known as Climate Resilient Development (CRD), is gaining international attention (IPCC 2022). Advancing research on CRD is an important task for future studies on the impacts of climate change and adaptation.

1.5 Introduction to the S-18 Research Project

1.5.1 Purpose and Structure of the S-18 Project

In line with the progress of Japan's climate change adaptation policies, a strategic research project was launched in 2020 under the Environment Research and Technology Development Fund. This is the S-18 research project, "Comprehensive Research on Projection of Climate Change Impacts and Evaluation of Adaptation." This project has a five-year plan running from FY2020 to FY2024, involving more than 170 researchers from 27 universities and research institutions. The project is expected to contribute to the next climate change impact assessment of the MOE, scheduled for publication in 2025, and to the promotion of adaptation efforts at the local level. High spatial resolution impact projection data are essential for the formulation of adaptation plans by local governments. In addition to expanding the range of assessments to include impacts on a wider variety of crops, the lives of the people, and urban life, the development of evaluation methods to support policy assessments is also expected.

In response to this background, the S-18 project set its overall goal as "the creation of the latest scientific information on impact projections and adaptation assessments to support Japan's climate change adaptation." Through the achievement of this goal, the project aims to: (1) contribute to the government's climate change impact assessment report and adaptation plan review scheduled for 2025, (2) assist local governments in identifying vulnerable areas and formulating and implementing adaptation plans, (3) contribute to international efforts such as the IPCC's seventh Assessment Report and the Paris Agreement, and (4) provide recommendations on how to create a resilient society in response to climate change (S-18 Project 2024).

As shown in Fig. 1.6, the S-18 project is structured into five themes aligned with the research objectives. The research topics for each theme are as follows:

- Theme 1: Development of Comprehensive Research Framework for Impact Projection and Evaluation of Adaptation
- Theme 2: Projection of Climate Change Impacts and Evaluation of Adaptation Options for Agriculture, Forestry and Fisheries
- Theme 3: Projection of Climate Change Impacts and Evaluation of Adaptation to Natural Disasters and Water Resources
- Theme 4: Projection of Climate Change Impacts on Quality of Life (QoL) of People and Their Associated Infrastructure and Local Industries and Evaluation of Adaptation Options
- Theme5: Development of Economic Assessment Methods for Impact of Climate Change and Adaptation Options

This structure covers six of the seven areas targeted by Japan's adaptation plan. The remaining area, the natural ecosystems sector, is studied by the Climate Change

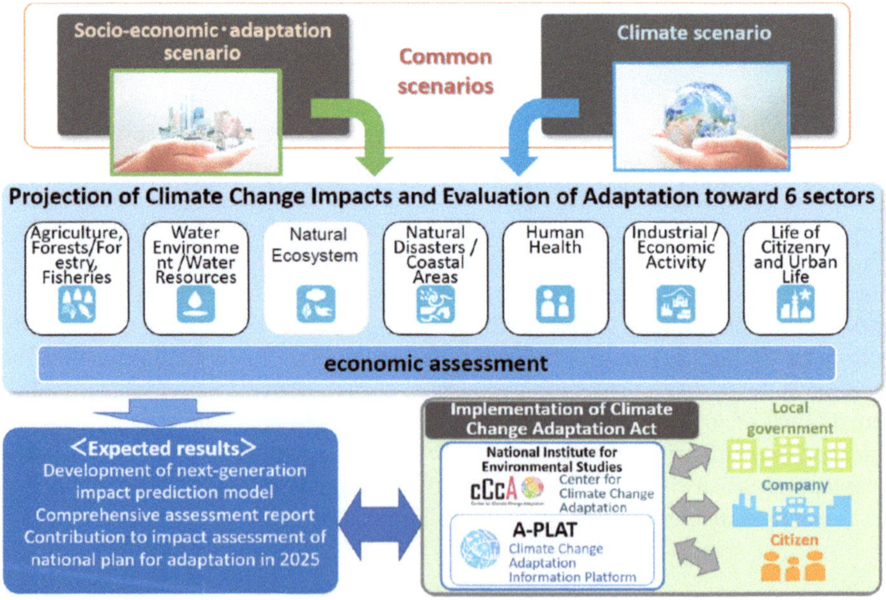

Fig. 1.6 Structure of the S-18 Project

Adaptation Center of the National Institute for Environmental Studies. Together, both collaborate to cover all sectors.

1.5.2 Research Framework of the S-18 Project

The S-18 project has established a common research framework. This framework combines (1) three levels of global warming using the GHGs concentration pathways (RCP 8.5, 4.5, 2.6), (2) five global climate models, (3) three socioeconomic scenarios based on Japanese SSPs, and (4) with or without adaptation measures, to predict future impacts under a variety of conditions (Fig. 1.7). The basic evaluation cases of this study include: (1) an evaluation focusing solely on changes in climate hazards, (2) an evaluation considering the synergistic effects of climate change and societal changes, and (3) an additional evaluation to assess the impact of higher population decline and aging scenarios.

The spatial resolution of the projection and evaluation in this project is 1 km × 1 km, based on Japan's Digital National Land Information System (Third mesh). All impact items are calculated on this unified grid, and the nationwide distribution is shown using GIS. In some cases, evaluations are aggregated by municipalities to indicate the distribution across municipalities. The time frame covers from 2020 to 2100, with particular focus on three time slices: near future (2030), mid-term (2050), and long-term (2090).

The common scenario assumptions for the research are set as follows:

1. Global Warming Levels (RCP)

From the five warming levels (RCPs) used in the IPCC Sixth Assessment Report (AR6), the following three levels were selected. The temperatures in parentheses

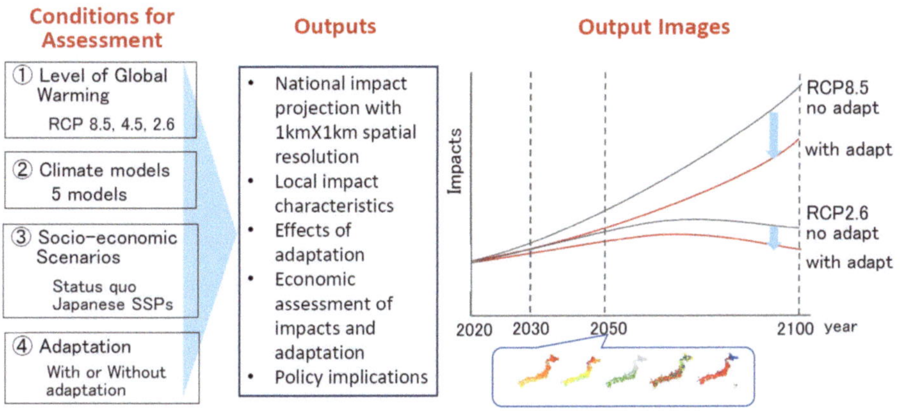

Fig. 1.7 Research framework of the S-18 Project

represent the projected global average temperature for each scenario by 2100. The meaning of each scenario is shown as follows:

- RCP8.5 (SSP5–8.5: +4.4 °C): Baseline scenario without mitigation measures, prior to the Paris Agreement.
- RCP4.5 (SSP2–4.5: +2.7 °C): Assumed to be similar to current national targets, regarded as the present Business as Usual (BaU).
- RCP2.6 (SSP1–2.6: +1.8 °C): Scenario aligned with the Paris Agreement's 2 °C target.
- Additionally, some studies have included an evaluation with RCP1.9 (SSP1–1.9: +1.4 °C).

2. Climate Scenarios

In collaboration with the National Institute for Environmental Studies (NIES), Japan Agency for Marine-Earth Science and Technology (JAMSTEC), and other organizations, a Climate Scenario Working Group was established for the S-18 Project. By downscaling atmospheric and oceanic projection data for each warming level, climate scenarios around Japan were developed. The global climate models (GCMs) used as the basis were selected from climate model groups in CMIP5 and CMIP6, choosing five models each with high reproducibility for around Japan. The spatial resolution of the land-based data is also based on the Third Mesh of Japan's Digital National Land Information (1 km × 1 km). Users can select the scenario that best fits their purposes. Detailed explanations of the climate scenarios are provided in Chap. 2 (Wakazuki et al. 2024).

3. Socioeconomic Scenarios

We created socioeconomic scenarios (changes in population, households, and land use distribution) up to 2100. These data are also available at the Third Mesh and municipal levels. The socioeconomic scenarios used in the S-18 project are based on the global SSP scenarios from the IPCC Sixth Assessment Report, adapted to reflect Japan's social conditions as the Japan-specific SSPs (Chen et al. 2020) and a scenario where population and land use remain fixed at current levels. In this study, we used SSP 1 (decarbonization and sustainability pathway), SSP 2 (mid-level pathway), and SSP 5 (fossil fuel-intensive development pathway). Detailed explanations of the socioeconomic scenarios are provided in Chap. 3 (Yoshikawa et al. 2024).

1.6 Structure of This Book

The S-18 project has conducted a nationwide impact assessment and published findings on the impact profiles across Japan and the effectiveness of adaptation measures. While the full scope of these results will be presented in a separate report, this book focuses on new research findings in each field. The structure of this book is as follows:

Part 1 introduces the concept of the S-18 project. Following this chapter, the next two chapters present the climate change scenarios and socioeconomic scenarios that served as the common foundation for the project. Part 2 focuses on research related to agriculture, forestry, and fisheries. It reports on the impacts and adaptation measures for a wide variety of crops, starting with Japan's staple food, rice, as well as studies on forestry, and wakame seaweed farming in the fisheries sector. Part 3 addresses the impacts on ecosystems and adaptation measures. The three reports are contributions from the research projects conducted at the NIES' Climate Change Adaptation Center. Part 4 deals with changes in natural disasters and adaptation measures. This section introduces the latest research on topics such as sea-level rise, coastal areas, river flooding, stakeholder disaster prevention behavior, and agricultural drought risks related to climate change. Part 5 covers the impacts on urban areas and daily life, using quality of life (QoL) as a comprehensive index, along with adaptation measures. Among the wide range of research topics in this field, we report on local governments' awareness of adaptation measures, community-level planning, urban material flows, urban transportation, and the health impacts and measures related to Dengue fever. Part 6 discusses the economic evaluation and climate change policies. The economic evaluation includes studies on farmers' adaptation and the economic impacts of sea-level rise. Furthermore, the final chapter discusses the climate change policy from the viewpoint of climate security.

While a wide variety of research is reported, each chapter is self-contained, allowing readers to select chapters of interest. We hope that the 24 reports will not only introduce Japan's experiences but also provide insights that may help various countries and regions identify and undertake similar efforts.

References

Chen H, Matsuhashi K, Takahashi K, Fujimori S, Honjo K, Gomi K (2020) Adapting global shared socio-economic pathways for national scenarios in Japan. Sustain Sci 15(3):985–1000

d4PDF (2021) Database for Policy Decision making for Future climate change (d4PDF). https://climate.mri-jma.go.jp/d4PDF/index_en.html

IPCC (2022) Summary for policymakers. In: Pörtner H-O, Roberts DC, Tignor M, Poloczanska ES, Mintenbeck K, Alegría A, Craig M, Langsdorf S, Löschke S, Möller V, Okem A, Rama B (eds) Climate Change 2022: Impacts, adaptation, and vulnerability. Cambridge University Press, pp 3–33. https://doi.org/10.1017/9781009325844.001

IPSS (2023) Population Projections for Japan (2023 revision): 2021 to 2070. https://www.ipss.go.jp/pp-zenkoku/e/zenkoku_e2023/pp2023e_Summary.pdf

JMA (2024) Climate Change Monitoring Report 2023, p 94.

MEXT, JMA (2020) Climate Change in Japan 2020: Report on Assessment of Observed/Projected Climate Change Relating to the Atmosphere, Land and Oceans, p 263. (in Japanese)

Mimura N (2022) Climate change adaptation and building social resilience. Trends in the Sciences, The Japan Science Support Foundation 27(2):59–63. (in Japanese)

Mimura N, Hijioka Y (2022) Climate change and natural disasters in Japan. In: Rosenzweig C, Parry M, Del Mar M (eds) Our warming planet. World Scientific, pp 498–525

MLIT (2020) River Basin disaster resilience and sustainability by all- Japan's new policy on water-related disaster risk reduction, Water and Disaster Management Bureau. MLIT., https://www.mlit.go.jp/river/kokusai/pdf/pdf21.pdf.

MOE, MEXT, MAFF, MLIT, JMA (2018) Synthesis Report on Observations, Projections and Impact Assessments of Climate Change, 2018: Climate Change in Japan and Its Impacts, p 130. (in Japanese).

S-18Project (2024) S-18Project HP. https://s-18ccap.jp/en/
Shion Y, Miyashita T, Yasuda T, Shimura T, Mori N (2024) Analysis of inundation risk change in the zero-meter zones of the three major bays in Japan under different climate change scenarios, JSCE (Coastal), in press. (in Japanese)
UNEP (2022) Emissions Gap Report 2022: The Closing Window-Climate crisis calls for rapid transformation of societies, p 101.
Wakazuki Y, Shiogama H, Ishizaki NN, Hayashi M (2024) Chapter 2 Climate change scenarios for impacts and adaptation research. In: Mimura N, Takewaka S (eds) Climate change impacts and adaptation strategies in Japan. Springer-Nature, Tokyo, pp 17-27
Yoshikawa S, Imamura K, Takahashi K, Matsuhashi K (2024) Chapter 3 development of socio-economic scenarios for climate change impacts and adaptation research. In: Mimura N, Takewaka S (eds) Climate change impacts and adaptation strategies in Japan. Springer-Nature, Tokyo, pp 29-41

Chapter 2
Climate Change Scenarios for Impacts and Adaptation Research

Yasutaka Wakazuki, Hideo Shiogama, Noriko N. Ishizaki, and Michiya Hayashi

Abstract This chapter explains the NIES2020 climate scenario used for climate change impact assessments in the S-18 Project. As impact projections at national and regional levels increasingly require high spatial resolution data, it has become essential to downscale climate scenarios to finer scales. In the S-18 Project, climate scenarios were downscaled to a 1-km grid covering all of Japan. Selecting the most appropriate climate models for the target region is also critical. This chapter outlines the concept, development process, and validation of NIES2020, designed to meet Japan's specific needs. Additionally, we address the "hot model issue" observed in some CMIP6 models and the challenges associated with the SSP3-7.0 scenario. Through these discussions, we emphasize the importance of active dialogue and collaboration between climate scenario developers, impact assessment researchers, and policy-making researchers in evaluating climate change impacts and formulating effective countermeasures.

Keywords Climate scenario data · High-resolution downscaling data · Selecting the most suitable climate models · Hot model issue · Distinctiveness of SSP3-7.0

2.1 Introduction

2.1.1 Design Policy for Climate Scenario Data

Future climate projections for the atmosphere, land surface, and oceans are essential for detailed impact assessments of climate change across various research fields at the national or regional scale. This projected information is referred to as a climate

Y. Wakazuki (✉)
Ibaraki University, Mito, Japan

H. Shiogama · N. N. Ishizaki · M. Hayashi
National Institute for Environmental Studies, Tsukuba, Japan

N. Mimura, S. Takewaka (eds.), *Climate Change Impacts and Adaptation Strategies in Japan*, https://doi.org/10.1007/978-981-96-2436-2_2

scenario. This chapter provides an overview of Japan's high-resolution climate scenario data, adopted in the S-18 Project, with a particular focus on atmospheric and land surface data.

Future climate is projected by numerical simulations using climate models. A climate model is a program that mathematically represents various processes of the Earth's surface, such as atmospheric elements, atmospheric circulation and cloud-precipitation processes, land surface processes, ocean circulation, ice sheet, and sea ice processes. Since the future concentration pathways of greenhouse gases and socioeconomic conditions are unknown, multiple scenarios are given as forcings for simulations. Then, the forced responses of the Earth's surface climate conditions are computed using climate models. In the Coupled Model Intercomparison Project Phase 6 (CMIP6), climate simulations were conducted under common conditions by research institutes worldwide, and the results were compared. The CMIP6 project utilized the SSPx-RCPy scenarios, which combine Representative Concentration Pathways (RCPs—scenarios depicting future greenhouse gas concentration levels) with Shared Socioeconomic Pathways (SSPs—scenarios illustrating future global socioeconomic conditions), hereafter referred to as SSPx-y. The variable x ranges from 1 to 5, reflecting population growth, economic development, technology, and resource use differences. The variable y, which can be 1.9, 2.6, 3.4, 4.5, 6.0, 7.0, or 8.5, represents the level of radiative forcing (the external force heating the Earth) and indicates the intensity of various factors driving climate change by the end of the twenty-first century. Based on these scenarios, future climate projections were generated using climate models. The S-18 Project employs climate model projections based on five main scenarios: SSP1–1.9, SSP1-2.6, SSP2-4.5, SSP3-7.0, and SSP5-8.5.

Climate scenario data should be as high-resolution as possible because impact assessments based on coarse climate scenarios may differ significantly from those derived from high-resolution scenarios. High-resolution data is required in Japan, where steep terrain and complex land-use patterns prevail. For the S-18 Project, the spatial resolution of the climate scenarios is set to approximately 1 km. This finer resolution allows for accurate assessments of climate change impacts in areas such as agriculture, urban environments, hydrological disasters, and water resources, as it captures the interplay between atmospheric conditions, complex terrain, and the distribution of farmland and cities. However, the raw climate model simulation data, which forms the basis of these scenarios, typically has a resolution of around 100 km, which is insufficient for direct use in the S-18 Project's impact assessments. To address this, 1-km resolution data, downscaled and bias-corrected through statistical methods (NIES2020; Ishizaki et al. 2022), was created and adopted as the common climate scenario.

2.1.2 NIES2020

In CMIP6, numerical experiments of climate models projecting future climate have been conducted at many research institutes worldwide. In NIES2020, five representative models were selected by Shiogama et al. (2021), considering the uncertainty in climate change projections. Shiogama et al. (2021) excluded models that have too large global average temperature changes compared to the observational data over the past decades (so-called hot models), considering the hot model issue discussed later. Five climate models were then selected to reasonably capturer the range of possible climate projection for eight variables, including temperature and precipitation, radiation, and surface wind speed, which are often used in impact assessments (Shiogama et al. 2021; Hayashi and Shiogama 2022) (Fig. 2.1, Table 2.1).

The previous Japanese reports, the "Climate Change Impact Assessment Report (Ministry of the Environment 2020)" and "Climate Change in Japan (MEXT and JMA 2020)," were based on analyses of CMIP5 climate model simulation data. The reports discussed the mitigation effects of greenhouse gas emissions by comparing the highest concentration scenario, RCP8.5, with the lowest scenario, RCP2.6, among the available choices. Therefore, for developing NIES2020, the scenarios of SSP5-8.5 and SSP1-2.6 were selected to compare with the results using the CMIP5-based climate scenarios. In addition, SSP2-4.5 was selected as an intermediate emissions scenario. Japan has also proposed ambitious climate policies aimed at limiting warming to 1.5°C above pre-industrial levels, such as the goal of achieving carbon neutrality by 2050. In light of this social context, SSP1-1.9 was considered to assess the potential benefits, assuming these policies are successful. Since the establishment of the Paris Agreement in 2015, the likelihood of greenhouse gas concentrations reaching the levels assumed in RCP8.5 has decreased due to mitigation efforts. As a result, it has become difficult to refer to RCP8.5 as a business-as-usual (BAU) scenario, as was done previously (e.g., Hausfather and Peters 2020). Given that SSP3-7.0 is one of the common scenarios in the Inter-Sectoral Impact Model Intercomparison Project Phase 3b (ISIMIP3b), NIES2020 also provides partial data under SSP3-7.0. However, as discussed later, it is essential to recognize the distinct characteristics of this scenario.

We introduce the method used in NIES2020 for generating high-resolution data (downscaling) from climate models. First, the output data from historical and future projection experiments of CMIP6 are interpolated to a 1-km grid. This data is then bias-corrected using the nonparametric cumulative distribution function-based downscaling method (CDFDM) proposed by Iizumi et al. (2010), applying daily data from 1900 to 2100. This approach employs statistical methods for downscaling climate models instead of physical and dynamical methods using numerical models (regional climate models). Although the ability to represent heavy precipitation events is relatively limited, it seems that we can use multiple climate models with a low computational cost to assess model uncertainty. The Agro-Meteorological Grid Square Data (Ohno et al. 2016) is used as a reference for bias correction for the

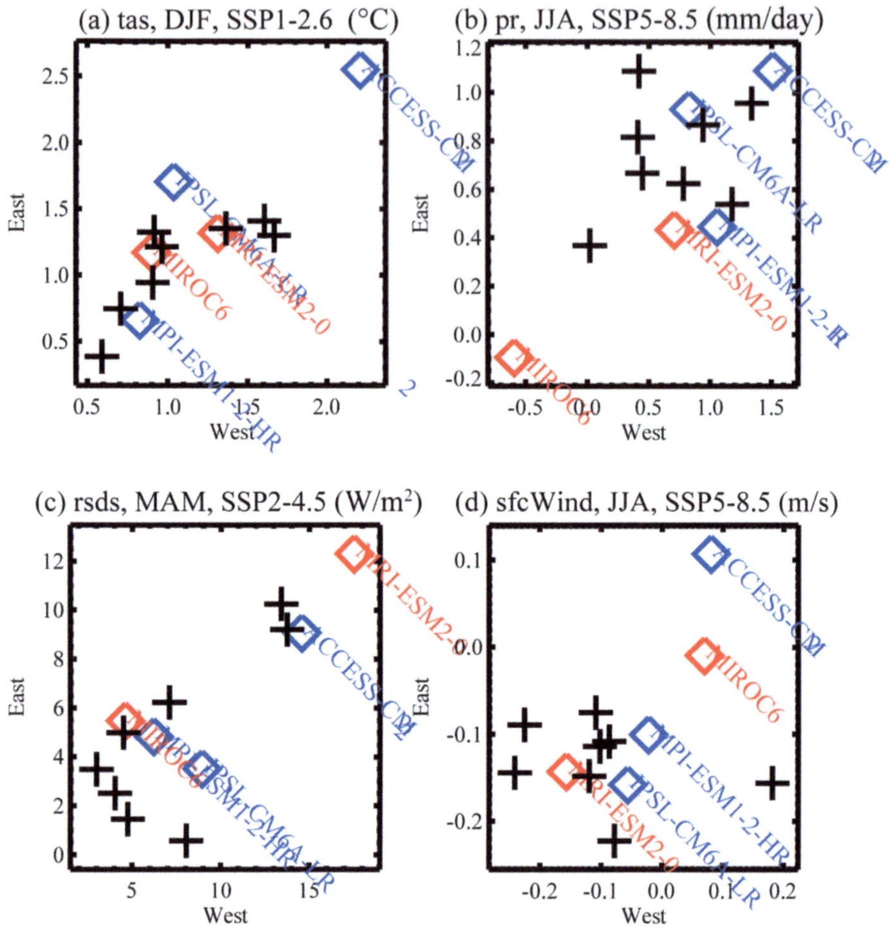

Fig. 2.1 Five representative models of NIES2020. (Source: Reproduced from Shiogama et al. 2021). Red and blue symbols indicate two Japanese models and the other three models, respectively, which were selected by Shiogama et al. (2021). Black symbols represent the other CMIP6 models. The horizontal and vertical axes show area-averaged changes for West Japan and East Japan, respectively, for 2080–2099 compared to 1995–2014. The variables displayed are: (**a**) surface air temperature (December–February mean for SSP1-2.6, °C), (**b**) precipitation (June–August mean for SSP5-8.5, mm/day), (**c**) downward shortwave radiation at the surface (March–May mean for SSP2-4.5, W/m^2), and (**d**) wind speed at the 10 m level (June–August mean for SSP5-8.5, m/s)

variables listed in Table 2.1. For some scenarios, cloud cover percentage and surface pressure are also corrected.

In the S-18 Project, climate change impacts are evaluated by comparing climate conditions during three future target periods—around 2030, around 2050, and at the end of the twenty-first century—with current climate conditions. Figure 2.2 shows the spatial distribution of climate change for temperature- and precipitation-related elements estimated in NIES2020. This map represents the average of three

Table 2.1 Configurations of NIES2020 used in S-18 Project

Climate Models	MIROC6, MRI-ESM2-0, ACCESS-CM2, IPSL-CM6A-LR, MPI-ESM1-2-HR
Elements	Mean, maximum, and minimum surface air temperature (°C; tas, tasmax, tasmax), Daily precipitation (mm/day; pr), Downward shortwave radiation at the surface (MJ/m²/day; rsds), Surface wind speed (m/s; sfcwind), Surface relative humidity (%, rhs), Downward longwave radiation at the surface (MJ/m²/day; rlds)
Future target periods	Around 2030 (2020–2040), Around 2050 (2040–2060), and The end of the twenty-first century (2080–2100)

Source: Created from the contents of Ishizaki et al. (2022). The climate models, elements, and future target periods adopted in NIES2020 are listed. The units and commonly used symbols for the elements are shown in parentheses

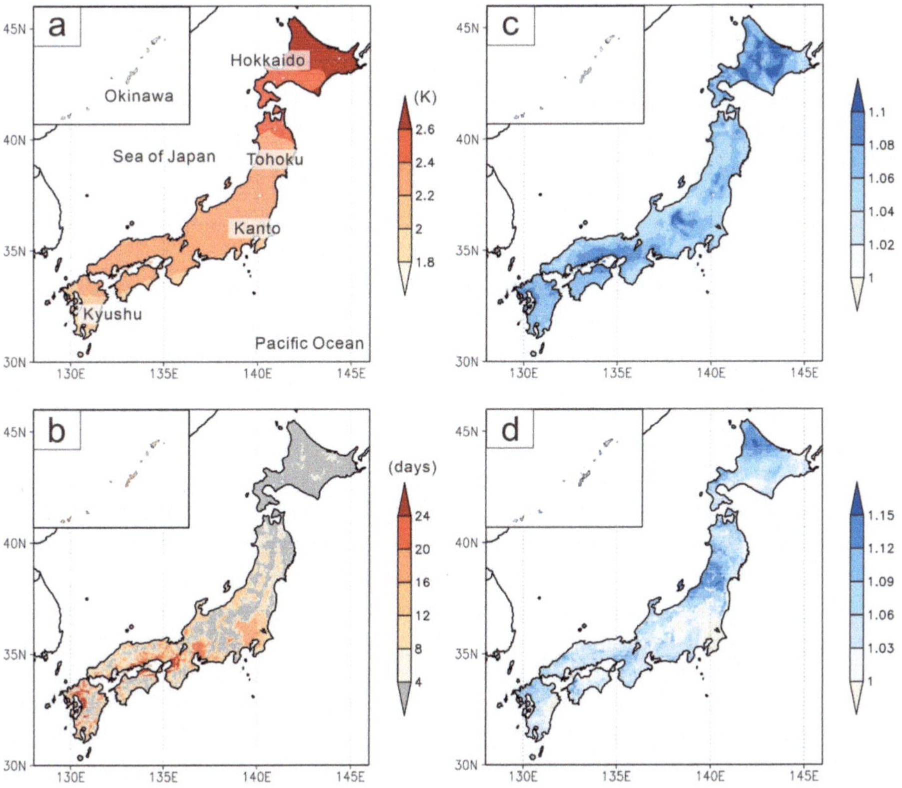

Fig. 2.2 High-resolution climate change signals of NIES2020. (Source: Created by the author). Spatial distributions of climate change signals derived from NIES2020, showing changes from 2010–2030 to 2080–2100, based on the average of five climate models under SSP1-2.6, SSP2-4.5, and SSP5-8.5 scenarios: (**a**) Difference in mean surface air temperature, (**b**) Difference in the number of days with maximum air temperatures of 35 °C or higher, (**c**) Ratio of change in precipitation, and (**d**) Ratio of change in maximum daily precipitation

scenarios and five GCMs. Warming is pronounced in northern Japan, with the number of extremely hot days (days with a maximum temperature of 35 °C or more) increasing in the plains from eastern to western Japan. Precipitation increases by about 5% overall, with the most significant increases in Hokkaido. Maximum daily precipitation tends to rise by about 10%, except in the Kanto region, with particularly notable increases on the Sea of Japan side of the Tohoku region and in Hokkaido.

2.2 Hot Model Issue

Due to significant discrepancies (uncertainties) in future projections among climate models, addressing these uncertainties is crucial. The Emergent Constraint (EC) approach is a promising method for reducing uncertainty in future projections. EC is a statistical technique that compares climate model simulations with past observational data and has been actively studied over the past 15 years (Knutti 2010; Shiogama et al. 2011; Hall et al. 2019). Compared to CMIP5, CMIP6 contains more "hot models," which project particularly large future global average temperature increases. However, recent studies have shown that hot models tend to overestimate global temperature trends from 1980 to the present, suggesting low reliability in projecting future temperatures. As a result, the upper bound of temperature projections should be adjusted downward (Tokarska et al. 2020). Consequently, the Sixth Assessment Report of the Intergovernmental Panel on Climate Change (IPCC) Working Group I (IPCC-AR6-WGI) presents a reduced uncertainty range for global average temperature projections rather than using the full range of CMIP6 models (Hausfather et al. 2022). Shiogama et al. (2022b) also revealed that hot models overestimated future global average precipitation increases.

Many impact assessment studies use hot models, but it remains to be seen if and how these models overestimate impacts (Hausfather et al. 2022). Shiogama et al. (2022a) evaluated the economic impacts (GDP reduction due to climate change) using a statistical model called an impact emulator (Takakura et al. 2021). This machine learning-based emulator can replicate economic impact assessments of complex models across nine sectors: agricultural productivity, undernourishment, heat-related excess mortality, cooling/heating demand, occupational health costs, hydropower generation, thermal power generation, fluvial flooding, and coastal inundation (Takakura et al. 2019). The economic impacts across these nine sectors were assessed by inputting temperature and precipitation projection data from CMIP5 and CMIP6 climate models into this impact emulator. By combining these results with the EC method of Shiogama et al. (2022b), it was demonstrated that hot models also overestimate future economic impacts (GDP loss). By reducing the weighting of hot models, the inter-model variance was decreased by 31%.

Shiogama et al. (2024) investigated if we can reduce uncertainties in regional projections of future changes in various meteorological variables and extreme event indices. Figure 2.3 illustrates how much the projection uncertainty (inter-model

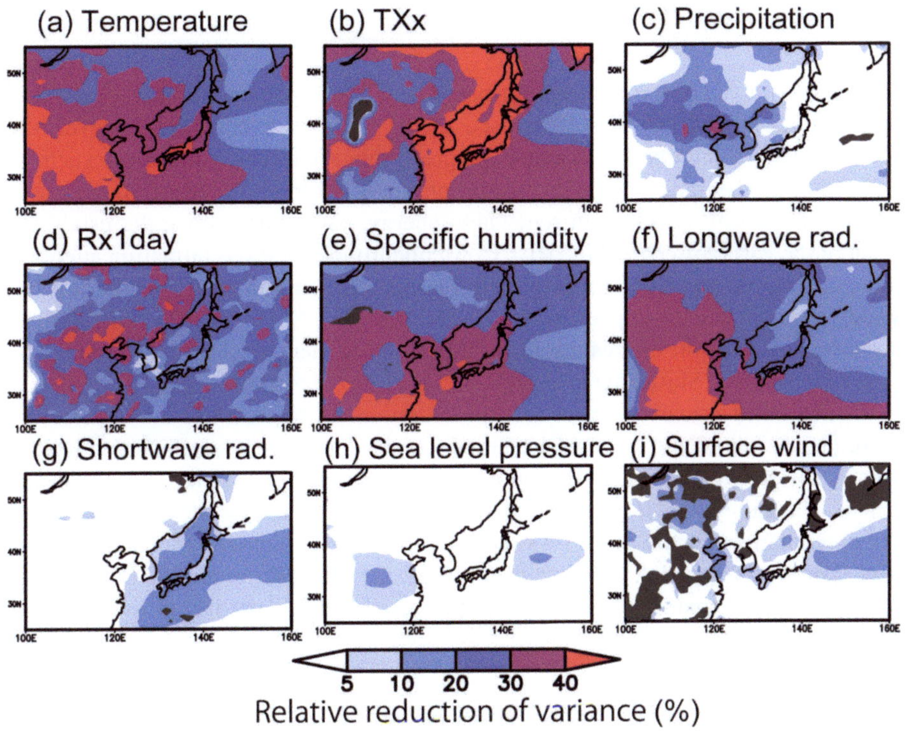

Fig. 2.3 Relative reduction of variance (%) owing to the emergent constraints based on the performance of the past global mean temperature trend simulations. (Source: Reproduced from Shiogama et al. 2024). Shading indicates relative reductions of inter-model variance of changes (2051–2100 compared to 1851–1900) in (**a**) temperature, (**b**) annual maximum daily maximum temperature, (**c**) precipitation, (**d**) annual maximum daily precipitation, (**e**) specific humidity, (**f**) downward longwave radiation, (**g**) downward shortwave radiation, (**h**) sea level pressure, and (**i**) surface wind speed, respectively

variance) of variables around Japan can be reduced using the EC method. It is possible to reduce the variance for temperature, annual maximum daily maximum temperature, annual maximum daily precipitation, humidity, longwave radiation, and shortwave radiation around Japan. In particular, the variance of the annual maximum daily maximum temperature is reduced by more than 40%. Additionally, along the Pacific coast of eastern Japan and Kyushu, the uncertainty in surface wind speed can be reduced by approximately 5%. These findings suggest that we have decreased the risk of overestimating the uncertainty range in impact assessment studies by excluding hot models in creating NIES2020. In other words, if a climate scenario includes hot models, there is a possibility of overestimating the uncertainty range in impact assessments for some sectors such as heat stress, floods, and renewable energy, which are critically influenced by these variables.

2.3 The Distinctiveness of SSP3-7.0

In previous impact assessment studies based on CMIP5 models, RCP8.5 (equivalent to SSP5-8.5 in CMIP6) was often used as the upper-end scenario. However, the likelihood of a significant increase in greenhouse gas concentrations, such as SSP5-8.5, has become low due to recent progress in emission reduction measures and technological advances in renewable energy, particularly since the establishment of the Paris Agreement in 2015. Consequently, SSP3-7.0 is attracting attention as an alternative upper-end scenario in impact assessment studies. However, it is not widely recognized by impact model researchers that SSP3-7.0 is a "distinctive" scenario. As a result, the results from impact assessment models may not be correctly interpreted (Shiogama et al. 2023).

Unlike other scenarios where aerosol emissions decrease, SSP3-7.0 assumes an extreme and unrealistic trend where aerosol emissions increase, reflecting a lack of significant progress in air pollution control measures in countries like India and China. The large amount of aerosol emissions in SSP3-7.0 have the effect of reducing precipitation. Figure 2.4 shows future projections of temperature and precipitation changes averaged across Japan in NIES2020 (Hayashi et al. 2024). After the mid-twenty-first century (around 2040), projected temperature increases are higher

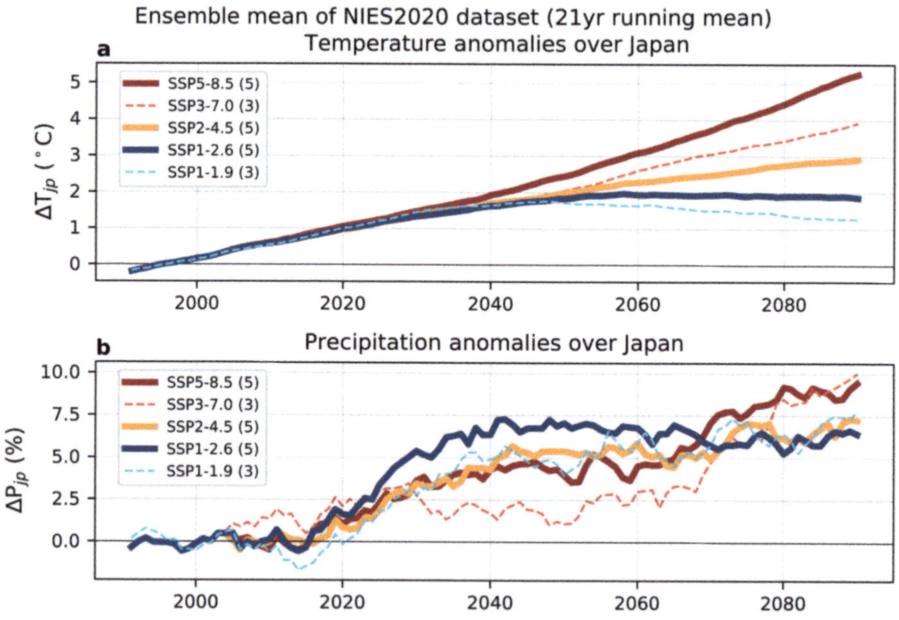

Fig. 2.4 Changes in temperature and precipitation averaged across Japan in NIES2020. (Source: Reproduced from Hayashi et al. 2024). Ensemble mean time series of the 21-year running mean (**a**) temperature (°C) and (**b**) precipitation (%) anomalies relative to 1981–2010 averaged across the 1-km mesh Japanese land grids in NIES2020. The ensemble sizes for each scenario are shown in the legend

under high-emission scenarios like SSP5-8.5 and lower under low-emission scenarios like SSP1-2.6. Although increases in precipitation are projected for all SSP scenarios, the increase in precipitation in the mid-twenty-first century is more pronounced under SSP1-2.6 than under SSP2-4.5 and SSP5-8.5 and is suppressed under SSP3-7.0.

These results suggest that temperature increases do not solely determine changes in Japan's precipitation. Under SSP1-2.6, the rapid reduction of fossil fuel consumption also drastically reduces aerosols, leading to increased precipitation. In contrast, under SSP3-7.0, precipitation hardly increases due to the extreme aerosol emissions. To our knowledge, no studies have examined the effects of this distinctiveness in the precipitation change projection under SSP3-7.0 on impact assessments. It is essential to be cautious when conducting impact assessments using SSP3-7.0, and if possible, it is desirable to compare the results with those of SSP5-8.5. Since the scenario equivalent to SSP5-8.5 (RCP8.5) was used as the upper-end scenario before the Paris Agreement, comparing the impact assessment results of SSP5-8.5 with lower-emission scenarios can also be beneficial in examining the effects of emission reduction efforts on climate change impacts after the Paris Agreement.

2.4 Conclusion

This chapter explained the NIES2020 climate scenario used for climate change impact assessments in the S-18 Project. As impact projections for national and regional levels increasingly require high spatial resolution, it has become essential to downscale climate scenarios to finer scales. In the S-18 Project, climate scenarios were downscaled to a 1-km grid that covers all of Japan. Selecting the most suitable climate models for the target region is also crucial. This chapter outlined the concept, creation process, and validation of NIES2020, which were developed in response to such demands for Japan. Furthermore, we discussed the "hot model issue" identified in some CMIP6 models and the issues with the SSP3-7.0 scenario. Through these discussions, we emphasize the importance of active dialogue and collaboration between climate scenario developers, impact researchers, and policy researchers in assessing the impacts of climate change and developing countermeasures. We aim to continue these efforts moving forward.

References

Hall A, Cox P, Huntingford C et al (2019) Progressing emergent constraints on future climate change. Nat Clim Chang 9:269–278. https://doi.org/10.1038/s41558-019-0436-6

Hausfather Z, Peters GP (2020) Emissions-the 'business as usual' story is misleading. Nature 577:618–620. https://doi.org/10.1038/d41586-020-00177-3

Hausfather Z, Marvel K, Schmidt GA et al (2022) Climate simulations: recognize the 'hot model' problem. Nature 605:26–29. https://www.nature.com/articles/d41586-022-01192-2

Hayashi M, Shiogama H (2022) Assessment of CMIP6-based future climate projections selected for impact studies in Japan. SOLA 18:96–103. https://doi.org/10.2151/sola.2022-016

Hayashi M, Ishizaki NN, Shiogama H, Wakazuki Y (2024) Scenario dependence of future precipitation changes across Japan in CMIP6. SOLA 20:207–216. https://doi.org/10.2151/sola.2024-028

Iizumi T, Nishimori M, Ishigooka Y, Yokozawa M (2010) Introduction to climate change scenario derived by statistical downscaling. J Agric Meteorol 66:131–143. https://doi.org/10.2480/agrmet.66.2.5. (in Japanese)

Ishizaki NN, Shiogama H, Hanasaki N, Takahashi K (2022) Development of CMIP6-based climate scenarios for Japan using statistical method and their applicability to heat-related impact studies. Earth Space Sci 9:e2022EA002451. https://doi.org/10.1029/2022EA002451

Knutti R (2010) The end of model democracy? Clim Chang 102:395–404. https://doi.org/10.1007/s10584-010-9800-2

Ministry of Education, Culture, Sports, Science and Technology (MEXT) and Japan Meteorological Agency (JMA) (2020) Climate change in Japan 2020 –Report on assessment of observed/projected climate change relating to the atmosphere, land and oceans. https://www.data.jma.go.jp/cpdinfo/ccj/index.html. Accessed 18 May 2023 (in Japanese)

Ministry of the Environment (2020) Assessment report on climate change impacts in Japan. https://www.env.go.jp/content/900516663.pdf. Accessed 18 May 2023 (in Japanese)

Ohno H, Sasaki K, Ohara G, Nakazono K (2016) Development of grid square air temperature and precipitation data compiled from observed, forecasted, and climatic normal data. Clim Biosph 16:71–79. https://doi.org/10.2480/cib.J-16-028. (in Japanese)

Shiogama H, Emori S, Hanasaki N et al (2011) Observational constraints indicate risk of drying in the Amazon basin. Nat Commun 2:253. https://doi.org/10.1038/ncomms1252

Shiogama H, Ishizaki NN, Hanasaki N et al (2021) Selecting CMIP6-based future climate scenarios for impact and adaptation studies. SOLA 17:57–62. https://doi.org/10.2151/sola.2021-009

Shiogama H, Takakura J, Takahashi K (2022a) Uncertainty constraints on economic impact assessments of climate change simulated by an impact emulator. Environ Res Lett 17:124028. https://doi.org/10.1088/1748-9326/aca68d

Shiogama H, Watanabe M, Kim H, Hirota N (2022b) Emergent constraints on future precipitation changes. Nature 602:612–616. https://doi.org/10.1038/s41586-021-04310-8

Shiogama H, Fujimori S, Hasegawa T et al (2023) The importance of recognizing the 'distinctiveness' of SSP3-7.0 for use in climate change impact assessments. Nat Clim Chang 13:1276–1278. https://doi.org/10.1038/s41558-023-01883-2

Shiogama H, Hayashi M, Hirota N, Ogura T (2024) Emergent constraints on future changes in several climate variables and extreme indices from global to regional scales. SOLA 20:122–129. https://doi.org/10.2151/sola.2024-017

Takakura J, Fujimori S, Hanasaki N et al (2019) Dependence of economic impacts of climate change on anthropogenically directed pathways. Nat Clim Chang 9:737–741. https://doi.org/10.1038/s41558-019-0578-6

Takakura J, Fujimori S, Takahashi K et al (2021) Reproducing complex simulations of economic impacts of climate change with lower-cost emulators. Geosci Model Dev 14:3121–3140. https://doi.org/10.5194/gmd-14-3121-2021

Tokarska KB, Stolpe MB, Sippe S et al (2020) Past warming trend constrains future warming in CMIP6 models. Sci Adv 6:eaaz9549. https://doi.org/10.1126/sciadv.aaz9549

Chapter 3
Development of Socioeconomic Scenarios for Climate Change Impacts and Adaptation Research

Sayaka Yoshikawa, Kohei Imamura, Kiyoshi Takahashi, and Keisuke Matsuhashi

Abstract Possible future scenarios of both climate and socioeconomic changes are commonly used to assess the impacts and adaptation measures of climate change. This chapter provides an overview of the development of the common socioeconomic scenarios in Japan. The data prepared for the S-18 common socioeconomic scenario are population, number of households, and land use (building site areas by use). The actual impacts of climate change occur alongside social changes in the population, number of households, and land use, including vacant houses. These data are important in many fields, such as natural disasters, biodiversity conservation, and CO_2 emission reduction, to formulate climate change impacts and adaptation strategies.

Keywords Socioeconomic scenario · Land use · Vacant houses · Building site areas by use · Japan SSPs

3.1 Introduction

In research on the impacts of and adaptation to climate change, scenarios that assume future climate and socioeconomic conditions are typically used in some form. The scenario is the input data for a model that assumes a possible future. There are various uncertain elements in assuming a possible future. It is difficult to

S. Yoshikawa (✉)
Nagasaki University, Nagasaki, Japan
e-mail: sayaka_yoshikawa@nagasaki-u.ac.jp

K. Imamura
Global and Local Environment Co-creation Institute, Ibaraki University, Mito, Japan

K. Takahashi · K. Matsuhashi
National Institute for Environmental Studies, Tsukuba, Japan

N. Mimura, S. Takewaka (eds.), *Climate Change Impacts and Adaptation Strategies in Japan*, https://doi.org/10.1007/978-981-96-2436-2_3

predict based on past data because the situation of technological progress, lifestyle, economic development, and policies of each country changes with the progress of society. Therefore, scenario development has primarily been conducted by using integrated assessment models. Shared Socioeconomic Pathways (SSPs), commonly used for global impact assessment, are typical socioeconomic scenarios (hereafter, Global SSPs). Global SSPs were developed as socioeconomic scenarios that can be used across fields in response to a request from the Intergovernmental Panel on Climate Change (IPCC) in 2007 for new scenario development to assess the impacts of climate change (Kriegler et al. 2012; O'Neill et al. 2012). However, Global SSPs are only data aggregated into five regions (population), 32 regions (GDP), or data by country, and are not suitable for detailed analysis and evaluation targeting the interior of each country. In addition, these Global SSPs may not fully reflect the future prospects based on the policy situation of each country.

The development of Japan SSPs was conducted considering these issues. Chen et al. 2020 created the only narrative scenario of Japan SSPs. Japan SSPs were partially changed from the narrative scenario of Global SSPs throughout various medium- and long-term plans, and measures in Japan were investigated. In Japan SSPs, the migration rate, birth rate, number of immigrants, etc., are set for each scenario, and age- and sex-specific population data for every five years from 2020 to 2100 are provided by selecting relevant data from the future population data of the National Institute of Population and Social Security Research (2017) (National Institute for Environmental Studies 2021).

In the S-18 project, we established common socio-economic scenarios based on the Japan SSPs which are necessary to project and assess impacts and adaptation to climate change. This chapter provides an overview of the development of the common socioeconomic scenarios in Japan.

3.2 Narrative Scenario Selection

In the S-18 project, three main RCPs (RCP8.5, RCP4.5, and RCP2.6) and one additional RCP (RCP1.9) were adopted as climate scenarios. The socioeconomic scenarios corresponding to these four RCPs followed the scenarios with the highest priority in O'Neill et al. (2016): SSP5, SSP2, and SSP1. Japan SSP3 shows that a large population decrease was adopted to clarify the impact of the degree of population decrease, since there is not much difference in the population data between Japan SSP1 and Japan SSP5. Additionally, a status quo scenario was set, assuming that the current social situation in 2015 will continue until 2100 to understand only the impact of climate change. In IPCC sixth assessment report (IPCC 2021), scenarios that integrated socioeconomic and climate scenarios, namely SSP5-8.5, SSP2-4.5, SSP1-2.6, and SSP1-1.9, were used. However, we decided that it is not always necessary to set the climate scenario and socioeconomic scenario together.

3.3 Creating Data That Constitutes Socioeconomic Scenarios

The data prepared for the S-18 common socioeconomic scenario were population, number of households, and land use (building site areas by use). As a first step, a demand survey was conducted targeting researchers in various fields who conducted impact assessments of the S-18 project. It was found that 80% of the teams need land use/land cover (hereafter, referred to as LULC), and 50% of the teams need population. In particular, there were requests such as "It necessary for the building site areas to be divided by use (industrial use, commercial use, residential use)" and "It necessary for distribution maps of household projection." These data were estimated based on the Japan MESH3 Boundaries (approximately 1 km × 1 km) or on a municipal basis according to the spatial resolution of the impact assessment.

There is a future projection of the number of households by the National Institute of Population and Social Security Research (2018, 2019); however, this only goes up to the 2040s by prefecture. It is necessary to estimate that up to 2100 (as detailed in Sect. 3.3.2). Although there are projection studies on vacancy rates (Ishikawa et al. 2016), housing floor area (Nakano 2014), building sites (Hanasaki et al. 2012, 2014; Yoshikawa et al. 2022), and artificial land use (Etoh and Onishi 2018), to our knowledge, no study has estimated building site areas by use using the Japan MESH3 Boundaries (approximately 1 km × 1 km). After dividing the building site areas by use in 2015 (the base year), we conducted a future projection of the building site areas by use from 2020 to 2100 every five years (Yoshikawa et al. 2024) (detailed in Sect. 3.3.3).

3.3.1 *Population*

We decided to use the National Institute for Environmental Studies (2021), which is an age- and sex-specific population estimate in accordance with Japan SSPs, as population data. In Japan SSP1, a high birth rate of 18% is maintained under the assumption that the birth rate will accelerate due to the improvement of the child-rearing environment through educational investment, but the total population in 2100 will be 73 million. The mortality and migration rates remained medium. On the other hand, in Japan SSP5, the labor market gradually opens, and international mobility increases, resulting in a net immigration of 250,000 foreigners by 2035. The total population was assumed to be 79 million by 2100, with medium birth and death rates. The total population decreased in the long term in both scenarios; however, the decrease was estimated to be greater in Japan SSP1 (Fig. 3.1). Approximately 2.7% under Japan SSP1 and 4.7% under Japan SSP5 municipalities of the 1683 municipalities increased from 2015 to 2100, with a population decrease in more than 1600 municipalities. Even in municipalities where the population increases, in many cases, it is not due to an increase in the birth population but due to an increase

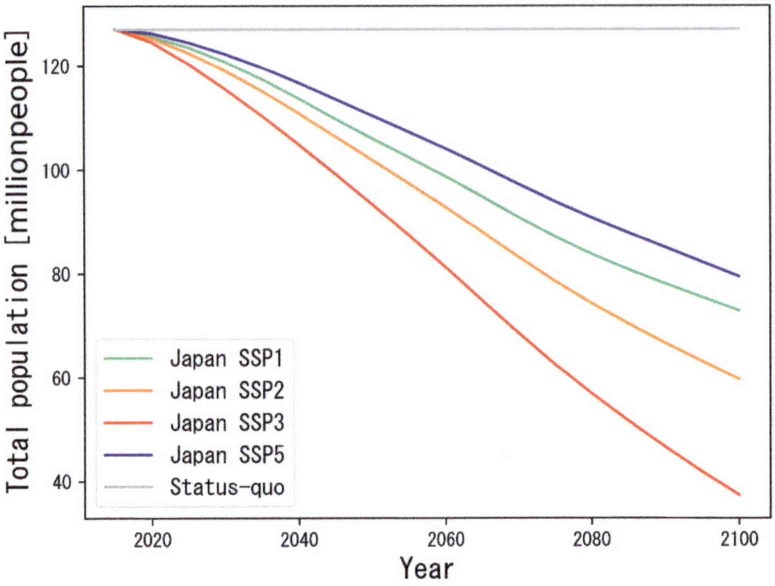

Fig. 3.1 Changes in population projections from 2015 to 2100 under Japan SSPs and status quo scenarios

in the population aged 65 and over. The concentration of the population in large metropolitan areas is projected to continue until 2100, with Japan SSP5 (64.2% of the total population) estimated to have a higher concentration than Japan SSP1 (49.8% of the total population).

3.3.2 Households

We estimated the future number of households for each scenario by multiplying the future population and the future household head rate for each Japan SSP (Yoshikawa et al. 2024). The future population is described in Sect. 3.3.1. The household head rates divided by gender, five-age class, and family type of the household head for each prefecture were obtained from the National Institute of Population and Social Security Research (2019). The estimated total number of households in each scenario is shown in Fig. 3.2. The total number of households will decrease from 53.2 million in 2015 to between 18.81 million and 36.55 million in 2100. As the method of multiplying the future population by the household head rate is used, the number of households will decrease sharply after 2030, similar to the trend in population decrease. The scenario with the highest decrease in total households was Japan SSP3, and the scenario with the least decrease was Japan SSP5. This trend was

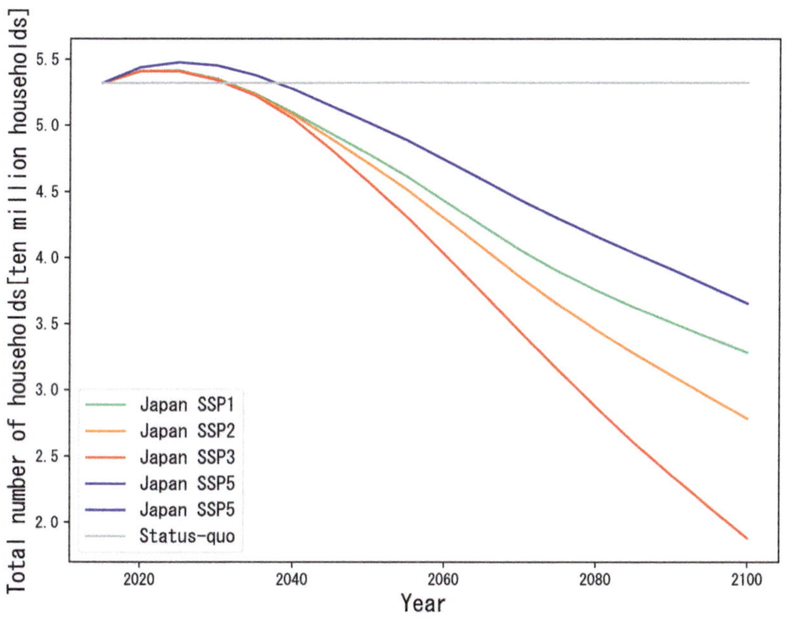

Fig. 3.2 Estimated changes in the total number of households under Japan SSPs. (Modified from Yoshikawa et al. 2024)

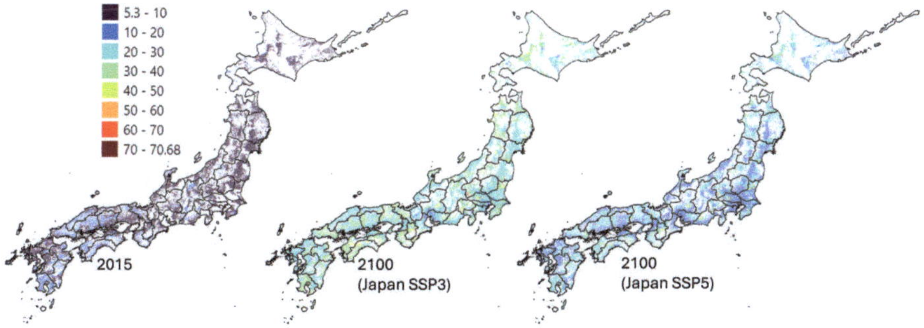

Fig. 3.3 Distribution of the proportion of households in which the household head is 85 years old or older to the total number of households by municipality in 2015 and 2100

similar to the estimated results for the total population for each scenario. Figure 3.3 shows the proportion of households in which the household head is 85 years old or older to the total number of households. In 2015, the average ratio of households aged 85 years or older to the total number of households in each municipality was 8.1%. The average value in 2100 is expected to be 28.7% under Japan SSP3 and 20.5% under Japan SSP5, which is 2.5 to 3.5 times that of 2015. This suggests that the aging of the Japanese society will progress further, and major changes are expected.

3.3.3 Land Use

In Japan, changes in LULC due to farmland expansion, urbanization, and economic development have occurred in various places following economic development after World War II (Himiyama 1999). Land Utilization Tertiary Mesh Data (Ministry of Land, Infrastructure, Transport and Tourism National Land Numerical Information 2023) have been developed since 1976 as a nationwide database of LULC in Japan. The Land Utilization Tertiary Mesh Data have been developed, focusing on areas dominated by humans, such as residential areas and transportation facilities identified based on topographic maps, and vegetation areas calculated from satellite image data such as Landsat, TERRA (Aster), and ALOS.

Future projections of LULC changes are important in many fields, such as natural disasters, biodiversity conservation, and carbon cycling, for formulating climate change impacts and adaptation strategies. In many previous studies on LULC projection in Japan, changes in population dynamics have been considered the main factor of LULC changes, and population changes have functioned as proxy variables for LULC changes (Hanasaki et al. 2012, 2014; Etoh and Onishi 2018; Yoshikawa et al. 2022). Therefore, spatially explicit population projections play an important role in predicting LULC changes (Shen et al. 2008).

In the LULC change estimation by Yoshikawa et al. (2024), building site areas, which are one of the land-use types in the "Land Utilization Tertiary Mesh Data" of the National Numerical Information, were targeted. First, building site areas were classified into four categories: industrial, commercial, residential, and others. For each, they used the developed estimation method to make future projections after 2020 based on 2015. The area changes in the industrial and commercial building site areas were estimated using population as a proxy variable. Residential building site areas were divided into residences with and without households using household projection. The areas of the residential building sites and former residential building sites without residents were then estimated. The spatial resolution of these estimates was a tertiary mesh (approximately 1 km × 1 km).

Based on past trend analysis, it was found that both industrial building site areas and commercial building site areas increased for less than 20 years after the start of the population decline in each prefecture. Based on this provision, the areas of both industrial and commercial building sites will increase until 2045 under Japan SSP5 and until 2035 under Japan SSP3. As a result, the total area of industrial building sites for each scenario was 1538 km^2 in 2015 and 1564–1572 km^2 in 2100. The total area of commercial building sites for each scenario was 2887 km^2 in 2015 and 3090–3115 km^2 in 2100.

The number of vacant houses for detached houses and apartment houses for the estimation of residential building site areas is shown in Fig. 3.4. The number of vacant houses, which is the difference between the number of existing houses and the number of households, tends to increase with population decline. The number of

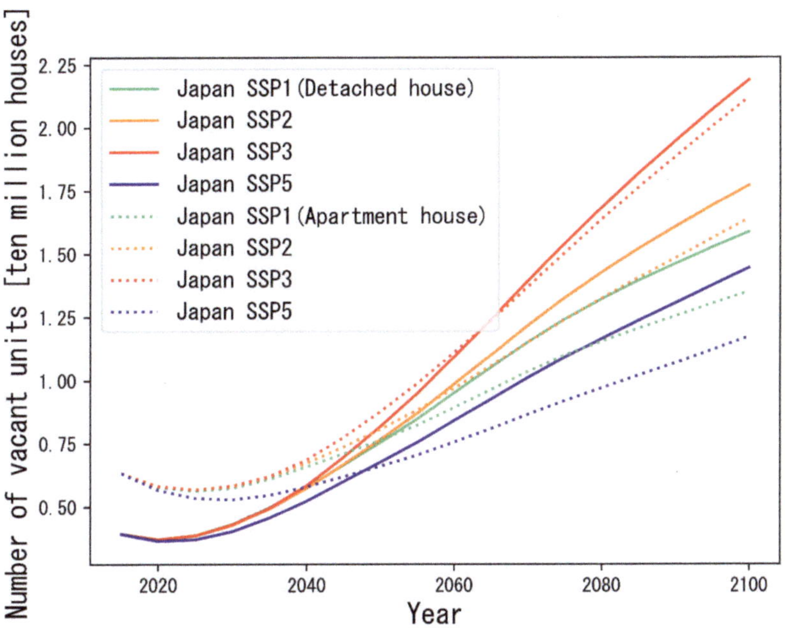

Fig. 3.4 Changes in the number of vacant units of detached houses and apartment houses under Japan SSPs. (Modified from Yoshikawa et al. (2024))

vacant houses were 3.93 million for detached houses and 6.34 million for apartment houses in 2015, and 14.45 to 21.88 million for detached houses and 11.72 to 21.19 million for apartment houses in 2100. Figure 3.5 shows the distribution of vacancy rates by municipality in Japan. The vacancy rate is the ratio of the number of unoccupied houses to the total number of existing houses. The vacancy rate in 2015 was 22.5%, on a national average. By 2100, the vacancy rates under Japan SSP3 and Japan SSP5 were 78.1%and 66.5%, Japan SSP5. The increase in the vacancy rate was noticeable in areas far from the major urban areas. If such a large number of vacant houses occur, there must be some policy intervention; however, no policy intervention was set in this estimate.

Figure 3.6 shows the changes in residential and former residential building site areas by scenario. The area covered by residential building sites was 8558 km^2 in 2015 and 3571–5704 km^2 in 2100. Compared to 2015, it would be 0.42 times under Japan SSP3 and 0.66 times under Japan SSP5 by 2100. These trends are almost the same as those of the decreasing rate of households. The area of the residential building sites was 1014 km^2 in 2020 and 4048–6180 km^2 in 2100. The areas of the residential building sites were the largest in Japan SSP3, where the rate of decrease in households was the largest. After 2080, former residential building site areas under Japan SSP3 would exceed the areas of the residential building sites.

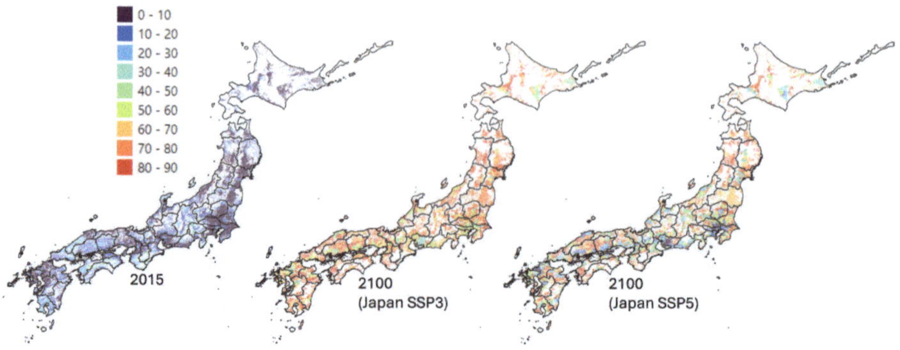

Fig. 3.5 Distribution of vacant unit rates by municipality in 2015 and 2100

3.4 Verification of the Data Created

3.4.1 Comparison of Building Site Areas by Use

The total evaluated land area of commercial, industrial, and residential land, which is subject to property tax, by prefecture (Local Taxation Bureau, Ministry of Internal Affairs and Communications 2016) was considered as the actual measurement. Property tax is levied on the owners of land, houses, or depreciated assets (these are called "fixed assets") as of January 1 every year, based on the price of these fixed assets. The total evaluated land area was taxable land designated according to the rules of Local Tax Law. The estimated areas of building sites for each use, which were the results of this study, were compared with the actual measurements by prefecture. Correlation coefficient between the actual measurements and the estimated values was 0.79 for commercial building site areas, 0.74 for industrial building site areas, and 0.99 for residential building site areas, respectively. It can be observed that the estimation of residential building site areas was accurate. The high accuracy of this stacking method based on the number of households is noteworthy. However, the estimated areas for industrial building sites were underestimated. This is because of the differences in the determination of building sites between the National Land Utilization Tertiary Mesh Data and the Industrial Land Polygon Data used as the basic data. The building site areas in the Land Utilization Tertiary Mesh Data are determined by satellite data; therefore, areas without buildings, such as parking lots and material storage areas, are no longer included in the industrial building site. It is important to note that the estimation of building site areas by use only represents the area where a building is located.

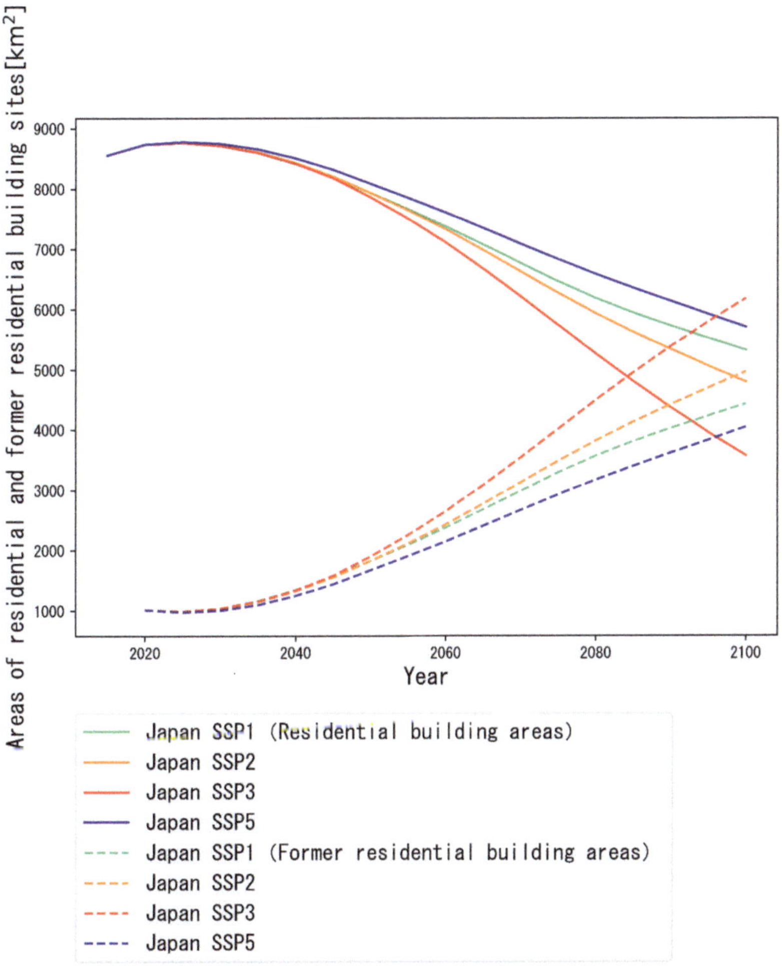

Fig. 3.6 Changes in residential and former residential building site areas under Japan SSPs. (Modified from Yoshikawa et al. (2024))

3.4.2 Comparison of Building Site Area

Figure 3.7 shows a comparison between Yoshikawa et al. (2022) and Yoshikawa et al. (2024). Yoshikawa et al. (2022) adopted a method that assumes that the building site area changes in proportion to the population for two Japan SSPs. Yoshikawa et al. (2024) divided the building site areas by use and estimated these areas by different methods. The results were divided into five categories: industrial, commercial, residential, other, and former residential building site. Yoshikawa et al. (2022) reported that the building site areas have decreased rapidly with a decrease in population. However Yoshikawa et al. (2024), expresseds a gradual decrease after

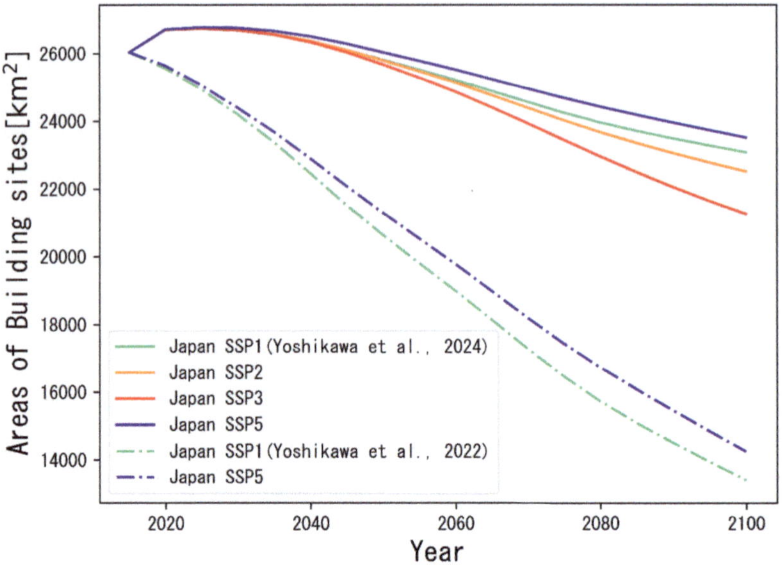

Fig. 3.7 Comparison of building site areas in the S-18 common socioeconomic scenario (Yoshikawa et al. 2024) with Yoshikawa et al. (2022)

2040 by accumulating the site area from the number of buildings. The building site areas of Yoshikawa et al. (2024) were considered more realistic estimates. In addition, the establishment of a new land use category called "Former building site areas" in Yoshikawa et al. (2024) was a significant improvement in reflecting reality. By setting this up, it became possible to express land use that transitioned from building site to other land use types. If the total estimated value of the building site area for each use exceeds the upper limit, the land use to be changed to the building site will vary depending on the scenario assumption and the assumption in each field. Therefore, we did not deliberately process the conversion of other land use types. This is a future challenge and includes the possibility of extending the data flexibly according to the research purpose.

3.4.3 Comparison with the 6th National Land Use Plan

In Japan, a national land use plan has been formulated that sets target values for LULC. The target values were set based on past performance and measures of related ministries and agencies. The latest version, the 6th National Land Use Plan, was formulated in July 2023 (Ministry of Land, Infrastructure, Transport, and Tourism 2023). We compared the LULC projections in Yoshikawa et al. (2024) with the national plan. The population and number of households in the national plan are assumed to be approximately 118 million and 53 million, respectively, in the target year 2033. The residential area is estimated to be 11,900 km^2, the industrial area is 1700 km^2, and the other residential areas are 6100 km^2. In the projection under Japan SSP1-SSP3, the number of households in 2030 is 53.42 to 53.48 million, the area for residential building sites is 8713 to 8724 km^2, and the area for industrial building sites is 1537 km^2. Compared to the increase or decrease ratio from 2020, the residential land of the national land use plan is 0.99 times, and the building site areas of Japan SSP1 to Japan SSP3 is approximately 0.98 times. The trends in the number of households and residential areas generally matched. Industrial building sites in Yoshikawa et al. (2024) are underestimated compared with the national plan. This is because there is a difference in the treatment of land without buildings, such as parking lots and material storage areas, for industrial building sites. While this is included in the government's estimate, it does not include the projection of Yoshikawa et al. (2024). For residential areas, there was almost no change from the target standard value of 2020 in the national plan, because the plan assumed that vacant houses are likely to be maintained in residential areas. This is different from the treatment of the projection in Yoshikawa et al. (2024), which dealt with former residential building site areas.

3.5 Future Challenge

The impact of climate change on each sector occurs amid major social changes in population, number of households, and land use including vacant houses. The development of socioeconomic scenarios that assume future social changes is very important for projecting the impact of climate change. Therefore, we have been developing S-18 common socioeconomic scenarios in accordance with Japan SSPs for climate change and adaptation assessment (Yoshikawa et al. 2022, 2024). Yoshikawa et al. (2024) estimated the number of households and the building site areas by use from 2015 to 2100 for the entire area of Japan in Japan MESH3 Boundaries (approximately 1 km × 1 km).

The number of households showed a high correlation with the population decline, and all scenarios showed a rapid decrease after 2030. For residential building site areas, Yoshikawa et al. (2024) distinguished between detached houses and apartment houses, making it possible to estimate the proportion of vacant houses. The method could be more accurate for estimating residential building site areas and former residential building site areas, which account for 62–72% of the total building site areas in 2100.

As of 2100, industrial, commercial, and other building site areas account for 28–38% of the total building site areas. Based on an analysis of past changes, it was modeled that industrial and commercial building site areas tend to increase in prefectures where the population has started to decrease for less than 20 years. The industrial and commercial building site areas have remained constant since 2040, because we assume that these areas will remain constant for 20 years after the start of the population decline in each prefecture. We set a simple assumption because long-term future trends are unknown. However, this approach may be unrealistic. To solve this problem, it is necessary to identify the driving factors and proxy variables related to the decrease or increase in industrial and commercial building site areas under a population decline. For example, an approach that utilizes economic factors and considers industrial and commercial structures based on research, such as Honjo et al. (2021), can be considered. In addition, future estimates are needed for land use categories other than building sites, such as farmlands and forests. These will be the future challenges.

Finally, the projection data of households and building site areas by use created by Yoshikawa et al. (2024) are available on the National Institute for Environmental Studies Climate Change Adaptation Information Platform (A-PLAT). The development of country-specific socioeconomic scenarios plays a very important role in the future impact and assessments of adaptation measures for climate change. We look forward to further sophistication of socioeconomic scenario development while considering the situation in each country.

References

Chen H, Matsuhashi K, Takahashi K, Fujimori S, Honjo K, Gomi K (2020) Adapting global shared socio-economic pathways for national scenarios in Japan. Sustain Sci 15(3):985–1000

Etoh N, Onishi A (2018) Estimation of future land use change in 109 water systems of Japan. J Jpn Soc Hydrol Water Resour 31(5):364–379

Hanasaki N, Takahashi K, Hijioka Y (2012) Climate and Socioeco-nomic scenarios for climate change impact and adaptation assessments in Japan. Environ Sci 25(3):223–236

Hanasaki N, Takahashi K, Hijioka Y, Kusaka H, Iizumi N, Ariga T, Matsuhashi K, Mimura N (2014) Climate, population, and land use scenarios for climate change impacts and adaptation polices assessments in Japan (Second edition). Environ Sci 27(6):362–373

Himiyama Y (1999) Land use/cover change in Japan: from the past to the future. Hydrol Process 12:1995–2001

Honjo K, Gomi K, Kanamori Y, Takahashi K, Matsuhashi K (2021) Long-term projections of economic growth in the 47 prefectures of Japan: an application of Japan shared socioeconomic pathways. Heliyon 7(3):e06412

Intergovernmental Panel on Climate Change (IPCC) (2021) Summary for policymakers. In Masson-Delmotte V, P Zhai, A Pirani, SL Connors, C Péan, S Berger, N Caud, Y Chen, L Goldfarb, MI 450 Gomis, M Huang, K Leitzell, E Lonnoy, JBR Matthews, TK Maycock, T Waterfield, O Yelekçi, R Yu, and B. Zhou (eds) Climate Change 2021: The Physical Science Basis. Contribution of Working Group I to the Sixth Assessment Report of the Intergovernmental Panel on Climate Change. Cambridge University Press, Cambridge, United Kingdom and New York, NY, USA, pp. 3–32

Ishikawa M, Matsuhashi K, Ariga T, Kanamori Y, Kurishima H (2016) Estimation of current and future distribution of vacant dwellings within municipalities: with a focus on the difference between the numbers of dwellings and households, Papers on city planning, 51(3)

Kriegler E, O'Neill BC, Hallegatte S, Kram T, Lempert RJ, Moss RH, Wilbanks T (2012) The need for and use of socioeconomic scenarios for climate change analysis: a new approach based on shared socio-economic pathways. Glob Environ Change 22:807–822

Local Taxation Bureau, Ministry of Internal Affairs and Communications (2016) Summary record of prices, etc. of fixed assets for fiscal year 2015

Ministry of Land, Infrastructure, Transport and Tourism (2023) The 6th National Land Use Plan (National Plan), https://www.mlit.go.jp/common/001100246.pdf
Nakano K (2014) A forecasting method for future occupied dwellings considering the change in household type. Environ Syst Res 70(4):68–77
National Institute for Environmental Studies (2021) Population scenarios by Japan SSPs Version 2., https://adaptationplatform.nies.go.jp/socioeconomic/population.html
National Institute of Population and Social Security Research (2018) Household Projections for Japan: 2015–2040, Outline of Results and Methods, https://www.ipss.go.jp/pp-ajsetai/e/hhprj2018/t-page_e.asp
National Institute of Population and Social Security Research (2019) Estimated Future Number of Households in Japan (by Prefecture), Research on Population Problems No. 343, ISSN1347-5428
O'Neill BC, Carter TR, Ebi KL, Edmonds J, Hallegatte S, Kemp-Benedict E, Kriegler E, Mearns L, Moss R, Riahi K, van Ruijven B, van Vuuren D (2012) Meeting report of the workshop on the nature and use of new socioeconomic pathways for climate change research. National Center for Atmospheric Research, Boulder
O'Neill BC, Tebaldi C, van Vuuren DP, Eyring V, Friedlingstein P, Hurtt G, Knutti R, Kriegler E, Lamarque J-F, Lowe J, Meehl GA, Moss R, Riahi K, Sanderson BM (2016) The Scenario Model Intercomparison Project (ScenarioMIP) for CMIP6. Geosci Model Dev 9:3461–3482
Shen Y, Oki T, Utsumi N, Kanae S, Hanasaki N (2008) Projection of future world water resources under SRES scenarios: water withdrawal. Hydrol Sci J 53:11–33
Statistics Bureau, Ministry of Internal Affairs and Communications (2017) Regional mesh statistics on the 2015 census
Yoshikawa S, Takahashi K, Wu W, Matsuhashi K, Mimura N (2022) Development of common socio-economic scenarios for climate change impact assessments in Japan. Geosci Model Dev Discuss:gmd-2022-169
Yoshikawa S, Imamura K, Yamasaki J, Nitanai R, Manabe R, Murayama A, Takahashi K, Matsuhashi K, Mimura N (2024) Estimation of future building area by use for data development associated with Japan SSPs. J JSCE (Global Environment Engineering) 12(2)., accepted

Part II
Agriculture, Forestry, and Fishery

Chapter 4
Coping with Climate Change: An Evaluation of Agricultural Impacts and Adaptation in Japan

Toshihiro Hasegawa, Yasushi Ishigooka, Kou Nakazono, Toshihiko Sugiura, and Hitomi Wakatsuki

Abstract Agriculture is highly susceptible to climate change, with Japan's diverse climatic conditions leading to region- and commodity-specific impacts. Since around 2000, rising temperatures and extreme heat have notably affected both field and horticultural crops. In field crops, heat exposure during critical periods for rice grain quality, especially in central Japan, has increased the appearance of chalky grains, reducing premium-grade rice. In fruit trees, heat-related disorders such as poor coloration and sunburn have frequently lowered produce quality and market value. As global warming progresses, these effects are projected to worsen, with suitable growing regions, particularly for horticultural crops, expected to shift northward or to higher altitudes, altering regional production dynamics. Adaptation strategies, including heat-tolerant cultivars and adjusted planting schedules, are under consideration, but their effectiveness in specific local contexts and under extreme weather conditions requires further study. Current research also remains insufficient to fully assess the feasibility of various adaptation options. Locally tailored strategies that account for environmental, economic, and social factors are crucial for effective climate adaptation in Japan's agriculture.

Keywords Chalky rice grains · Crop quality losses · Heat-tolerant cultivars · Poor coloration · Shifting plating times · Suitable growing areas · Sunburn

---

T. Hasegawa (✉) · Y. Ishigooka · H. Wakatsuki
National Agriculture and Food Research Organization (NARO), Tsukuba, Japan
e-mail: thase@affrc.go.jp

K. Nakazono
Central Region Agricultural Research Center, National Agriculture and Food Research Organization, Tsukuba, Japan

T. Sugiura
National Agriculture and Food Research Organization, Tsukuba, Japan

N. Mimura, S. Takewaka (eds.), *Climate Change Impacts and Adaptation Strategies in Japan*, https://doi.org/10.1007/978-981-96-2436-2_4

4.1 Introduction

Agriculture has been and will continue to be one of the industries most affected by ongoing and future changes in climate. The latest Intergovernmental Panel for Climate Change report (IPCC AR6) confirmed that climate change has hindered agricultural productivity over the past 50 years. Weather extremes linked to climate change have triggered food production shocks, placing millions at risk of acute food insecurity (Bezner Kerr et al. 2022). Globally, each degree of additional warming increases risks of food insecurity, threatening livelihoods and quality of life. While this global assessment emphasizes the urgency of addressing climate change, it also highlights that no universal adaptation strategy will suffice, given the varying impacts, risks, and resources across regions. Country- and region-specific impacts and adaptation options, therefore, require special attention.

Japan spans a wide latitudinal range (20 to 45°N) and experiences diverse climatic conditions, leading to regionally distinct agricultural practices. Consequently, the impacts of climate change and the strategies to mitigate these impacts are highly variable depending on the region and the crops. Studies assessing climate change impacts on Japan's agriculture began in the early 1990s, shortly after the IPCC initiated its assessments (Horie 2019). These assessments have been regularly updated using the latest climate scenarios and crop models to project impacts for the mid-to-late twenty-first century. The overall risks are projected to rise, but they remain highly region- and commodity-specific, requiring careful examination.

In Japan, since around 2000, rising temperatures and an increasing frequency of extreme heat events have been observed in various crops, including rice and major fruit crops (Sugiura et al. 2012). This trend has intensified, with record-high temperatures becoming more common, especially during the summer. Adaptation strategies are being implemented, but the options and their effectiveness vary by region. In this chapter, we first review recent updates on the observed impacts of climate change on Japan's agriculture. We then cover projected impacts, highlighting key risks for major commodities based on the latest research. Finally, we explore practical strategies to cope with climate change and discuss gaps and challenges in aligning adaptation measures with climate development trajectories.

4.2 Observed Impacts of Climate Change

In Japan, region- and sector-specific impacts were also evident in the previous National Assessment Report on Climate Change Impacts (MOE 2020). However, since then, record-breaking summer temperatures have been recorded for two consecutive years in 2023 and 2024, increasing the risks of heat-related impacts in the agricultural sector.

Heat-related disorders in various crops have already been recognized in all prefectures in Japan since the early 2000s (Sugiura et al. 2012). Corresponding to the emerging impacts of climate change, the Ministry of Agriculture, Forestry, and Fisheries, Japan (MAFF), has started to collect distribution of the impacts across 47 prefectures in Japan and to annually release reports to show heat-related damages that occurred each year and prefecture (MAFF 2022). We summarized this report for the past ten years (2012–2021) and confirmed that the effects on some commodities, including staples, fruits, vegetables, and livestock species, are already widespread and/or have been increasing in the past ten years (Table 4.1).

4.2.1 Field Crops

For rice, the most notable heat-related disorder is the occurrence of chalky grains, which results from abnormal starch accumulation during the grain-filling period (about 20 days after the heading stage or the appearance of ears). Chalky grains degraded the appearance and milling quality of rice and were reported in as many as 36 prefectures in 2019. The number of affected prefectures has been rising over the past decade (Table 4.1).

To assess the exposure to heat during the 20-day period after heading, critical for chalky grain formation, we derived the Mean Excess Temperature (MET26), by averaging the excess daily air temperature above 26 °C during this period. The MET26 index was derived from national 1-km mesh meteorological data for 23 years (2001–2023) and the heading dates reported by MAFF. The estimated values for each 1-km grid were then weighted by the paddy area ratio (Hasegawa et al. 2024).

The MET26 index averaged across the nation has significantly increased over this period, with regional variations in the rate of increase and the overall MET26 levels (Fig. 4.1a). The index was particularly high in the Kanto, Hokuriku, and Kinki regions. In 2023, Japan's hottest summer on record, the MET26 index greatly exceeded previous levels in regions like Tohoku and Hokuriku, with all four prefectures in Hokuriku recording temperatures above 3 °C. In the Kanto, Tokai, and Kinki regions, the MET26 index in 2023 aligned with previous trends, indicating that record-high summer temperatures are becoming the norm. Higher MET26 values had a strong negative impact on the percentage of first-grade rice aggregated for the whole country: a 1 °C increase in MET26 led to a 15-point decrease in the percentage of first-grade rice, confirming the negative impacts of recent warming on rice quality (Fig. 4.1b).

For wheat and soybeans, heat-related disorders have not been as widely reported (Table 4.1). However, in 2010, high temperatures during the flowering-to-maturity period caused an estimated loss of 11 billion JPY (about 78 million USD) in Hokkaido due to reduced yields (Hokkaido Sapporo District Meteorological Observatory 2010; Nishio et al. 2011). Similar reductions in single grain weight were observed in multiple sowing experiments in Kyushu when temperatures

Table 4.1 The number of prefectures reporting heat-related disorders in crops and animals in Japan

Crops/Animals	Types of disorder	Climatic drivers	Consequences	Number of prefectures Mean (Maximum-Minimum) 2012–2016	2017–2021
Rice	Appearance of chalky grains	High temperature during the grain-filling period	Reduced appearance quality/price	24(17–29)	31(23–36)
Wheat & Barley	Late frost damages	Warmer winter advancing crop stage increasing chances of exposure to frost damage during the panicle development	Reduced quality/yield	3 (2–4)	4(1–6)
Legumes	Green stem disorder	Heat/drought during the reproductive phase	Reduced yield	4 (1–8)	2 (1–5)
Grape	Poor skin color	High temperature in summer	Reduced yield/ quality	13(6–18)	21(20–25)
Apple	Poor skin color	High temperatures from apple coloring to harvest (August–November)	Reduced quality	7(4–11)	8(6–10)
	Sunburn	High temperatures during the fruit growth (July–September)	Reduced quality/yield	6(6–7)	6(5–7)
Satsuma mandarin	Peel puffing	High temperature and excessive soil moisture during the fruit growth period	Reduced quality, storability	9(5–14)	11(9–13)
	Sunburn	High temperature and excessive soil moisture during the fruit growth period	Reduced yield and quality	4(2–6)	8(5–11)
Japanese pear	Poor germination	High temperature during the dormancy-germination period.	Reduced yield and quality	5 (2–8)	7(4–10)
Tomato	Poor fruit setting	High temperature during the growing season (May–September)	Reduced yield/ quality	19(13–27)	15(12–17)

(continued)

Table 4.1 (continued)

Crops/Animals	Types of disorder	Climatic drivers	Consequences	Number of prefectures Mean (Maximum-Minimum)	
				2012–2016	2017–2021
Strawberry	Delayed flower initiation	High temperature during flower initiation (July–November)	Delayed harvest, reduced yield/ quality	10 (6–15)	10(3–15)
Spinach	Poor growth due to heat	High temperature during the summer growing season	Reduced yield/ quality	6(4–7)	6 (4–8)
Onion	Poor growth including tip burn	High temperatures during the growing season	Reduced yield/ quality	11(8–14)	10(8–12)
Chrysanthemum	Advanced or delayed flowering	High temperature during the growing season (April–December)	Missed shipping opportunities	12(7–14)	19(17–23)
Dairy cattle	Poor milk yields	High temperatures in summer (June–October)	Reduced yield/ quality	15(13–18)	16(14–19)
	Mortality	High temperature and humidity in summer (July–September)	Reduced production	10(6–14)	14(12–17)
	Poor reproduction	High temperature in summer (June–October)	Reduced productivity	11(9–16)	9(7–11)
Beef cattle	Poor weight gain	High temperature in summer (July–September)	Reduced yield	10 (8–14)	8(7–8)
	Mortality	High temperature in summer (July–September)	Reduced productivity	6(4–8)	10 (9–12)
	Poor reproduction	High temperature in summer (July–September)	Reduced productivity	5(4–6)	4(3–6)
Swine	Poor weight gain	High temperature in summer (July–September)	Reduced yield	8(5–10)	7(5–9)
	Mortality	High temperature and humidity in summer (July–September)	Reduced productivity	6(5–10)	10(9–12)

(continued)

Table 4.1 (continued)

Crops/Animals	Types of disorder	Climatic drivers	Consequences	Number of prefectures Mean (Maximum-Minimum)	
				2012–2016	2017–2021
	Poor reproduction	High temperature in summer (July–September)	Reduced productivity	9(8–11)	7 (5–9)
Layer hen	Decline in spawning rate and egg weight	High temperature and humidity in summer (July–September)	Reduced productivity	12(10–14)	10(9–11)
	Mortality	High temperature in summer (July–September)	Reduced productivity	11(9–12)	14 (13–17)
Broiler chicken	Decline in growth rate	High temperature and humidity in summer (July–September)	Reduced productivity	9(6–10)	4(3–5)
	Mortality	High temperature in summer (July–September)	Reduced productivity	8(5–11)	12(10–14)

Data summarized from annuals reports by MAFF (2022) for the ten-year period from 2012 to 2021, divided into two five-year periods. The values represent the annual number of prefectures (out of 47) reporting the disorder averaged over each five-year period The maximum and minimum numbers of reporting prefectures for each period are shown in parentheses

exceeded 15 °C for 30 days after flowering (Fukushima 2012). In the Kanto region, Minoda et al. (2015) analyzed 45 years of data from Saitama Prefecture and found that high temperatures from November to April prior to the flowering stage reduced grain yield by limiting grain number. Additionally, warmer winters in the Kanto and Tokai regions have advanced the booting stage, increasing the risk of frost exposure during the critical developmental stage of early spike (ear) development (Nakazono 2013; Morisaki et al. 2018).

In soybeans, chamber experiments showed that single grain weight decreases almost linearly with higher temperatures over a wide range of growing-season temperatures, from about 20 to 30 °C (Kumagai and Sameshima 2014; Tacarindua et al. 2013). A recent study confirmed that higher temperatures during the late seed-filling period can lead to green stem disorders, where pods mature but stems and leaves remain green, reducing harvest efficiency and increasing the prevalence of stained grains (Yamazaki and Kawasaki 2023), necessitating the development of effective countermeasures.

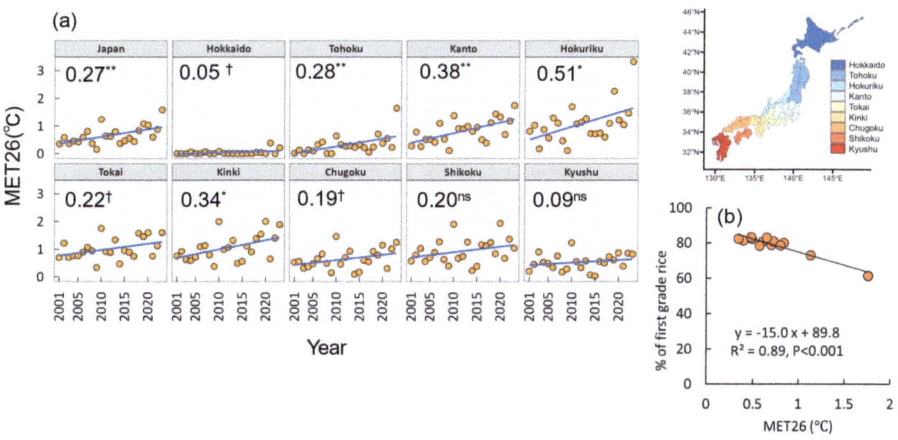

Fig. 4.1 Mean Excess Temperature (MET26) and first-grade rice. (**a**) MET26 from 2001 to 2023 in Japan's nine regions. (**b**) Relationship between the percentage of first-grade rice and MET26. The MET26 index represents the heat exposure during the 20-day period, critical for rice grain development, derived by averaging the excess daily air temperature above 26 °C. Values for each year were derived from national 1-km mesh meteorological data and heading dates reported by MAFF. When aggregated at the regional and national level, the values were weighted by the paddy area ratio in each grid. The value in each panel in (**a**) represents the regression slope (°C per decade) followed by its statistical significance. **, *, and † indicate that the slope is statistically significant at $P < 0.01$, $P < 0.05$, and $P < 0.1$, respectively, while n.s. denotes not significant

4.2.2 Fruits, Vegetables, and Flowers

Major fruit crops such as grapes, apples, and satsuma mandarins have been significantly impacted by rising summer temperatures in nearly all major producing prefectures. Common issues include poor skin coloration and sunburn. Warmer winters have also reduced productivity by hindering proper germination due to insufficient dormancy. These damages have been frequently reported since the early 2000s, but an increasing number of prefectures have recently reported problems with grapes and Japanese pears, indicating that these issues are becoming more widespread (Table 4.1).

For black grape cultivars such as Kyoho, Pione, and Suzuka, Sugiura et al. (2018) found that skin color ratings, measured on a scale from 0 (green) to 12 (black), decreased linearly with rising temperatures—by approximately 1.1 units per °C—contributing to the recent increase in heat-related quality losses. High summer temperatures also influence acidity in grapes (Sugiura et al. 2020), which has different implications depending on their intended use. For table grapes, reduced acidity can be advantageous, especially in cooler regions where overly sour grapes are undesirable. However, for wine grapes, a certain level of acidity is important. Rising temperatures necessitate careful management to maintain optimal acidity as grapes mature faster under warmer conditions.

In Japanese pears, higher winter temperatures have led to flowering disorders, particularly in lower-latitude regions (Tominaga et al. 2022). Like many deciduous fruit trees, Japanese pear trees require a specific amount of cold exposure, known as chilling hours, during their dormancy period to ensure proper flowering. For example, the popular pear variety "Kosui" needs 750 hours below 6 °C, or 1160 hours at 9 °C, to break dormancy. Recent warming trends have prolonged the dormancy period and delayed dormancy break. A lack of chilling during the dormancy period also causes flowering disorder, making pear production unstable (Tominaga et al. 2022).

Among vegetables and flowers, chrysanthemum cultivation has been increasingly affected by unexpected and unseasonable flowering times (Table 4.1). Market demand for specific flowers change dramatically, and unexpected flowering time due to warmer and variable weather has complicated the shipping timing, making the adjustment of flowering time difficult and risky.

4.2.3 *Livestock*

The impacts of recent climate change have also become evident across major livestock species (Table 4.1). Dairy cattle have been particularly vulnerable, experiencing reduced milk yields, poor reproduction rates, and increased mortality due to heat stress. Across major species, the number of prefectures reporting livestock mortality has risen by nearly 50% from the 2012–2016 period to 2017–2021, averaged across different species. This surge in mortality highlights the increasing intensity of heat stress and serious consequences for the productivity.

4.3 Projected Impacts and Adaptation Measures

Nationwide projections of crop yields or suitable cropping areas have so far been limited to a few major crops and fruit trees. However, various adaptation measures have already been planned or implemented. This section introduces recent quantitative evaluations of future climate impacts and adaptation strategies.

4.3.1 *Field Crops*

Rice has been extensively studied, with impact assessments regularly updated to incorporate new climate scenarios and crop simulation models (Ishigooka et al. 2021). Studies on the effects of climate change, including elevated atmospheric CO_2 on grain yields, have yielded consistent results since the early projections. Climate change is generally expected to benefit rice production in northern regions by reducing cold injury occurrence and enhancing photosynthesis, biomass, and yields due to elevated CO_2. Using the most recent climate data from NIES2020 (MIROC-6) (Ishizaki et al. 2022), we reran the rice simulation model reported by Ishigooka et al. (2021) under three emission scenarios and summarized the results in relation to global warming levels (GWL) from the preindustrial period (1850–1900).

National bulk rice production, assuming the current planted area remains constant throughout the twenty-first century, is projected to remain relatively stable up to a GWL of 3 °C (Fig. 4.2a), beyond which production falls sharply as GWL increases. Regionally, however, significant impacts are projected at lower GWLs: the proportion of areas experiencing yield losses of 10% or more is projected to rise sharply from 2 °C GWL, reaching 80% at around 4 °C GWL (Fig. 4.2b). Furthermore, high-quality rice production with chalky grain percentages below 30% is projected to be more severely affected by global warming. Negative impacts are expected even at a small GWL increase from the current level of 1.1 °C, with a 10% decline at 2 °C GWL and a 20–30% decline at 3 °C GWL (Fig. 4.2c).

Adaptation efforts are underway. Regional shifts in planting time to avoid heat exposure during critical stages have been implemented over the past 20 years, particularly in central and western Japan. The aim is to push back the critical 20-day period after the heading stage to avoid mid-summer heat. However, the benefits of later planting have been reduced or nullified in recent years due to rising temperatures, which have shortened the time from transplanting to heading, leading to little or no delay in the heading date, except in Kyushu. Ishigooka et al. (2021) noted that shifting planting time is an effective strategy to secure yields, though it involves trade-offs between quantity and quality. Later planting generally improves quality but can reduce quantity, while earlier planting can increase yields but compromise quality. The direction of change varies by region; currently, warm regions benefit from earlier planting, while later planting is favored in higher-latitude or higher-altitude areas (Ishigooka et al. 2021).

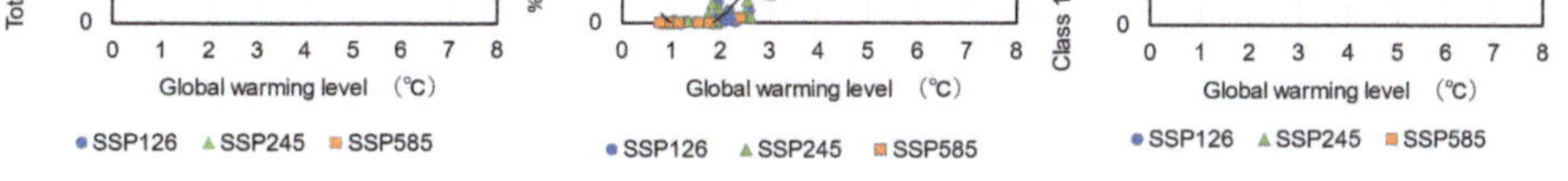

Fig. 4.2 Nationally aggregated projected impacts on rice production under three emission scenarios (SSP1-2.6: low, SSP2-4.5: medium, SSP5-8.5: very high) as a function of global warming levels from the preindustrial level. (**a**) Total production relative to that of 2001–2020; (**b**) percentage of area with a yield loss of 10% or more; (**c**) High-quality grain yield (yield with less than 30% chalky grains) relative to that of 2001–2020. (Recalculated from Ishigooka et al. (2021))

Heat-tolerant cultivars, which can maintain high appearance quality under elevated temperatures during grain-filling, are considered a key adaptation strategy (Ishimaru et al. 2016). In 2017, heat tolerance levels for cultivars were standardized by assigning check cultivars to different tolerance ranks. Currently, these ranks range from "weak" to "tolerant" across five levels (Sato 2017).

Recently, Wakatsuki et al. (2024) developed an empirical model based on approximately 1300 field observations from across Japan, quantifying the effects of key climatic factors such as MET26, relative humidity, solar radiation, and cultivar heat tolerance ranks. The analysis revealed that at an MET26 of 2 °C (equivalent to 28 °C), the percentage of chalky grains was 45% for cultivars with "weak" heat tolerance, 24% for "medium" cultivars, and 11% for "tolerant" cultivars. As temperatures rise to around 30 °C (MET26 of 4 °C), the percentage of chalky grains increases, even in "moderately tolerant" and "tolerant" cultivars. However, the benefits of heat-tolerant cultivars become more pronounced at this higher temperature. For instance, at an MET26 of 4 °C, a one-rank improvement in heat tolerance from "medium" to "moderately tolerant" could reduce the percentage of chalky grains by over 10 percentage points.

In wheat, adaptation strategies focus on addressing two major challenges associated with the accelerated development, such as early floral development caused by warmer winters. First, earlier onset of floral development increases the risk of frost damage, which typically occurs between February and April. Second, accelerated development shortens the time required for producing sufficient spikes and grain numbers. A potential solution to mitigate these negative effects is to use cultivars that maintain an extended growth period, even in warmer winter conditions. Vernalization requirements (chilling requirements) and photoperiod sensitivity are key traits influencing the earliness of wheat cultivars. Recent studies have shown that introducing genes to enhance photoperiod sensitivity or vernalization requirements can extend the period before floral development under high temperatures during early growth stages (Matsuyama et al. 2015, 2017, 2024). Sawada et al. (2019) demonstrated that increasing vernalization requirements successfully prolonged vegetative growth and boosted biomass accumulation under warm conditions after sowing. Adjusting sowing times in conjunction with these genotypes could be an effective adaptive response to warming. However, to fully realize the potential of these cultivars, management practices—such as fertilizer timing and application rates—must also be optimized based on specific cultivar–environment combinations (Kawakita et al. 2021).

Another method to delay the early onset of spike formation is a traditional Japanese practice known as ground rolling. Initially used during the early vegetative stage in winter to control excessive vegetative growth, recent studies have shown that ground rolling is also effective in slowing down accelerated plant development during warm winters. Experimental evidence from the Tokai region demonstrated that ground rolling at the one- or two-leaf stage can successfully delay floral initiation and development by about 5 days (Mizumoto et al. 2022). This delay reduces the risk of frost exposure, which, when avoided, can significantly increase grain yields (Mizumoto et al. 2023). However, the effectiveness of this method has yet to be tested across different regions and climates in Japan.

4.3.2 Fruits and Vegetables

Perennial crops, such as fruit trees, face significant challenges from climate change due to their reliance on specific climatic conditions, which restrict their growth to particular regions (Sugiura 2019). Early studies have already shown that a large proportion of current growing regions for major temperate fruits in Japan, like apples and grapes, will become climatically unsuitable. As a result, these suitable growing regions are projected to shift to higher latitudes and altitudes (Sugiura et al. 2014, 2019b; 2024). A more recent study using the latest climate scenarios from the IPCC AR6 (IPCC 2021) highlights a stark contrast between the outcomes under different emission scenarios. Under a very high emission scenario (SSP5-8.5), all current areas suitable for satsuma mandarin cultivation will become unsuitable, while 80% will remain viable under a low-emission scenario (SSP1-2.6), underscoring the critical importance of limiting global warming (Sugiura et al. 2024).

Adaptation technologies vary depending on the crop and the specific climatic disorders it faces, but one straightforward solution is to avoid heat exposure. Sunburn can be effectively mitigated using shading materials to block sunlight, particularly from the sides, or by covering individual fruits with paper bags (Sugiura 2019). For black grapes, poor coloration caused by summer heat can be minimized by advancing the fruit pigmentation stage to cooler periods in the year through the use of plastic covers with side films in the spring (Sugiura et al. 2019b). This method helps maintain skin color even in southern Japan until the end of the century under RCP4.5, where warming impacts on coloration are projected to be significant. Another approach to enhance coloration in black grapes is the application of abscisic acid solution at the start of pigmentation, which increases anthocyanin content in the skin and improves coloration, even under high temperatures (Sugiura et al. 2019a).

Replacing cultivars is a more difficult adaptation strategy for fruit farmers compared to annual crops, as fruit trees remain commercially viable for many years, and switching to new cultivars represents a long-term investment. However, some cultivars have been developed to reduce the negative impacts of warming. For example, certain apple and black grape cultivars have been bred to improve coloration under high temperatures, while yellow cultivars, which are not affected by poor coloration, are also considered a viable adaptive option (Sugiura 2019). A superior black grape cultivar is projected to retain its dark color under an intermediate emission scenario (RCP4.5) until the end of the century (Sugiura et al. 2019b).

Introducing subtropical fruit species presents an adaptation opportunity by leveraging warmer conditions in Japan, potentially replacing crops like satsuma mandarin. For instance, tankan, a subtropical citrus, has been identified as a suitable replacement for satsuma mandarin in low-lying coastal regions where the climatic conditions for the latter are expected to deteriorate (Sugiura et al. 2014). A recent study compared the thermal suitability of satsuma mandarin and avocado under three emission scenarios for the twenty-first century (Sugiura et al. 2024). Avocado-suitable growing regions are projected to increase from around 20% during the baseline period to 70–90% by mid-century, depending on the scenario (Fig. 4.3). Under a very high emission scenario (SSP5-8.5), all areas would become suitable for avocado but too warm for satsuma mandarin by the late century. While these simulations only consider temperature conditions, they indicate a dramatic shift in thermal suitability for fruit trees by the mid-to-late century under a high-emission scenario.

Many vegetables, especially leafy ones, have shorter life cycles, which may allow them to avoid excessive heat by adjusting planting times and choosing suitable growing regions. Another key feature of vegetables is their freshness, making them unsuitable for long-term storage. Year-round vegetable production is made possible through regional collaboration, utilizing Japan's wide range of latitudes and altitudes, known as the relay production regime (Sato 2022). For example, lettuce, which thrives in cooler climates, is supplied to the Tokyo market from high-altitude regions like Nagano and Gunma in the summer (June–September), from Ibaraki in the fall, and from warmer regions such as Shizuoka and Hyogo in mid-winter (Sato 2022).

Climate change is likely to affect the growth duration of vegetables, disrupting this coordination and potentially leading to supply issues. Ohara and Okada (2020a)

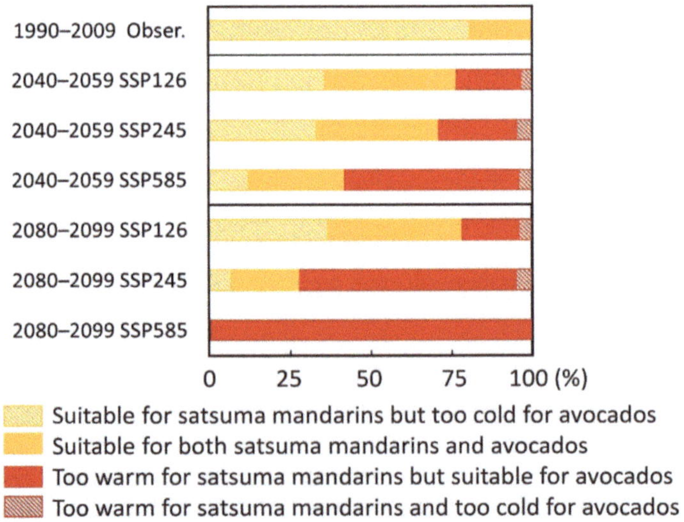

Fig. 4.3 Subsequent changes in suitable locations for satsuma mandarins in Japan under observed temperatures for 1990–2009. Each change was projected by using temperature based on MRI-ESM2-0 under SSP1-2.6 (SSP126), SSP2-4.5 (SSP245), and SSP5-8.5 (SSP585) for 2040–2059 and 2080–2099. The area of suitable locations for satsuma mandarins for 1990–2009 is considered 100%. (Source: Sugiura et al. (2024))

developed a model to estimate climatically determined growth durations for five vegetable crops—cabbage, daikon radish, lettuce, and Chinese cabbage—under field conditions. Using this model, they projected changes in monthly harvest areas during three future periods in the twenty-first century under the RCP8.5 scenario (high emissions), assuming planting times remain unchanged. They also estimated the suitable growing seasons across four periods using projected temperatures (Ohara and Okada 2020b). Significant impacts on harvest areas are expected as early as 2026–2045, particularly for crops like lettuce, cabbage, Chinese cabbage, and daikon radish. For example, harvests of these crops are projected to increase in November and December but decline sharply in February and March due to shortened growth durations due to temperature rise in cool regions. Carrot harvests are predicted to show more fluctuation, as both summer-autumn harvests in cool regions and winter-spring harvests in warm regions will likely shorten, leading to frequent supply gaps.

By the late century (2081–2100), suitable growing regions for crops like Chinese cabbage, carrots, and daikon radish are expected to decline. However, projected temperatures are not expected to severely hamper climatic suitability in the 2041–2060 period for any of the five vegetable species studied despite an average temperature rise of 2.7 °C. It should be noted, however, that the projections in this study do not account for potential production losses due to extreme heat or drought, which are expected to increase in frequency and severity as global temperatures rise. More frequent heat-related disorders could disrupt the stability of vegetable supply, and these projections highlight the need for careful adjustments in planting times across different growing regions.

4.4 Challenges and Conclusions

This brief review highlights the urgent need for action in response to climate change. It confirms that climate change impacts in Japan are not a distant concern but a present reality, with the potential to become more widespread and severe as global warming levels (GWLs) increase. However, many of the projections in simulation studies tend to overlook the immediate effects of extreme weather events, which are already disrupting food supply chains. It is essential to track environmental changes in real-time and promptly assess both the impacts and associated risks. Detailed regional monitoring, combined with fine-scale weather data, is critical for improving predictive accuracy. Strengthened efforts to establish comprehensive monitoring and modeling systems are vital for understanding the changing environmental conditions and their impacts on food supply, quality, and access.

A recent global review of climate change adaptation studies using crop simulations (Wakatsuki et al. 2023) revealed that current research explores only a fraction of the adaptation strategies available for the future. The study highlighted the need for simulations to consider a wider range of variables when evaluating climate impacts and adaptation options. As noted by Ishigooka et al. (2021), adaptation strategies can differ significantly based on whether the focus is on quantity or quality outcomes. Nearly all adaptation measures require additional costs, labor, and resources, factors often overlooked in current studies. Moreover, it is essential to account for the environmental dimensions of adaptive responses to ensure their sustainability. For example, introducing new crop species may take advantage of changing climate conditions, but the potential environmental consequences, such as unsustainable exploitation of regional resources, must be carefully evaluated to avoid maladaptation.

Effective adaptation strategies must be tailored to the specific characteristics of each region, requiring a deep understanding of how climate change impacts vary across locations. Adaptation measures should be optimized at the regional level. Furthermore, it is important to recognize that the actors involved in adaptation efforts vary depending on the context. While many adaptation technologies are farm-based, some transformative measures, such as those seen in vegetable relay production systems, require cross-regional coordination. Identifying who leads these transboundary adaptation efforts requires careful socioeconomic consideration, a factor often underexplored in current assessments. To facilitate these actions, scientific evaluation of adaptation effectiveness and feasibility must be enhanced. This includes developing methods for quantitatively assessing the real-world impacts of adaptation measures and continuously monitoring their impacts, success, and practicality.

References

Bezner Kerr R, Hasegawa T, Lasco R et al (2022) In: Pörtner HO, Roberts DC, Tignor M et al (eds) Food, fibre, and other ecosystem products. Cambridge University Press (CUP), Cambridge/New York, pp 713–906. https://doi.org/10.1017/9781009325844.007

Fukushima A (2012) Effects of global warming on growth and yield of wheat in Japan: data analysis of field experiments. Jpn J Crop Sci 81:83–88. https://doi.org/10.1626/jcs.81.83

Hasegawa T, Wakatsuki H, Yoshimoto M et al (2024) Trends in ripening temperature and percentage of first-grade rice in Japan from 2001 to 2023, and the effectiveness of heat-tolerant cultivars. Abstracts of the Meeting of the Crop Science Society Japan, 258:49. https://doi.org/10.14829/jcsproc.258.0_49

Hokkaido Sapporo District Meteorological Observatory (2010) Hokkaido Agricultural Weather Disaster Bulletin [Online]. Available: https://www.data.jma.go.jp/sapporo/bosai/past_kishou/pdf/nougyou_saigai_h22_01.pdf. Accessed 19 Sept 2024

Horie T (2019) Global warming and rice production in Asia: Modeling, impact prediction and adaptation. Proc Jpn Acad, Ser B 95:211–245. https://doi.org/10.2183/pjab.95.016

IPCC (2021) Summary for policymakers. In: Masson-Delmotte V, Zhai P, Pirani A et al (eds) Climate change 2021: The physical science basis. Contribution of working group I to the sixth assessment report of the intergovernmental panel on climate change. Cambridge University Press, Cambridge/New York, pp 3-33. https://doi.org/10.1017/9781009325844.001

Ishigooka Y, Hasegawa T, Kuwagata T et al (2021) Revision of estimates of climate change impacts on rice yield and quality in Japan by considering the combined effects of temperature and CO_2 concentration. J Agric Meteorol 77:139–149. https://doi.org/10.2480/agrmet.D-20-00038

Ishimaru T, Hirabayashi H, Sasaki K et al (2016) Breeding efforts to mitigate damage by heat stress to spikelet sterility and grain quality. Plant Prod Sci 19:12–21. https://doi.org/10.1080/1343943X.2015.1128113

Ishizaki NN, Shiogama H, Hanasaki N et al (2022) Development of CMIP6-based climate scenarios for japan using statistical method and their applicability to impact studies. Earth Space Sci 9:1–12. https://doi.org/10.1029/2022EA002451

Kawakita S, Ishikawa N, Takahashi H et al (2021) Interactions of cultivar, sowing date, and growing environment differentially alter wheat phenology under climate warming. Agron J 113:4982–4992. https://doi.org/10.1002/agj2.20911

Kumagai E, Sameshima R (2014) Genotypic differences in soybean yield responses to increasing temperature in a cool climate are related to maturity group. Agric For Meteorol 198:265–272. https://doi.org/10.1016/j.agrformet.2014.08.016

MAFF (2022) Global Warming Impact Study Report [Online]. Tokyo. Available: https://www.maff.go.jp/j/seisan/kankyo/ondanka/report.html. Accessed 19 Sept 2024

Matsuyama H, Fujita M, Masako S et al (2015) Growth and yield properties of near-isogenic wheat lines carrying different photoperiodic response genes. Plant Prod Sci 18:57–68. https://doi.org/10.1626/pps.18.57

Matsuyama H, Seki M, Shimazaki Y et al (2017) Differences in growth properties between near-isogenic wheat lines carrying different types of vrn-d1 in the genetic background of two early cultivars. Jpn J Crop Sci 86:311–318. https://doi.org/10.1626/jcs.86.311

Matsuyama H, Sawada H, Fujita M et al (2024) Growth and yield-related traits of near-isogenic wheat lines carrying different alleles at the Vrn-D1 locus. Plant Prod Sci 00:1–11. https://doi.org/10.1080/1343943X.2024.2363547

Minoda T, Kobayashi K, Hirasawa T (2015) Effects of climatic changes on growth, yield, and yield components of winter wheat cultivar'Norin 61'across 45 years: Analysis of an experimental record in an upland test field of Saitama Prefecture, Japan. Jpn J Crop Sci 84:285–294. https://doi.org/10.1626/jcs.84.285

Mizumoto A, Tanio M, Watanabe K et al (2022) Ground rolling delays apical development and reduces frost injury in early-sown spring wheat. Plant Prod Sci 25:434–439. https://doi.org/10.1080/1343943X.2022.2136097

Mizumoto A, Tanio M, Nakazono K et al (2023) Early ground rolling is highly effective in delaying spikelet initiation in early-sown spring wheat. Plant Prod Sci 26:402–410. https://doi.org/10.1080/1343943X.2023.2251183

MOE (2020) Climate Change Impact Assessment Report, Japan [Online]. Tokyo. Available: https://www.env.go.jp/content/000120415.pdf. Accessed 19 Sept 2024

Morisaki K, Yamashita Y, Hiraiwa K et al (2018) Frost damages on wheat harved in 2016 and 2017 in Aichi Prefecture, Japan, and factors associated with them. Res Bull Aichi Agric Res Center 66:63–66

Nakazono K. (2013). Modeling the phenological development of wheat for optimal planning and harvest. Ph.D. Agricultural Sci Ph.D., Kyoto University

Nishio Z, Ito M, Tabiki T et al (2011) II. Analysis of reduction in yield of wheat (Triticum aestivum L.) due to high summer temperatures in Hokkaido in 2010. Miscellaneous Publication of the NARO Hokkaido Agricultural Research Center 69:1–7. https://www.naro.go.jp/publicity_report/publication/archive/files/cryo_siryo_02_komugi.pdf

Ohara G, Okada K (2020a) An attempt to identify the production situations for major open-field vegetables. Clim Biosphere 20:67–75. https://doi.org/10.2480/cib.J-20-047

Ohara G, Okada K (2020b) An impact evaluation of global warming on market arrival based on the identification of production situations for major open-field vegetables in Japan. Clim Biosphere 20:107–116. https://doi.org/10.2480/cib.J-20-046
Sato H (2017) Selection of standard varieties for ripening performance under high temperatures in paddy rice across Japan, excluding Hokkaido. Availabile: https://www.naro.go.jp/project/results/4th_laboratory/nics/2017/17_038.html. Accessed 19 Sept 2024
Sato F (2022) Challenges of relay shipping of open field vegetables and the significance of growth forecasting. J Agric Soc Japan 1686:6–21. https://www.dainihon-noukai.or.jp/library/61bc776bb57d7bf14fe411a4/63450ef44ed620a7226146b4.pdf
Sawada H, Matsuyama H, Matsunaka H et al (2019) Evaluation of dry matter production and yield in early-sown wheat using near-isogenic lines for the vernalization locus Vrn-D1. Plant Prod Sci 22:275–284. https://doi.org/10.1080/1343943X.2018.1563495
Sugiura T (2019) Three climate change adaptation strategies for fruit production. In: Shirato Y and Hasebe A (eds) Climate smart agriculture for the small-scale farmers in the Asian and Pacific Region, National Agriculture and Food Research Organization, Tsukuba, Japan. pp 277–292. https://www.naro.go.jp/english/laboratory/niaes/fftc-marco_book2019.pdf
Sugiura T, Sumida H, Yokoyama S et al (2012) Overview of recent effects of global warming on agricultural production in Japan. Jpn Agric Res Quart 46:7–13. https://doi.org/10.6090/jarq.46.7
Sugiura T, Sakamoto D, Koshita Y et al (2014) Predicted changes in locations suitable for Tankan cultivation due to global warming in Japan. J Jpn Soc Horticult Sci 83:117–121. https://doi.org/10.2503/jjshs1.CH-094
Sugiura T, Shiraishi M, Konno S et al (2018) Prediction of skin coloration of grape berries from air temperature. Horticult J 87:18–25. https://doi.org/10.2503/hortj.OKD-061
Sugiura H, Azuma A, Nishimura R et al (2019a) Effect of spray treatment with a new liquid formulation containing Abscisic Acid (S-ABA) on the fruit quality of 'Pione' grape at the onset of pigmentation. Horticult Res (Japan) 18:81–87. https://doi.org/10.2503/hrj.18.81
Sugiura T, Shiraishi M, Konno S et al (2019b) Assessment of deterioration in skin color of table grape berries due to climate change and effects of two adaptation measures. J Agric Meteorol 75:67–75. https://doi.org/10.2480/agrmet.D-18-00032
Sugiura T, Sato A, Shiraishi M et al (2020) Prediction of acid concentration in wine and table grape berries from air temperature. The Horticulture Journal 89:208–215. https://doi.org/10.2503/hortj.UTD-141
Sugiura T, Sugiura H, Konno S et al (2024) Assessing the expansion of suitable locations for avocado cultivation due to climate change in Japan and its suitability as a substitute for satsuma mandarins. J Agric Meteorol, advpub 80:111. https://doi.org/10.2480/agrmet.D-24-00017
Tacarindua CRP, Shiraiwa T, Homma K et al (2013) The effects of increased temperature on crop growth and yield of soybean grown in a temperature gradient chamber. Field Crop Res 154:74–81. https://doi.org/10.1016/j.fcr.2013.07.021
Tominaga A, Ito A, Sugiura T et al (2022) How is global warming affecting fruit tree blooming? "Flowering (Dormancy) Disorder" in Japanese Pear (Pyrus pyrifolia) as a case study. Front Plant Sci 12:1–17. https://doi.org/10.3389/fpls.2021.787638
Wakatsuki H, Ju H, Nelson GC et al (2023) Research trends and gaps in climate change impacts and adaptation potentials in major crops. Curr Opin Environ Sustain 60:101249–101249. https://doi.org/10.1016/j.cosust.2022.101249
Wakatsuki H, Takimoto T, Ishigooka Y et al (2024) Effectiveness of heat tolerance rice cultivars in preserving grain appearance quality under high temperatures in Japan—a meta-analysis. Field Crop Res 310:109303–109303. https://doi.org/10.1016/j.fcr.2024.109303
Yamazaki R, Kawasaki Y (2023) Effect of high temperature during the late seed filling period on green stem disorder in soybean. Field Crop Res 302:109092–109092. https://doi.org/10.1016/j.fcr.2023.109092

Chapter 5
Learn and Predict from Data: Statistical Analysis of Climate Change Impacts on Crop Production

Gen Sakurai

Abstract This chapter presents the results of crop yield predictions under future climate conditions in Japan. For nine crops, including soybeans, rice, and wheat, the author developed models using statistical methods (generalized additive model) based on historical meteorological and crop yield data, and predicted future yields at the municipal level across Japan. While process-based crop models are commonly used to predict crop yields, this chapter introduces a method that leverages crop yield data spanning both low and high latitudes to make predictions while also addressing the challenges of extrapolation. The chapter also explains the statistical methods used to quantify the potential effects of adaptation measures against climate change.

Keywords Crop Yield · Statistical model · Generalized additive model · Uncertainty · Extrapolation

5.1 Introduction

Plants are one of the entities most affected by global warming. The life activities of plants are carried out by a vast combination of chemical reactions, and as long as they are chemical reactions, each reaction fundamentally increases exponentially with an increase in temperature. On the other hand, each reaction catalyzed by enzymes, which are proteins, decreases in reaction speed at high temperatures because the proteins become inactive (Somero 1978). Plants, which cannot actively control their own body temperature, grow while strongly affected by daily temperature changes. In fact, the growth rate and photosynthesis rate of crops are clearly functions of temperature (Jones 2013). It goes without saying that crops, in

G. Sakurai (✉)
National Agriculture and Food Research Organization, Tsukuba, Japan
e-mail: sakuraigen@naro.affrc.go.jp

N. Mimura, S. Takewaka (eds.), *Climate Change Impacts and Adaptation Strategies in Japan*, https://doi.org/10.1007/978-981-96-2436-2_5

particular, are directly connected to human society and are directly affected by changes in the external environment, making the effects of global warming more likely to manifest.

5.2 General Climate Change Impact Assessment in the Agricultural Field

In studies evaluating the productivity of crops under future climate change, an approach based on process-based models is often taken. In process-based models for crops, photosynthesis rates and growth rates are calculated from daily average temperature, solar radiation, humidity, etc., and the process of plants accumulating biomass and growing is mathematically modeled. Many studies have reported the relationship between past weather and crop productivity (Pörtner et al. 2022), and the impact of climate change on crop productivity has been verified using process-based models. In particular, research on major crops such as corn, soybeans, wheat, and rice is active, and there is concern about a decrease in crop productivity near the equator (Pörtner et al. 2022). Also, many model comparison studies have verified the characteristics of these model predictions (for example, Müller et al. 2021). In Japan, the impact of climate change on crop yield has been studied using a process-based model for rice (Masutomi et al. 2009; Ishigooka et al. 2020, 2021), and while there is a possibility that crop yield itself may increase due to an increase in temperature, there is concern that quality may decrease.

One advantage of using a process-based model is its detail. Process-based models are developed by combining not only field data but also laboratory data under various temperature conditions, making them suitable for verifying the impact of future warming (Asseng et al. 2015). However, their complex model structures increase the number of parameters required for each crop, limiting the types of crops for which process-based crop models have been developed. Also, because numerous equations are inherent, the uncertainty of each equation affects the uncertainty of yield predictions in a chain reaction. Therefore, when comparing the results of multiple process-based models, the prediction results usually vary to some extent depending on the model (Rozenzweig et al. 2014; Martre et al. 2015; Müller et al. 2019; Schewe et al. 2019).

5.3 Predicting the Impact of Climate Change Using Statistical Methods

On the other hand, the approach using statistical models is a method of abstractly modeling and analyzing the relationship from statistical information about crop yields and past weather data. Because it simplifies the relationship between crop yield and weather factors, it has the advantage of being applicable to any type of crop as long as there is data. In fact, many studies have been conducted using the statistical model approach in climate change impact assessment research, not only for major crops such as corn, rice, soybeans, and wheat, but also for other crops such as coffee beans (for example, Schlenker and Roberts 2009; Lobell et al. 2011; Hawkins et al. 2013; Kath et al. 2020). However, data-driven methods are rarely used to predict the impact of future climate change. This is because the climate factors predicted by Global Climate Models (GCMs), especially temperature, sometimes exceed the range of the training data, increasing uncertainty in terms of prediction robustness. In other words, there is always the problem of extrapolation (although this problem must be considered for the training data used to construct each equation in process-based models, it is rarely discussed).

To solve this extrapolation problem, it is necessary that the range of prediction data does not exceed the range of training data. For this purpose, it is necessary to analyze using data that spans a wide range of latitudes from low to high in the training data. Japan is a long, narrow island nation that belongs to a wide range of climate zones, from the subtropical climate of Okinawa to the subarctic climate of Hokkaido. Also, the Ministry of Agriculture, Forestry and Fisheries has been organizing long-term crop yield statistical data for various crops for each municipality. Using these data, it is possible to analyze the relationship between weather and crop yield for various types of crops, and the wide temperature range covered by the data has the potential to clear the problem of extrapolation in prediction. Data-driven research that takes advantage of the abundance of crop yield data in Japan has a significant advantage and has the potential to make unified predictions of the impact of climate change on dozens of types of crops.

In the S-18 project, we are leveraging the advantages of data on these crops, targeting over 50 crop species, and using uniform statistical methods to analyze the relationship between weather and crop yield uniformly, and conducting analysis on the productivity of Japan's crops in the future. In this chapter, I will illustrate some of the research results on rice, wheat, and beans, which are important items for Japan, to give an overview of the project research. Also, as shown in the analysis results of this study below, we can discuss the potential for adaptation by using statistical models appropriately. Through this chapter, I hope to understand the future potential of Japanese crops and deepen our understanding of the possibilities of future climate change impact prediction research using statistical models.

5.4 Method: Data

The crops targeted in this chapter are rice, wheat, two-row barley, six-row barley, naked barley, soybeans, green beans, fava beans, and azuki beans. The annual crop yield data by municipality from 1993 to 2020 was obtained from the Ministry of Agriculture, Forestry and Fisheries' crop statistics (https://www.estat.go.jp). These statistics show the crop production and harvested area by municipality. The yield obtained by dividing the crop production by the harvested area was used as the dependent variable in the statistical model. Past weather data was used from the Agro-Meteorological Grid Square Data, NARO (Ohno et al. 2016). This dataset is composed of weather data with a spatial resolution of about 1 km, interpolated based on AMeDAS observation data. When averaging the weather data uniformly in each municipality, weather values from nonagricultural areas such as forests and urban areas are mixed in, so only the temperature and solar radiation in agricultural land were weighted and averaged according to their area to represent the average weather of each municipality. The boundary data of the municipalities and land use information (agricultural land, forest, building area, etc.) were obtained from the National Land Numerical Information Download Service of the Ministry of Land, Infrastructure, Transport and Tourism (https://nlftp.mlit.go.jp/index.html). For the crop calendar, the Ministry of Agriculture, Forestry and Fisheries' published characteristics table of recommended varieties of rice and wheat was used. For beans, the Ministry of Agriculture, Forestry and Fisheries' published crop statistics were used.

For future weather data, we used NIES2020 (Ishigooka et al. 2021) from the National Institute for Environmental Studies. This dataset is a bias-corrected version of five global climate models (GCM) (ACESS-CM2, IPSL-CM6A-LR, MIROC6, MPI-ESM1-2-HR, MRI-ESM2-0) from CMIP6 at a spatial resolution of 1 km. Future land use scenarios were obtained from the SSP scenario constructed by Yoshikawa et al. (2022). This scenario provides future land use scenarios for each 1 km grid cell, similar to the National Land Numerical Information Download Service.

5.5 Method: Statistical Model

The following generalized additive model was used as a statistical model for past crop yields and meteorological factors. In the generalized additive model, a spline function was used as a nonlinear function.

$$\log\left(y_{i,t}\right) = f_{T_1}\left(T_{1,i,t}\right) + f_{T_2}\left(T_{2,i,t}\right) + f_S\left(S_{i,t}\right) + \alpha t + b_j + \epsilon_{i,t}, \tag{5.1}$$

$$\log\left(y_{i,t}\right) = h_{T_1}\left(T_{1,i,t}, C_i\right) + h_{T_2}\left(T_{2,i,t}, C_i\right) + h_S\left(S_{i,t}, C_i\right) + \alpha t + b_j + \epsilon_{i,t}, \tag{5.2}$$

Here, $y_{i,t}$ is the crop yield in year t of municipality i, $T_{1,i,t}$ and $T_{2,i,t}$ represent the average temperature during the early and late stages of the growing period, and $S_{i,t}$ represents the average solar radiation. Here, the potential crop yield b_j is in prefecture j, which is assumed to change with each. Also, crop yield has a trend, and that trend is represented by the term . $\epsilon_{i,t}$ is assumed to follow a normal distribution. f_{T_1} and f_{T_2} , f_S represent spline functions for average temperature (early and late stages) and solar radiation, respectively. The reason why the potential crop yield is calculated for each prefecture (assuming b_j) is that in Japan, recommendations for farming methods are generally made at the prefecture level. The term C_i in Eq. (5.2) represents the average weather of all years in municipality i (i.e., the average temperature of that region). Therefore, the function h_{T_1} and h_{T_2} and h_S each have two variables, allowing for different yield responses to average temperature and solar radiation for each average weather condition. The differences in what the statistical models of Eqs. (5.1) and (5.2) imply will be explained in Sect. 5.9 below. The analysis was performed using R (R Core Team 2024). The generalized additive model was performed using the gam function of the mgcv package. In the gam function, the settings other than the formula used the default settings.

5.6 Method: Future Prediction

For the functions of the above Eqs. (5.1) and (5.2), future weather values from each global climate model (Ishizaki et al. 2022) were substituted, and future yield predictions were made. For the year t in the term αt, it was set to 2020 to make the factors of yield increase and decrease only by weather factors. Here, for the municipalities i where future predictions are made, only municipalities that meet the following three conditions for future weather values were evaluated to consider the uncertainty of predictions due to extrapolation.

$$\Sigma_{t\in\tau}\Sigma_{g\in\gamma}1_{[T_{1\min},T_{1\max}]}\left(T'_{1,i,t,g}\right)<0.7N, \tag{5.5}$$

$$\Sigma_{t\in\tau}\Sigma_{g\in\gamma}1_{[T_{2\min},T_{2\max}]}\left(T'_{2,i,t,g}\right)<0.7N, \tag{5.6}$$

$$\Sigma_{t\in\tau}\Sigma_{g\in\gamma}1_{[S_{\min},S_{\max}]}\left(S'_{i,t,g}\right)<0.7N, \tag{5.7}$$

here $T'_{1,i,t,g}$, $T'_{2,i,t,g}$, and $S'_{i,t,g}$ is the average temperature during the first half of the growing season and the second half of the growing season and average solar radiation during the growing season predicted by the global climate model g, τ is thetarget year in the future, γ is the target global climate model, $1_{[T_{1\min},T_{1\max}]}\left(T'_{1,i,t,g}\right)$ and $1_{[T_{2\min},T_{2\max}]}\left(T'_{2,i,t,g}\right)$, $1_{[S_{\min},S_{\max}]}\left(S'_{i,t,g}\right)$ are indicator functions, and $T_{1\min}$ and $T_{2\min}$, $S_{\min}$are the lowest average temperatures and average solar radiation in the training data, $T_{1\max}$ and $T_{2\max}$, $S_{\max}$ represent the highest average temperatures and average solar radiation in the training data. N is the product of the global climate model and the target number of years. In other words, if the extrapolation ratio of future meteorological values used in future yield predictions does not exceed 30%, the degree of extrapolation is arbitrarily determined to be low. In addition, in the case of prediction by the additive model, it is possible to estimate the uncertainty of the estimated parameters, so in this analysis, the probability of yield reduction was calculated considering the error of estimation by the additive model and the variance of prediction by the global climate model, and the probability value was also presented.

5.7 Results

In the S-18 project mentioned in the previous section, predictions for more than 50 crops, including the nine crops introduced in this chapter, have been calculated for the years 2020 to 2100 for multiple socioeconomic scenarios (SSP). However, in this chapter, I have only illustrated the results for SSP5–8.5 from 2041 to 2060. Figure 5.1 shows a map of the predicted changes in crop yields using Eq. (5.1) for all nine crops. For azuki beans, naked barley, rice, soybeans, two-row barley, and wheat, yields are predicted to decrease in the low-altitude areas of western Japan and the Kanto region and increase in the Tohoku region. Also, for green beans and naked barley, a slight decrease is predicted in areas north of the Tohoku region. In particular, naked barley and soybeans are predicted to have a larger decrease in yield compared to other crops. On the other hand, green beans and six-row barley are predicted to increase in yield nationwide. Figure 5.2 shows the probability of decreases in yields predicted by Eq. (5.1) for all nine crops. Looking at the average increase or decrease (Fig. 5.1), a large decrease in yield for naked barley was predicted, but the uncertainty was high, and the probability of a decrease was not particularly high, ranging from 50% to 80% (the closer to 50%, the higher the uncertainty and the lower the accuracy of the prediction). On the other hand, while the degree of decrease in yields in western Japan and the Kanto region for azuki beans is not high, the accuracy is high, and there were many areas where a decrease in yields of 90% probability or more was predicted. Areas in western Japan and the Kanto region were also predicted to have decreases in yields of 90% probability or more for soybeans. On the other hand, areas with a high probability of increases in yields (i.e., areas with a near 0% probability of decreases in yields) were predicted to be nationwide for green beans and six-row barley, and the Tohoku region for wheat.

5.8 Advantages and Challenges of Statistical Models

The advantage of using statistical models is that they can output predictions as probability values. In the case of the additive model used this time, if the number of data is small and the error of the yield response curve to weather is large, the error of the predicted yield will also be large. Therefore, it can explicitly showcase where, even if a large decrease in yield is predicted on average for the rate of change in yield, as in the case of naked barley, the accuracy is not particularly high. In predictions

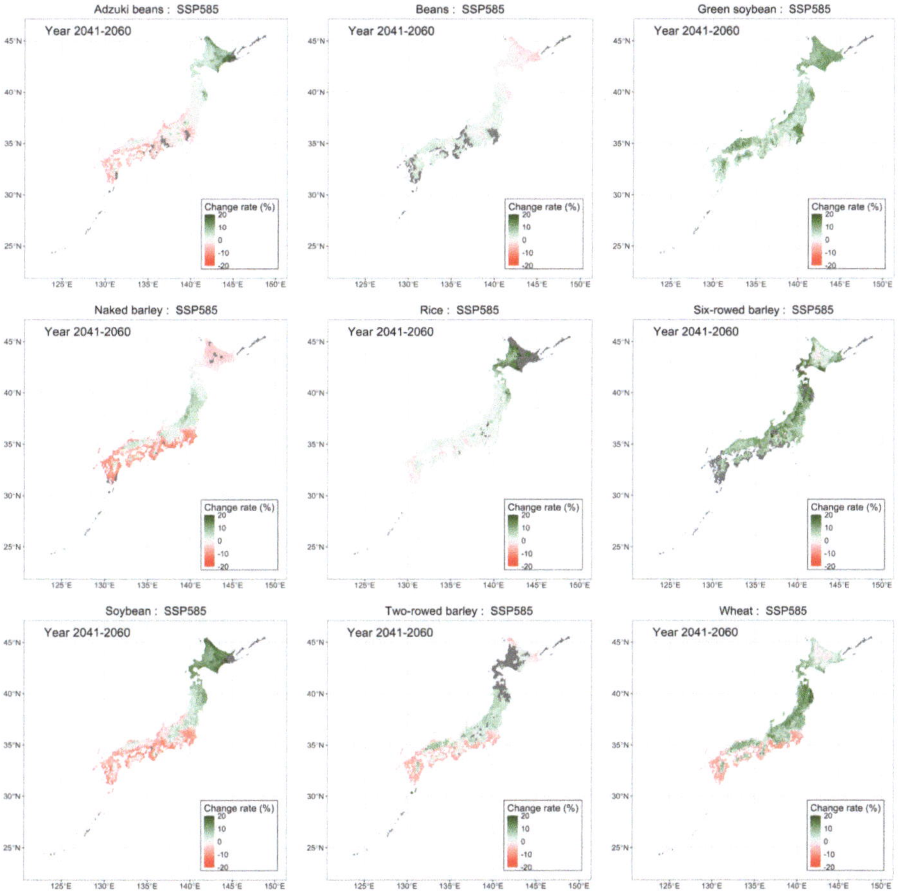

Fig. 5.1 Projected average yields from 2041 to 2060 for nine crops, expressed as percent change (%) from the 2018–2020 average. The average of five GCMs was used. The SSP585 was used for the scenarios. For these predictions, Eq. (5.1) was used

made by process-based models, evaluations that take into account uncertainty are rarely made, despite the large prediction error. Although future yield predictions using statistical methods are rare, I believe that the advantage of being able to evaluate uncertainty by utilizing the advantages of statistical methods should be reconsidered. However, statistical methods are likely to overlook important meteorological factors for crops because they have far fewer explanatory variables compared to process-based models. For example, there should be many crops that cannot be cultivated in the northernmost part of Hokkaido due to extremely low temperatures in early spring, but if average temperature and average solar radiation during the growing period are used as explanatory variables, this extreme temporary low temperature may be overlooked, posing a risk of calculating a certain yield. There is also a serious problem that it does not include effects such as the fertilization effect

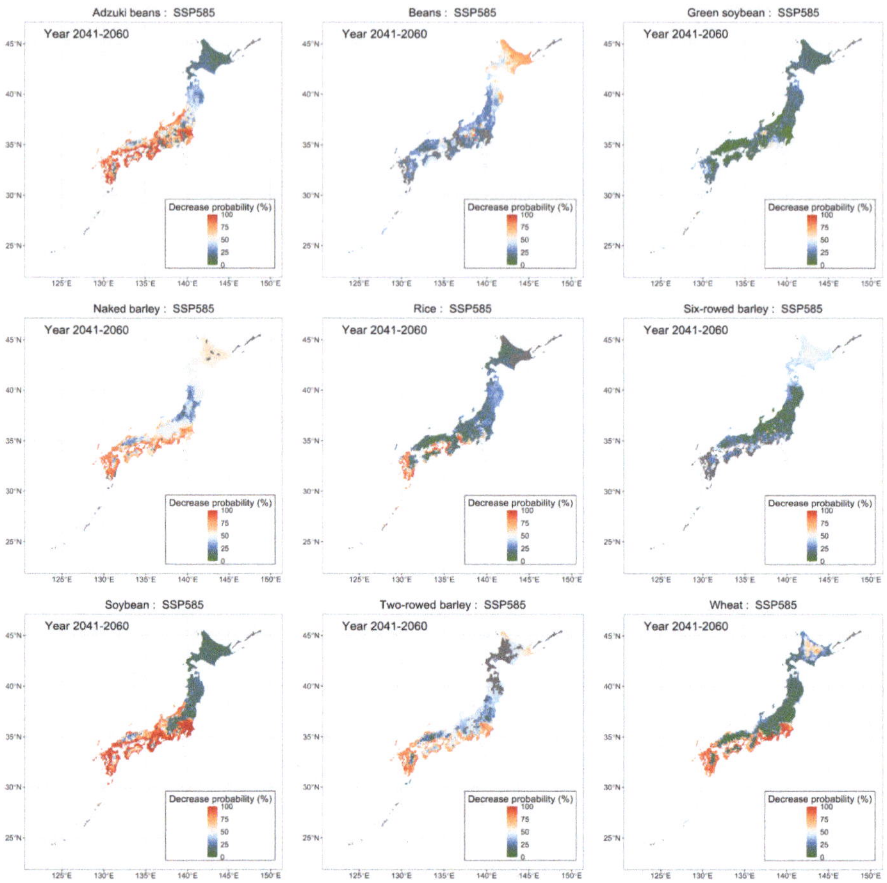

Fig. 5.2 Projected decrease probability (%) from 2041 to 2060 for nine crops. The average of five GCMs was used. The SSP585 was used for the scenarios. For these predictions, Eq. (5.1) was used

of carbon dioxide. Considering both the problems of predictions using statistical models and the problems of predictions using process-based models, it is desirable to predict future crop yields from multiple angles using both methods.

5.9 Effect of Adaptation: Differences in the Meaning of Different Statistical Models

Figure 5.3 shows the predicted probability of crop yield reduction for all nine crops predicted by Eq. (5.2). It can be seen that the degree of probability of increase or decrease is much larger than that predicted by Eq. (5.1). Particularly, the patterns of increase and decrease for beans and two-row barley are different. The difference between Eqs. (5.1) and (5.2) is that Eq. (5.1) assumes that the yield response curve to weather is the same in all regions, while Eq. (5.2) allows the response curve to be slightly different depending on the region. The characteristic that the response curve is slightly different for each region in Eq. (5.2) assumes that response curves are tailored to the farming methods and varieties of those regions. In other words, the prediction of Eq. (5.2) is a prediction in the case of changing weather conditions, with the farming methods and varieties of that region remaining the same.

For Eq. (5.1), although it assumes that the response curve does not change in any region, this can also be interpreted as a response curve that includes differences in farming methods and varieties in each region. In other words, for example, when the temperature zone changes, it can be interpreted as predicting the response in the case of not only meteorological factors but also indirectly transitioning farming methods and varieties to that temperature zone.

From the above, the prediction of Eq. (5.1) can be interpreted as a prediction that takes into account some changes in farming methods and varieties, that is, adaptation, and the prediction of Eq. (5.2) can be interpreted as a prediction with the current farming methods and varieties, that is, without considering adaptation (Sakurai et al. 2021). As shown in this study, by using statistical models appropriately, it is possible to estimate the potential for adaptation. I believe that, by evaluating the impact of future climate change using statistical models, we can gain more insights into not only the impact assessment but also the adaptation assessment of climate change in agriculture.

5.10 Conclusion

In this analysis, I presented the probability of future crop yield reduction for all nine crops (Fig. 5.2), along with the potential changes in future crop yields (Fig. 5.1). I also presented yield predictions for both adapted and nonadapted future weather conditions (Fig. 5.3). Currently, there is a shift from rice paddies to wheat and soybeans in Japan, but compared to rice, wheat and soybeans are predicted to have a high probability of reduced yields in low-altitude areas of western Japan. It indicates that soybeans may not be suitable for conversion from rice paddies in the future climate conditions, so it may be necessary to reconsider improvements in varieties and conversion to other field crops in future climate conditions. This study suggested that for other crops, regions where an increase in yield is expected and regions where a decrease in yield is feared may be divided. To adapt to climate change, measures for suitable land and crops anticipating future climate conditions are needed. For this purpose, using statistical methods should have one of the critical roles.

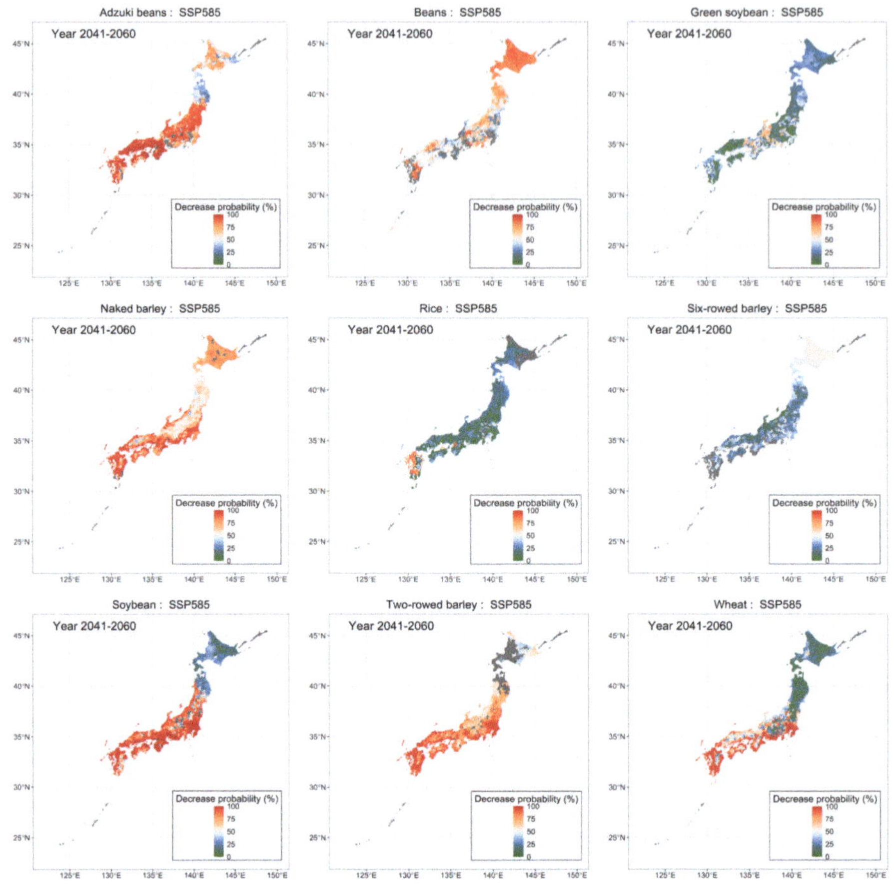

Fig. 5.3 Projected decrease probability (%) from 2041 to 2060 for nine crops. The average of five GCMs was used. The SSP585 was used for the scenarios. For these predictions, Eq. (5.2) was used

References

Asseng S, Ewert F, Martre P et al (2015) Rising temperatures reduce global wheat production. Nat Clim Chang 5(2):143–147

Hawkins E, Fricker TE, Challinor AJ et al (2013) Increasing influence of heat stress on French maize yields from the 1960s to the 2030s. Glob Change Biol 19(3):937–947

Ishigooka Y, Hasegawa T, Kuwagata T, Nishimori M (2020) Evaluation of the most appropriate spatial resolution of input data for assessing the impact of climate change on rice productivity in Japan. J Agric Meteorol 76(2):61–68

Ishigooka Y, Hasegawa T, Kuwagata T et al (2021) Revision of estimates of climate change impacts on rice yield and quality in Japan by considering the combined effects of temperature and CO2 concentration. J Agric Meteorol 77(2):139–149

Ishizaki N, Shiogama H, Hanasaki N, Takahashi K (2022) Development of CMIP6-based climate scenarios for Japan using statistical method and their applicability to heat-related impact studies. Earth Space Sci 9:e2022EA002451. https://doi.org/10.1029/2022EA002451

Jones HG (2013) Plants and microclimate: a quantitative approach to environmental plant physiology. Cambridge University Press

Kath J, Byrareddy VM, Craparo A et al (2020) Not so robust: Robusta coffee production is highly sensitive to temperature. Glob Change Biol 26(6):3677–3688

Lobell DB, Schlenker W, Costa-Roberts J (2011) Climate trends and global crop production since 1980. Science 333(6042):616–620

Martre P, Wallach D, Asseng S et al (2015) Multimodel ensembles of wheat growth: many models are better than one. Glob Change Biol 21(2):911–925

Masutomi Y, Takahashi K, Harasawa H, Matsuoka Y (2009) Impact assessment of climate change on rice production in Asia in comprehensive consideration of process/parameter uncertainty in general circulation models. Agric Ecosyst Environ 131(3–4):281–291

Müller C, Elliott J, Kelly D et al (2019) The global gridded crop model intercomparison phase 1 simulation dataset. Sci Data 6(1):1–22

Müller C, Franke J, Jägermeyr J et al (2021) Exploring uncertainties in global crop yield projections in a large ensemble of crop models and cmip5 and cmip6 climate scenarios. Environ Res Lett 16(3):034040

Ohno H, Sasaki K, Ohara G, Nakazono K (2016) Development of grid square air temperature and precipitation data compiled from observed, forecasted, and climatic normal data. Clim Bios 16:71–79

Pörtner HO, Roberts DC, Adams H et al (2022) Climate change 2022: impacts, adaptation and vulnerability. IPCC Sixth Assessment Report:37–118

R Core Team (2024) R: a language and environment for statistical computing. R Foundation for Statistical Computing, Vienna. https://www.R-project.org/

Rozenzweig C, Elliott J, Deryng D et al (2014) Assessing agricultural risks of climate change in the 21st century in a global gridded crop model intercomparison. Proc Natl Acad Sci USA 111(9):3268–3273

Sakurai G, Ishitsuka N, Okabe N (2021) Studies in climate change impact assessment. Jpn. J Appl Stat 50(2–3):55–74. (Japanese)

Schewe J, Gosling SN, Reyer C et al (2019) State-of-the-art global models underestimate impacts from climate extremes. Nat Commun 10:1005

Schlenker W, Roberts MJ (2009) Nonlinear temperature effects indicate severe damages to US crop yields under climate change. Proc Natl Acad Sci USA 106(37):15594–15598

Somero GN (1978) Temperature adaptation of enzymes: biological optimization through structure-function compromises. Annu Rev Ecol Syst 9:1–29

Yoshikawa S, Takahashi K, Wu W, Matsuhashi K, Mimura N (2022) Development of common socio-economic scenarios for climate change impact assessments in Japan. Geosci Model Dev. https://doi.org/10.5194/gmd-2022-169

Chapter 6
Projection of Climate Change Impacts and Evaluation of Adaptation Options for Forestry

Yasumasa Hirata, Jumpei Toriyama, Tokuko Ujino-Ihara, Katsuhiro Nakao, Wataru Murakami, Haruka Tsunetaka, Tomohiro Nishizono, Shoji Hashimoto, Kentaro Uchiyama, and Hideki Mori

Abstract Forests perform numerous vital functions that are closely linked to human life. However, there are growing concerns that the worsening climate crisis may significantly diminish the role of forests in mitigating and adapting to climate change. In this chapter, using the Biome-BGC process model, we predict the impact of climate change on the growth of Japanese cedar plantations by estimating net primary production. We also assess the vulnerability of different genetic groups of Japanese cedar to climate change. Based on these assessments, we propose adaptation measures and provide guidelines for selecting the most appropriate genetic groups. Further, we present an adaptation strategy that accounts for growth predictions and mountain disaster risks by developing a method to evaluate suitable and unsuitable forestry areas at local fine scales based on Japanese cedar growth patterns and the frequency of disaster risks.

Keywords Biome-BGC · Mountain disaster risk · Genetic group · Growth prediction · Japanese cedar · Net primary production

---

Y. Hirata (✉) · T. Ujino-Ihara · W. Murakami · H. Tsunetaka · T. Nishizono · S. Hashimoto
K. Uchiyama · H. Mori
Forestry and Forest Products Research Institute, Tsukuba, Japan
e-mail: hirat09@affrc.go.jp

J. Toriyama
Kyushu Research Center, Forestry and Forest Products Research Institute, Kumamoto, Japan

K. Nakao
Kansai Research Center, Forestry and Forest Products Research Institute, Kyoto, Japan

N. Mimura, S. Takewaka (eds.), *Climate Change Impacts and Adaptation Strategies in Japan*, https://doi.org/10.1007/978-981-96-2436-2_6

6.1 Introduction

Approximately 70% of Japan's land is covered by mountains and hills, with the majority being forested. Of these forests, about 40% are plantations. The main tree species planted in these plantations are cedar, cypress, and pine, with cedar accounting for 44% of the total. In the context of climate change, forests play an important role in mitigating climate change by absorbing and sequestering carbon dioxide from the atmosphere.

Furthermore, forests play an important role in climate change adaptation. In Japan, plains cover only about 90,000 square kilometers, roughly one-fourth of the country's land area. With about half of Japan's population concentrated in these areas, urban development has extended to the foothills of mountains in regional cities, exacerbating damage from natural disasters, especially in conjunction with recent extreme weather patterns. Forests provide various functions closely related to human life. Among these, the functions of preventing landslides, conserving soil, and protecting water resources are particularly important for reducing the impact of natural disasters.

However, there is concern that the escalating climate crisis may significantly reduce the role forests play in climate change mitigation and adaptation. Rising temperatures and changes in precipitation patterns are expected to accelerate drying, negatively impacting the growth of plantations (Shigenaga et al. 2005; Mitsuda 2018). The rise in temperature, on the other hand, also leads to increased productivity. Since growing conditions vary by region in our country, the effects differ depending on the area (Toriyama et al. 2018). Considering that cedar has been planted in high-altitude and cold regions where it is not naturally suited, further research is necessary to make accurate predictions about the impact of warming on the production of cedar plantation forests, taking into account these differences in planting environments. Additionally, the increasing frequency of large typhoons and intense short-term rainfall may weaken the functions of forests in preventing landslides, conserving soil, and other related functions, leading to an increased risk of mountain disasters. Regarding mountain disasters, an increase in the frequency of heavy rains with either extremely high rainfall intensity or total precipitation, or both, along with the expansion of regions frequently experiencing heavy rains and the broader geographic extent of individual heavy rains, will lead to more frequent landslides, debris flows, and slope failures, as well as larger-scale and changing patterns of such disasters (Saito and Matsuyama 2012; Tsunetaka 2021).

Therefore, this chapter predicts the impact of climate change, including heat and drought stress, on the growth of cedar plantations, evaluates the vulnerability of cedar genetic groups to climate change, and provides guidelines for selecting genetic groups from the perspective of plantation growth. It also describes adaptation measures that consider growth predictions and mountain disaster risks in response to increasingly severe heavy rainfall disasters.

6.2 The Impact of Climate Change on the Growth of Japanese Cedar and the Effectiveness of Adaptation Measures

6.2.1 *Impact on the Growth of Japanese Cedar Plantations*

Climate change is expected to affect the growth of Japanese cedar plantations. To evaluate the impact of climate change on the growth of Japanese cedar plantations, a model was developed to estimate the Net Primary Production (NPP) of Japanese cedar based on the process-based model, Biome-BGC, which simulates plant photosynthesis, respiration, and organ growth. The Biome-BGC model was developed by the Numerical Terradynamic Simulation Group at the University of Montana for the purpose of simulating the storage and flux of carbon, nitrogen, water, and energy. Biome-BGC was initially parameterized to estimate NPP for Japanese cedar plantations at 40 years of age (Toriyama et al. 2021). Subsequently, it was revised to encompass 60 and 80 years of age and to account for the distinctions among Japanese cedar varieties. The parameterized Biome-BGC model encompasses the entirety of Japan, with the exception of Hokkaido, where Japanese cedar plantations are scarce. The baseline year is 2000 and the spatial resolution is 1 km meshes.

The results of predicting the future NPP of Japanese cedar plantations using this model revealed significant differences depending on the climate scenarios (RCP) and climate models (GCM). Under a high-emission scenario (RCP8.5), fewer areas showed a decline in NPP compared to the low-emission scenario (RCP2.6), likely due to the higher CO_2 concentrations enhancing photosynthesis. Additionally, it was estimated that the NPP would generally increase in eastern Japan, while it will decrease in a wide area of western Japan under the RCP2.6 scenario (Fig. 6.1).

The differences in NPP of Japanese cedar plantations between 2000s and 2090s (RCP2.6, MIROC5) have also been predicted. This implies the case for the 80-year-old stand.

6.2.2 *Effectiveness of Adaptation Measures for Japanese Cedar Plantations*

The effectiveness of adaptation measures was also examined by the process-based model. The climate scenarios used had a spatial resolution of 1 km grids, utilizing five climate models from the second version of the Shared Socioeconomic Pathways (SSP). Two target periods were considered: the current climate (1995–2014) and future climate projections (2081–2100). The rotation periods were set at 40 years and 80 years, and the increase in accumulated wood biomass due to the shorter 40-year rotation was evaluated as an adaptation measure (sRot scenario in Fig. 6.2). Furthermore, changing the genetic type of Japanese cedar was also tested as an

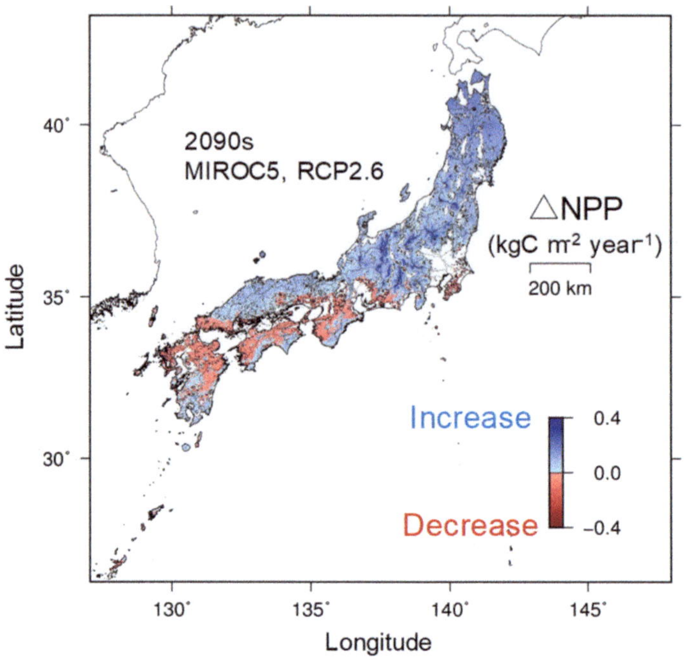

Fig. 6.1 Predicted future change in Net Primary Production (NPP) of Japanese cedar plantations

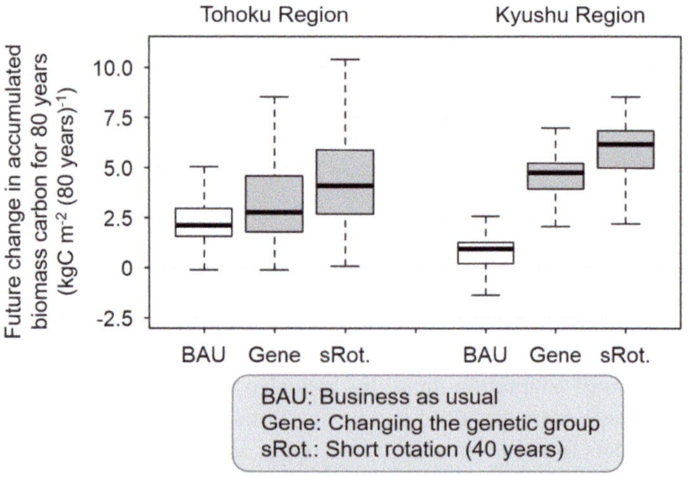

Fig. 6.2 Effect of adaptation measures in Japanese cedar growth model. The differences in accumulated biomass carbon of Japanese cedar plantations between 1995–2014 and 2081–2100 (SSP1-2.6) are predicted as a measure for the future change in forest growth. The positive value indicates a future increase in biomass carbon.

adaptation measure (Gene scenario in Fig. 6.2). The genetic groups of Japanese cedar were divided into the Pacific Ocean (PO) group and the Japan Sea (JS) group. In general, the PO group has a statistically higher rate of decline in timber volume growth after 40 years compared to the JS group, a difference that was factored into the analysis (Nishizono et al. 2014). In this study, the accumulated wood biomass was calculated based on the current and modified distribution of PO and JS types.

The nationwide model simulation of Japanese cedar growth revealed that the effectiveness of adaptation measures varied by region. In the Kyushu region, where accumulated biomass for 80 years of Japanese cedar plantations is expected to show both increases and decreases, the effectiveness of adaptation measures was found to be higher. Conversely, in the Tohoku region, where a widespread increase in growth is predicted, the effectiveness of these measures appeared to be limited (Fig. 6.2). The reason for this may be that, in the Kyushu region, the increase in biomass carbon slows more after 40 years, making the positive effects of harvesting and replanting more significant (e.g., sRot scenario in Fig. 6.2).

6.3 Genetic Vulnerability of Japanese Cedar to Climate Change

6.3.1 Environmental Adaptation of Japanese Cedar

Understanding how tree species respond to climate change is a crucial issue related to the stability of forest ecosystems. In widely distributed tree species like Japanese cedar (*Cryptomeria japonica*), which grow in a wide range of environments, specific genotypes have often become fixed in different regions over long periods, adapted to their local environment. This phenomenon is known as "local adaptation" in the context of environmental adaptation.

Local adaptation allows species to distribute across diverse environments, but it also leads to differences in how the same tree species respond to climate change depending on the region where they are grown. This complicates the understanding and prediction of tree species' responses to climate change. An evaluation of the environmental adaptation range of the PO group and JS group of Japanese cedar revealed that the two groups have different distributions along environmental axes represented by precipitation, suggesting that they are subject to different selective pressures from the environment (Fig. 6.3).

The vertical axis represents density (a measure of the relative frequency of genotypes), while the horizontal axis (PC2) represents an environmental gradient primarily explained by precipitation variables.

The foundation of these local adaptations lies in genetic variations driven by environmental factors. With the rapid advancements in genome analysis techniques in recent years, it has become possible to elucidate the genetic variations associated with local adaptation and to identify the genotypes and regional strains that are

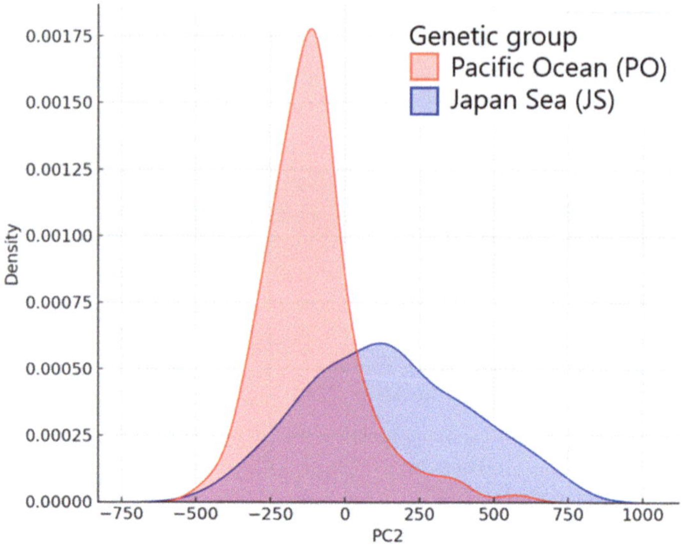

Fig. 6.3 Density distribution of the PO and JS genetic groups along the PC2 axis

advantageous for growth and survival under future climate conditions. In this section, we first extracted genetic variations related to climate and growth from the entire genome using natural Japanese cedar collection from across Japan planted in the provenance trial sites. Next, we employed spatial modeling techniques to elucidate the distribution of these adaptive genotypes across geographic regions. Finally, using multiple climate models we evaluated the potential disruption of current local adaptations under future climate scenarios, quantified as genetic offset on geographic maps.

6.3.2 Evaluation of Vulnerability of Japanese Cedar Under Future Climate Conditions

Through the analysis of over 30,000 genetic variations (single nucleotide polymorphisms, SNPs) obtained from 254 individuals from 20 populations encompassing the entire distribution range of Japanese cedar in Japan, we extracted SNPs significantly correlated with climate values utilizing a latent factor mixed model (LFMM). This model helps account for both environmental factors and population structure, allowing for more accurate identification of SNPs associated with environmental adaptation. The climate variables analyzed included the bioclimatic variables (bio1-19) from World Clim v2.0 (averages from 1970–2000). We subsequently modeled the distribution of locally adapted genotypes along environmental gradients and the extent to which local adaptation might be disrupted under future climate conditions using Gradient Forest method. This method models the relationship between genetic variation and environmental gradients, while incorporating spatial structure through the use of principal coordinates of neighbor matrices (PCNM). PCNM helps account for spatial autocorrelation, ensuring that geographic proximity is considered when assessing genetic adaptation. Additionally, we estimated the climate variables that significantly contribute to environmental adaptation. The climate models used were MIROC6 and MRI-ESM2, with socioeconomic scenarios ssp1-2.6 and ssp5-8.5. Genetic offset, which represents the degree of genetic mismatch or maladaptation due to changes in the environment, was calculated for regional strains in 2030, 2050, and 2090, for estimating regions with increased vulnerability.

The results of this projection showed that average winter temperature, summer precipitation during the warmest quarter, and annual precipitation contribute significantly to Japanese cedar's environmental adaptation (Fig. 6.4). A comparison of genetic structures derived from the full set of genetic variations and those specifically linked to environmental adaptation revealed substantial differences in regions like Kyushu, Shikoku, and the Pacific side of Japan, as well as the entire Japan Sea coast, suggesting strong local adaptation in these areas. In predictions of genetic

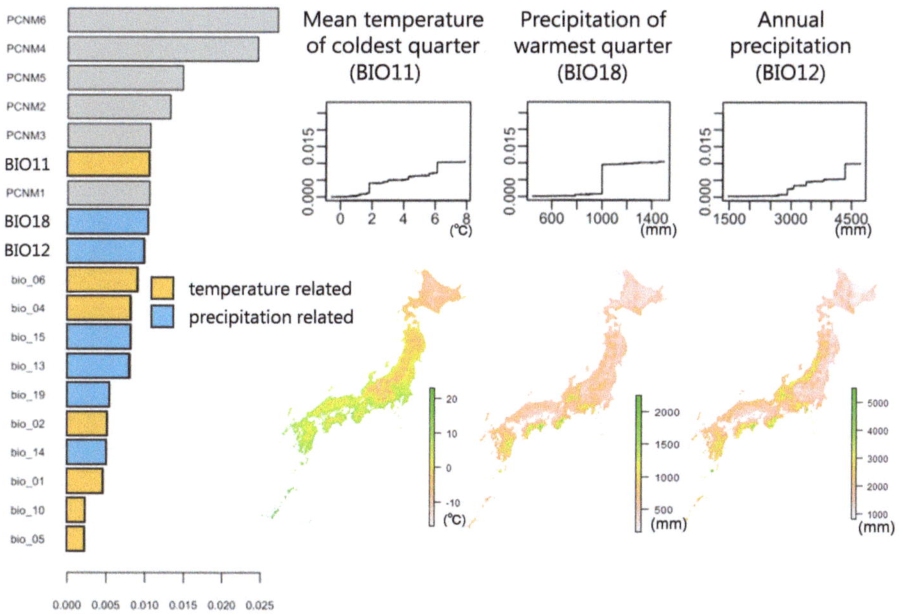

Fig. 6.4 Modeling the adaptive genetic variation of Japanese cedar in environmental space using Gradient Forest analysis. Left: Importance of 13 climate variables. Top right: Genotypic changes along environmental gradients for the top 3 climate variables. Bottom right: Distribution of current climate values.

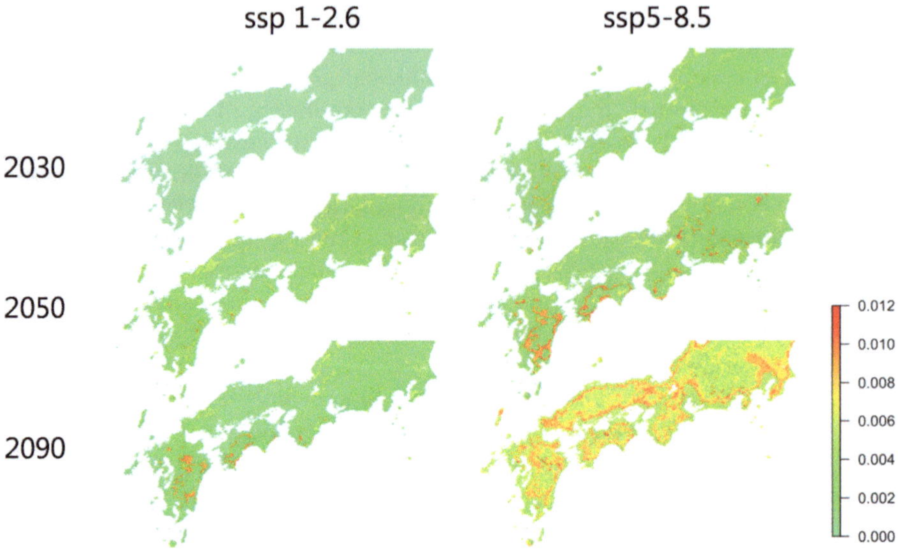

Fig. 6.5 Predicted genetic offset of Japanese cedar in 2030, 2050, and 2090 based on Gradient Forest analysis. The degree of genetic vulnerability under future climates is color-coded, estimated using the MIROC6 climate model, and both low (SSP1-2.6) and high (SSP5-8.5) emission scenarios.

offset using the MIROC6 climate model, under the low-emission scenario (SSP1-2.6), genetic offset increased in some inland areas of Kyushu and Shikoku and parts of the Tohoku region by 2050. On the other hand, under the high-emission scenario (SSP5-8.5), genetic offset was predicted to increase widely across lowland areas throughout the distribution range from 2050 and 2090 (Fig. 6.5). Similarly, predictions using the MRI-ESM2 climate model showed differences in the pattern and extent of genetic offset compared to MIROC6, emphasizing the variability in outcomes depending on the climate model used. These discrepancies are likely due to regional variations in predicted climate values (such as precipitation) between climate models. A principal component analysis of current environmental variables of the distribution range showed that winter precipitation most significantly affects the distribution of the PO and JS groups, consistent with the results from Gradient Forest modeling.

These results indicate that local adaptation to temperature and precipitation occurs in Japanese cedar, and as these factors change due to climate change, the existing adaptation patterns are likely to be disrupted, leading to decreased fitness in certain regions and genetic groups. Special attention should be given to regions predicted to have high genetic offset in common between the two climate models, as they may face increased vulnerability to future climate change.

6.4 Evaluation of Forestry Scenarios and Guidelines for Selecting Genetic Groups

6.4.1 Evaluation of Forestry Scenarios

In Japan, the area of plantations for timber production is expected to decrease in the future due to a decline in reforestation rates caused by a declining and aging population. On the other hand, the carbon sequestration function of forests is expected to mitigate future climate change. Therefore, we developed forestry adaptation scenario options that anticipate a reduction in the area of Japanese cedar plantations in the future. Specifically, for an area equivalent to approximately 4.3 million hectares nationwide, we recreated the age class structure and resource accumulation of Japanese cedar plantations as of 2010, and based on the assumptions of harvesting rates (5% and 15% over 10 years) and reforestation rates (25% and 75%) into the future, we developed four forestry scenarios: (1) high cover of cedar, (2) high activeness for regeneration of cedar, (3) low activeness for regeneration of cedar, and (4) low cover of cedar. These scenarios were input into the carbon cycle model Biome-BGC to estimate net ecosystem productivity (corresponding to forest carbon sequestration) for 2050 and 2090. Climate scenarios were evaluated using five climate models from the second edition of the Common Scenario Framework under the CMIP6, focusing on SSP1-2.6 and SSP5-8.5 scenarios.

All four forestry scenarios for Japanese cedar plantations showed a decline in net ecosystem productivity in 2050 and 2090 compared to 2010, to varying degrees. The two main factors were the expansion of broadleaf secondary forests and the aging of the plantations. In other words, under realistic harvesting speeds and reforestation rates, the area of Japanese cedar plantations will decrease, and the remaining plantations will age over the next few decades. Broadleaf secondary forests that replace the plantations have lower net ecosystem productivity than the plantations, and Japanese cedar itself also shows a decline in net ecosystem productivity due to aging. As a result, net ecosystem productivity in 2050 (SSP1-2.6) is estimated to decrease by about 20% on average compared to 2010 (Fig. 6.6).

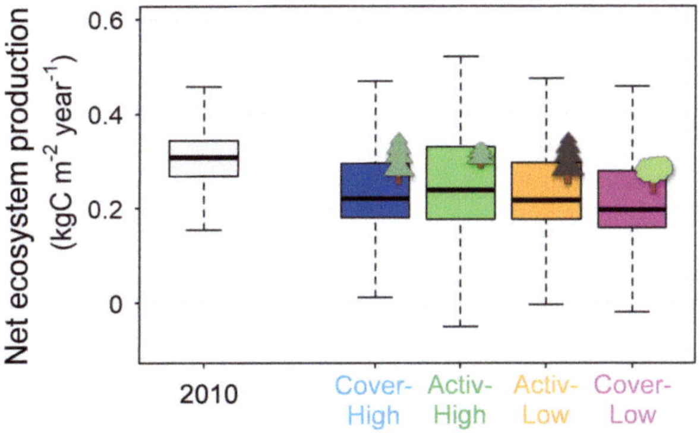

Fig. 6.6 Prediction of net ecosystem production in 2050 under four management scenarios of Japanese cedar plantation

$N = 49{,}440$ (number of 1 km^2 mesh); Results for SSP1-2.6 from the average of five climate models. The higher the net ecosystem production, the higher the forest carbon sequestration. The bars on the right show the results for 2050 for the four forest management scenarios. Cover-High and Cover-Low refer to maintaining high and low cedar cover, respectively; Activ-High and Activ-Low refer to maintaining high and low cedar replanting activity, respectively.

The second outcome of the plantation model simulation was that taking forestry options tailored to each region can effectively mitigate the nationwide decline in net ecosystem productivity. Specifically, in warm forestry regions dominated by the PO group (e.g., Kyushu), active reforestation with short rotation periods (i.e., high activeness scenario) is considered the only way to maintain relatively high forest growth and carbon sequestration. In contrast, in relatively cool regions dominated by the JS group (e.g., the Japan Sea side of the Tohoku region), passive reforestation with longer rotation periods (i.e., low activeness scenario) is considered a viable option. The differences between these regions reflect not only climate differences but also the different growth patterns of the genetic groups distributed there. In certain areas within the PO group distribution, including the Pacific side of the Tohoku region, where a decline in genetic fitness is predicted, the introduction of the JS group, which has broader environmental adaptability (Fig. 6.3), should be considered. On the other hand, introducing the PO group into regions where the JS group is planted is still considered difficult from the perspective of the genetic disturbance risk to remaining natural Japanese cedar forests.

6.4.2 Guidelines for Selecting Genetic Groups of Japanese Cedar Under Climate Change

Considering the predictions of genetic offsets under future climate conditions in the growth forecast, a selection guideline was developed based on two axes: the currently planted genetic group and the population issues due to future depopulation and aging.

In forestry regions dominated by the PO group:

1. In municipalities where significant population decline and aging are expected, the current plantation should be maintained or guided toward broadleaf forests.
2. In municipalities where depopulation and aging are limited, continue the plantation forestry with active reforestation and shorter rotation periods. In municipalities capable of maintaining higher reforestation rates, consider lineage or species conversion based on genetic offset results.

Meanwhile, in forestry regions dominated by the JS group:

3. In regions with significant population decline and aging, maintain the current the plantation.
4. In regions with limited population issues, continue the plantation forestry, including the possibility of extending rotation periods, as a provisional guideline package.

6.5 Evaluation of Adaptation Measures Considering Growth Prediction and Mountain Disaster Risk

6.5.1 Disaster Prevention Functions of Forests

Mountain disasters are influenced by factors such as geology, topography, slope, soil, and vegetation. Among these, vegetation is the factor where human intervention can play a role. In particular, maintaining forests in a healthy state in mountainous areas can reduce the factors that contribute to mountain disasters. However, the presence of these factors alone does not cause mountain disasters; they occur when combined with triggers such as rainfall or earthquakes.

Forests serve various functions, including timber production, biodiversity conservation, and global environmental protection. Among these, the functions of preventing landslides and conserving soil contribute to disaster prevention and mitigation against natural disasters. Forest root systems help prevent soil collapse, while trees and vegetation cover the ground, reducing soil erosion caused by rain. The development of root systems and undergrowth enhances these disaster prevention effects. While forests clearly have a role in preventing collapses, their ability to do so has limits. For example, even in forested areas, landslides can occur, and it is known that forest cover cannot always prevent collapses during extreme rainfall events. In this section, we will discuss a model that evaluates adaptation measures considering the risk of mountain disasters under climate change, combined with predictions of growth in Japanese cedar forests.

6.5.2 Regional Growth Prediction of Japanese Cedar Under Climate Change

Mountain disasters are localized events, so understanding the regional growth of Japanese cedar under climate change is crucial for considering adaptation measures that account for growth predictions and mountain disaster risks. Therefore, a model was developed to predict the crown height growth of Japanese cedar, an indicator of productivity, using high-resolution (25 m grid) data obtained from airborne LiDAR in Gujo City, Gifu Prefecture. The model employed the machine learning technique known as Random Forest, correlating tree height data with factors such as tree age and environmental conditions. The results demonstrated that the crown height growth of Japanese cedar could be quantitatively predicted with high accuracy based on factors like tree age, climate, and topography. Furthermore, using the constructed model, canopy height was predicted for years 20 to 100, and the grid within

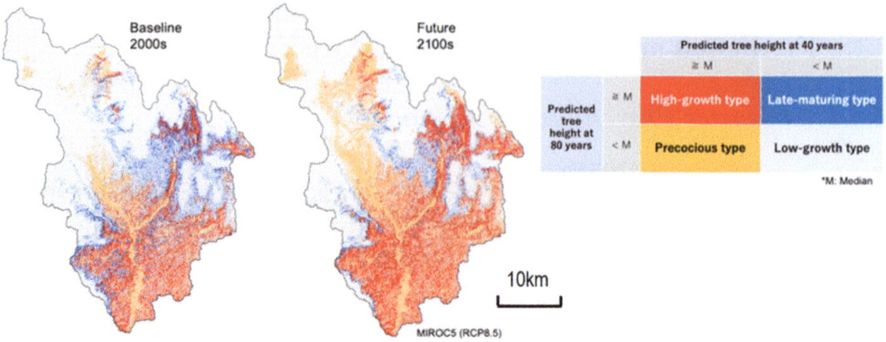

Fig. 6.7 Prediction of the impact of climate change on the tree height growth of Japanese cedar using the Common Scenario Version 1

the area was classified into four growth types (high-growth, late-maturing, early-maturing, and normal). As a result, it was revealed that in the study area, Gujo City, these types were spatially distributed according to climate conditions (Fig. 6.7).

Under future climate conditions, the area of high-growth types (red) and precocious types (orange) is expected to increase, and tree height growth of Japanese cedar is predicted to improve in Gujo, Gifu. It is considered preferable to choose short-rotation strategies as an adaptation measure.

6.5.3 Prediction of Rainfall Events Triggering Mountain Disasters

To predict mountain disaster risk, rainfall patterns from ten major landslide events, including the July 2012 Northern Kyushu Rainstorm and the July 2018 Heavy Rainfall, were compared. The analysis revealed that the risk of landslides substantially increases when the average hourly rainfall in the 72 hours preceding the rainfall peak reaches or exceeds the 100-year return level for the region (Fig. 6.8). The landslide risk prediction model is an empirical model that uses the 100-year return-level rainfall for a region as a threshold to determine future rainfall events with a high landslide risk. By applying future rainfall prediction data from shared climate scenarios, rainfall events exceeding the threshold were identified as rainfall events that potentially trigger landslides, and it was found that the frequency and location of such events would vary depending on the degree of climate change and the climate model used for rainfall estimation up to 2100.

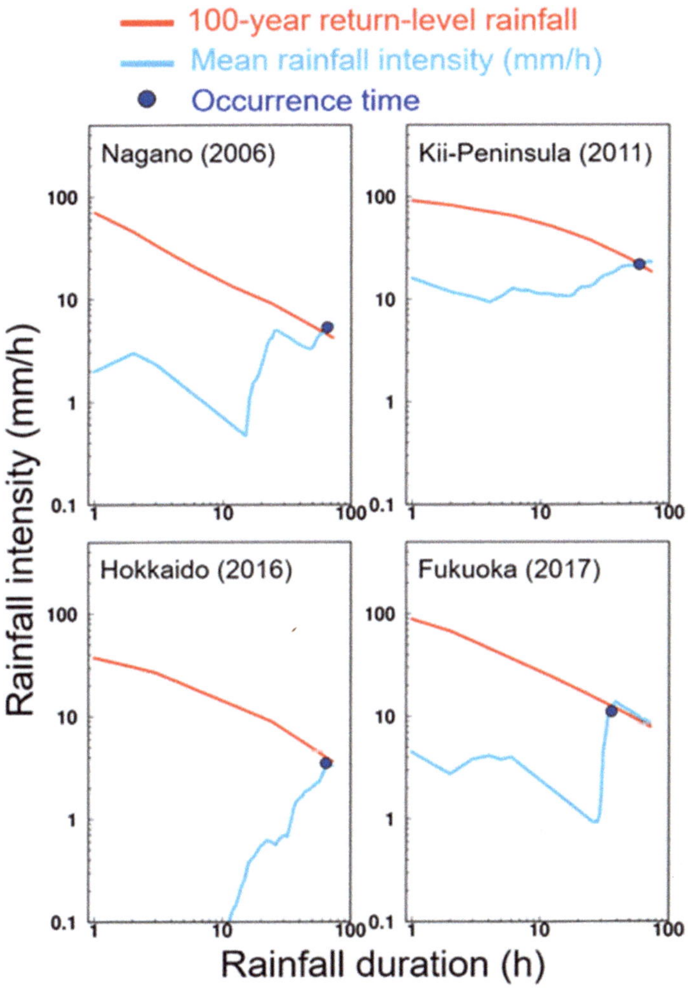

Fig. 6.8 Relation between changes in average hourly rainfall and the timing of landslide occurrences. (Reproduced from Tsunetaka (2021))

In all the cases investigated, it was found that landslides occurred around the point where the light blue line, representing the average hourly rainfall, intersects with the red line, which indicates the rainfall amount with a 100-year return level for that region. To compare among cases, the horizontal axis represents the 3-day (72-hour) period counted backward from the time when the 72-hour rainfall reached its maximum value.

6.5.4 Future Projections of Mountain Disaster Risk and Japanese Cedar Growth for Adaptation Strategies

For the model region, the growth of Japanese cedar, assuming a certain base year for planting, was predicted using the cedar crown height growth model (Fig. 6.9, upper side). The future climate scenario used the MIROC5 model, considered highly reproducible around Japan, with the most pronounced temperature rise scenario, RCP8.5. By estimating Japanese cedar crown height growth and the frequency of landslide risks based on future climate scenarios in the model region, it was found that topographical conditions (TWI: an index indicating soil moisture) had the most significant influence on cedar crown height growth, compared to parameters like tree age and climate conditions (temperature and precipitation). The prediction revealed that, due to spatial differences in these parameters, growth patterns would also differ spatially, with some areas expected to maintain high growth and others expected to see relatively low growth in the future. This suggested a significant spatial variation in cedar growth, differing from predictions under the current climate conditions (BAU scenario), highlighting the need for forestry adaptation measures considering future climate-induced growth changes.

Tree height growth was estimated using the Planted tree-height growth model. This model forecasts tree height using tree height data obtained from airborne laser scanning as the response variable and climate, topography, and forest age as explanatory variables. The estimation accuracy in the target area was sufficient. A model of the frequency of heavy rainfall events that cause landslides, calculated based on historical observation data, was used for landslide risk prediction.

The cumulative rainfall over three days up to 2100 was calculated using the same future climate scenario and rainfall prediction data. When this predicted cumulative rainfall reached the 100-year return-level rainfall under the current climate

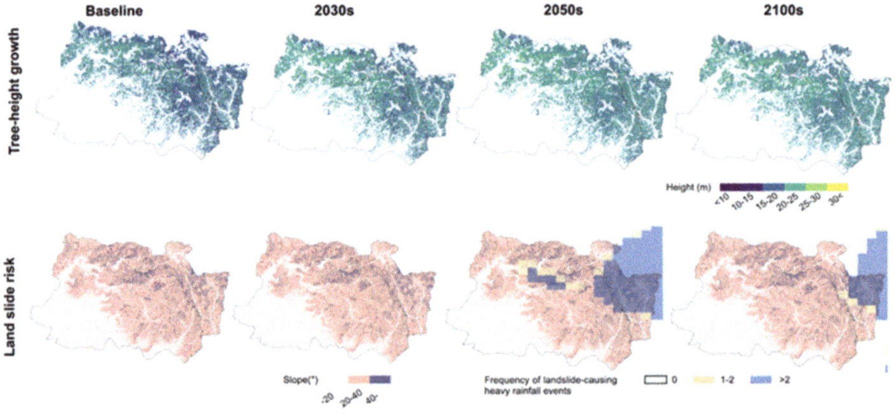

Fig. 6.9 Distribution of Japanese cedar tree height growth (upper side) and landslide risk (lower side) under future climate scenarios

conditions for the target region, the corresponding rainfall was identified as rainfall that potentially triggers landslides. The frequency of such rainfall events with the landslide risk was aggregated and mapped for each grid (about 1 km^2) (Fig. 6.9, lower side).

In the landslide risk frequency analysis, based on rainfall data from the future climate scenario (RCP8.5 MIROC5), no rainfall with the landslide risk (three-day cumulative rainfall exceeding the 100-year return-level rainfall) was predicted for the entire target region in the 2030s (2021–2040) (Fig. 6.9, lower side). However, in the 2050s (2041–2060), rainfall with a landslide risk was predicted to occur in the northeastern part of the model region. The spatial distribution of landslide risk differed from that of high cedar growth areas, allowing for the identification of regions where high growth and low landslide-related disaster risk overlap.

The results for cedar growth and landslide risk were calculated for the 2030s (2021–2040), 2050s (2041–2060), and 2100 s (2081–2100) to examine changes due to global warming in detail (Fig. 6.9). By integrating these results on GIS, a map was created to evaluate suitable and unsuitable areas for forestry in the target region, considering cedar growth patterns and landslide-risk frequencies (Fig. 6.10). Based on this integrated map, it is possible to zone areas suitable for safe and productive forestry, adapting to climate change, from the present to the future.

Based on the quality of growth predicted by the Japanese cedar tree height growth model, the level of landslide risk, and the slope gradient, the target area was divided into four categories.

Using GIS, the cedar crown height growth prediction model and landslide risk map were integrated. The model region was divided into four zones based on growth type, landslide risk level, and slope gradient (Fig. 6.10). In the northern region, areas were predicted to be suitable for forestry production under future climate

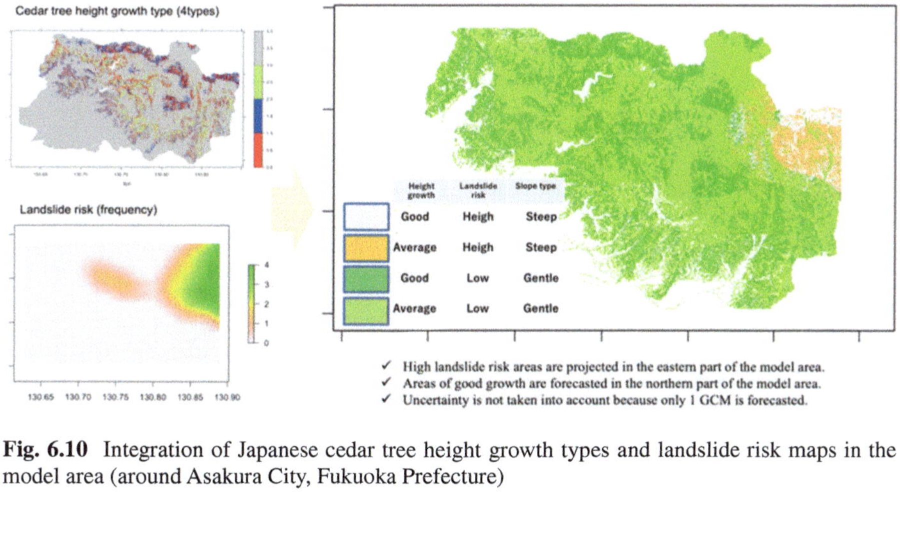

Fig. 6.10 Integration of Japanese cedar tree height growth types and landslide risk maps in the model area (around Asakura City, Fukuoka Prefecture)

conditions, showing high growth and low landslide risk. In contrast, in the eastern region, areas with moderate growth but high landslide risk were identified, requiring careful consideration for future land use.

6.6 Conclusions

In this chapter, using the process-based model, it was predicted that the net primary production (NPP) of Japanese cedar would decrease in many areas of western Japan, while it would generally increase in eastern Japan. Regarding adaptation measures for the plantations, such as shortening rotation periods and switching genetic groups in response to climate change, it was suggested that these measures would be more effective in the Kyushu region, whereas their effectiveness might be limited in the Tohoku region. Additionally, as an adaptation strategy that takes into account both growth projections and the risk of mountain disasters, a method was proposed to evaluate suitable and unsuitable areas for forestry on a grid-by-grid basis, considering sugi growth patterns and disaster risk frequency.

Up to this point, the challenge remains that if the current Japanese cedar plantations do not meet both the size suitable for timber use and the profitability for forest managers, they are unlikely to be harvested, making it difficult for forest owners to adopt adaptation measures. Additionally, the decline in labor force due to an aging population is expected to lead to reduced timber supply capacity and domestic demand, requiring adaptation measures that consider not only forestry but also changes in social structure.

References

Mitsuda Y (2018) Evaluating the effects of climate change on the potential site productivity of sugi (*Cryptomeria japonica*) planted forests in Kyushu Island, Japan. J For Plan 22(2):47–53. https://doi.org/10.20659/jfp.22.2_47

Nishizono T, Kitahara F, Iehara T, Mitsuda Y (2014) Geographical variation in age-height relationships for dominant trees in Japanese cedar (*Cryptomeria japonica* D. Don) forests in Japan. J For Res 19:305–316. https://link.springer.com/article/10.1007/s10310-013-0416-z

Saito H, Matsuyama H (2012) Catastrophic landslide disasters triggered by record-breaking rainfall in Japan : their accurate detection with normalized soil water index in the Kii peninsula for the year 2011. SOLA 8:81–84

Shigenaga H, Matsumoto Y, Taoda H, Takahashi M (2005) The potential effect of climate change on the transpiration of Sugi (*Cryptomeria japonica* D. Don) plantations in Japan. J Agric Meteo 60(5):451–456. https://doi.org/10.2480/agrmet.451

Toriyama J, Hashimoto S, Shimizu T, Sawano S, Osone Y, Lehtonen A (2018) Mapping of the productivity in Japanese cedar plantation in Kyushu region using a process-based model. Kyushu J For Res 71:33–37

Toriyama J, Hashimoto S, Osone Y, Yamashita N, Tsurita T, Shimizu T, Saitoh TM, Sawano S, Lehtonen A, Ishizuka S (2021) Estimating spatial variation in the effects of climate change on the net primary production of Japanese cedar plantations based on modeled carbon dynamics. PLoS One 16(2):e0247165. https://doi.org/10.1371/journal.pone.0247165

Tsunetaka H (2021) Comparison of the return period for landslide-triggering rainfall events in Japan based on standardization of the rainfall period. Earth Surf Process Landf 46:2984–2998. https://doi.org/10.1002/esp.5228

Chapter 7
Impact on Brown Macroalga *Undaria pinnatifida* Farming Under Changing Ocean Climate

Shigeho Kakehi, Goh Onitsuka, and Hideaki Kidokoro

Abstract To evaluate the effects of climate change on large brown seaweed Undaria pinnatifida, sporophyte growth was projected in the Sanriku coastal area (SCA) and the Seto Inland Sea (SIS), the major Undaria cultivation areas in Japan. We developed a growth model specific to each SCA and SIS to reproduce the total length (TL) of Undaria sporophytes in each area. Using this growth model, we projected the growth of sporophytes applying the environmental conditions projected by climate models in RCP scenarios. Geographical variations in TL were projected to occur in the future. Under RCP8.5, a decrease of –50% in TL is projected south of 39° N in the SCA and a decrease of –40% in TL is projected in the western SIS, indicating a large-scale reduction in U. pinnatifida yield in the future. Duration of cultivation is projected to decrease in the SCA, suggesting that the cultivation schedule will change significantly. Simulations of climate change adaptation using a high-temperature-tolerant cultivar in the SIS revealed that yield losses could be mitigated.

Keywords Undaria pinnatifida · Growth model · Total length · Future projection · Sanriku coastal area · Seto Inland Sea

7.1 Introduction

The large brown seaweed *Undaria pinnatifida* (wakame) is a commercially important macroalga mainly cultivated in East Asia (Yamanaka and Akiyama 1993). Global annual production of *U. pinnatifida* is approximately 2.8 million t in wet

---

The original version of the chapter has been revised. A correction to this chapter can be found at https://doi.org/10.1007/978-981-96-2436-2_25

S. Kakehi (✉)
Japan Fisheries Research and Education Agency, Shiogama, Miyagi, Japan
e-mail: kakehi@affrc.go.jp

G. Onitsuka
Japan Fisheries Research and Education Agency, Hatsukaichi, Hiroshima, Japan

H. Kidokoro
Japan Fisheries Research and Education Agency, Niigata, Japan

N. Mimura, S. Takewaka (eds.), *Climate Change Impacts and Adaptation Strategies in Japan*, https://doi.org/10.1007/978-981-96-2436-2_7

weight, accounting for 8.0% of the total seaweed production in 2020, and more than 99% of the total *Undaria* is cultivated (FAO 2022). This species is mainly used as food (Yamanaka and Akiyama 1993); however, it can also be used to extract bioactive substances and produce bioethanol (Murata et al. 2002; Kim et al. 2013). Seaweeds, including *U. pinnatifida*, play important roles in mitigating climate change by fixing greenhouse gases and reducing ocean acidification (e.g., Duarte et al. 2017).

U. pinnatifida production in Japan was ca. 43,000 t in 2021, and cultivation of this species is an important part of the seaweed farming industry (Ministry of Agriculture, Forestry and Fisheries 2023). This species is cultivated in two major areas in Japan: the Sanriku coastal area (SCA) and the Seto Inland Sea (SIS) (Fig. 7.1), accounting for 74% and 17% of the total domestic production in 2021,

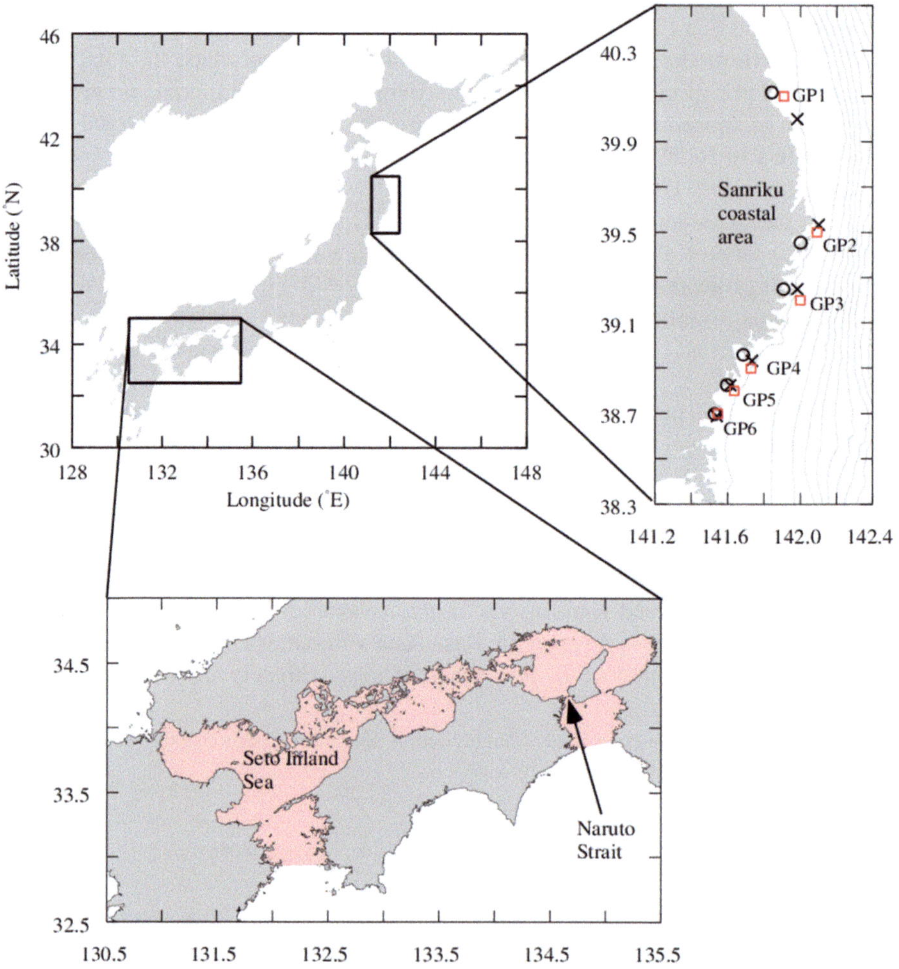

Fig. 7.1 Locations of the Sanriku coastal area (SCA) and the Seto Inland Sea (SIS). Open circles and crosses indicate stations where temperature and nitrate concentration was measured by public fisheries research institutes, respectively, and open squares indicate grid points representing the FORP-NP10 (GP 1–6) closest to the stations of public temperature and nitrate concentration datasets in the SCA

respectively (Ministry of Agriculture, Forestry and Fisheries 2023). The SCA, located on the Pacific coast of northeastern Japan, consists of many small ria-type bays that open to the North Pacific. *U. pinnatifida* produced in this area is known as "Sanriku wakame." The SIS, located in western Japan, is the largest semi-enclosed coastal sea in Japan, and encompasses over 700 small islands. Naruto Strait is one of the primary cultivation area (Fig. 7.1).

U. pinnatifida cultivation is conducted using longline ropes approx. 1 m below the sea surface where germinated sporophytes are installed (Fig. 7.2). The cultivation season for *U. pinnatifida* sporophytes extends from autumn to spring, when the temperature and nutrient concentrations are suitable for growth. The temperature range for suitable growth of *U. pinnatifida* is 5–20 °C (Epstein and Smale 2017) and the growth rate decreases at temperatures above 20 °C (Gao et al. 2013). A study reported that nitrate concentrations of ≥20 μg L^{-1} are suitable for the growth of germinated *Undaria* sporophytes (Kakehi et al. 2018). Nutrient depletion in seawater induces discoloration of sporophytes and diminishes their commercial value (Endo et al. 2017).

The effects of ocean warming on seaweed communities have been reported worldwide (Harley et al. 2012). Increased ocean temperatures in the SCA have been reported by the Japan Meteorological Agency and Kakehi et al. (2021). Higashi et al. (2020), using a numerical model that projects environmental conditions in the

(a)

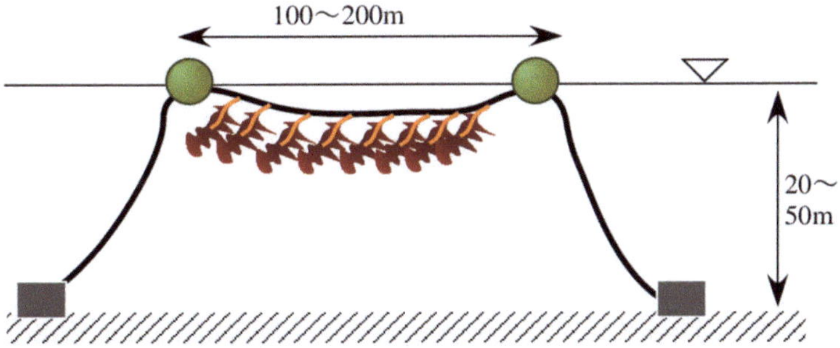

(b)

Fig. 7.2 Cultivation of *Undaria pinnatifida*. (**a**) Cultivation facility and (**b**) photo of cultivated sporophytes in the SCA

2090s in the Representative Concentration Pathway 8.5 (RCP8.5) scenario, a scenario of comparatively high greenhouse gas emissions without climate change mitigation efforts, showed that the winter water temperature in the 2090s in the eastern area of the SIS will rise by approximately 3 °C compared to that seen in the 1990s. Thus, the environmental conditions with respect to *U. pinnatifida* are projected to change. This might induce changes in the growth of the species and its potential for climate change mitigation.

The Comprehensive Research on the Projection of Climate Change Impacts and Evaluation of Adaptation Project (S-18 Project) (Mimura 2025) emphasizes the need to assess the impacts of climate change on fisheries and aquaculture and explore potential adaptation strategies. In this project, we evaluated the effects of climate change on *U. pinnatifida* growth in the SCA and SIS. We developed an individual-based growth model specific to each SCA and SIS to reproduce the total length of *Undaria* sporophytes in each area. Using this growth model, we projected the growth of *Undaria* sporophytes using the environmental conditions projected by climate models in RCP scenarios, and compared them with the current data.

7.2 Materials & Methods

7.2.1 Characteristics of Two Major Cultivation Areas

In this study, we evaluated the effects of climate change on *U. pinnatifida* growth in the SCA and SIS. The SCA is characterized by relatively low temperature and high nutrient environments. The cultivation of *U. pinnatifida* in the area is mainly controlled by nutrient concentrations (Kakehi et al. 2018). On the contrary, in the SIS, the cultivation environments are characterized by relatively high temperature and low nutrient concentrations, and temperature is a critical factor that controls *U. pinnatifida* growth (Onitsuka et al. 2024). Due to these environmental differences, we developed an individual-based growth model specific to each cultivation area. For the SIS, because high water temperature causes delays in the start of cultivation, we considered climate change adaptation using a high-temperature-tolerant cultivar, which is already being developed in the area.

7.2.2 Individual-Based Growth Model for U. pinnatifida *Sporophytes in the SCA*

Kakehi et al. (2024) developed an individual-based growth model for *U. pinnatifida* sporophytes and assessed the possible effects of climate change on its growth in the SCA. Temporal evolution of the total length (*TL*) of cultured *U. pinnatifida* was formulated as the temporal change in the total length due to photosynthesis and respiration. The photosynthesis rate was calculated based on temperature, nitrogen concentration, and light intensity, according to Kitadai and Kadowaki (2004). The respiration rate was defined as a function of temperature according to Sato et al. (2021). Unknown parameters included in the formulae were determined using case studies to reproduce the observed total length. Details are shown in Kakehi et al. (2024).

7.2.3 Projection of U. pinnatifida *Sporophyte Growth in the SCA*

When projecting the growth of *U. pinnatifida* sporophytes using the environmental conditions projected by climate models, the bias correction for the model outputs is essential. To correct the bias of the climate models by observed data, we used datasets observed by public fishery research institutes on environmental conditions along the coast of the SCA. Temperature and nitrate concentration data by the Iwate and Miyagi Fishery Research Institutes were available for the six points shown in Fig. 7.1. The 5-year average daily temperature at the points was estimated from the downloaded daily temperature data measured using a fixed-station buoy. The 5-year average daily nitrate concentration at the points was estimated using the downloaded monthly observed data through 5-year-averaging and linear interpolation. Daily downward solar radiation data obtained from the NCEP/DOE AMIP-II reanalysis (NCEP2) was averaged over 5 years to determine light intensity at the points.

To project potential *Undaria* sporophyte growth, we conducted our growth model calculation using the future environmental conditions by FORP-NP10 version 4, environmental datasets projected by climate models with a horizontal resolution of 10 km (Nishikawa et al. 2021). From the datasets, outputs by two climate models were provided: MIROC5 and MRI-CGCM3. Three experiments were conducted for each model: historical (1960–2005), RCP2.6 (2006–2100), and RCP8.5 (2006–2100) scenarios. In RCP2.6 scenario, low greenhouse gas emissions are required under stringent climate change mitigation policies. The grid point (GP) values for the environmental conditions (water temperature, nitrate concentration, and downward solar radiation) adjacent to the observation stations were extracted from the datasets (GP 1–6 in Fig. 7.1). Biases between the observed and modeled values were corrected by adding the model anomaly (projecting (RCP) calculation minus historical calculation) to the observation-based 5-year average environmental conditions following Yara et al. (2011).

The beginning of *U. pinnatifida* cultivation (i.e., installation of germinated sporophytes) for our projected calculations was set when the nitrate concentration exceeded 20 μg L^{-1} at temperatures <20 °C. The initial *TL* was set to be 1 cm. The end of cultivation was set when nitrate concentrations decreased to <20 μg L^{-1} or temperatures were > 20 °C. Daily *TL* was calculated using our growth model using the environmental conditions projected by two climate models with two scenarios. Using these results, we estimated *TL* at the end of cultivation (final *TL*), the duration of cultivation, and average temperatures during cultivation, and compared the values between the 2090s and the 2010s.

7.2.4 *Individual-Based Growth Model for* U. pinnatifida *Sporophytes in the SIS*

Onitsuka et al. (2024) developed an individual-based growth model for *U. pinnatifida* sporophytes and assessed the effects of projected climate change on its growth in the SIS. The model incorporates factors, including water temperature, dissolved inorganic nitrogen (DIN), current speed, and cultivation density. Additionally, the erosion at the blade tip was considered based on previous studies that simulated brown macroalgae (Broch and Slagstad 2012). Total length (*TL*) data collected from field observations conducted by Takenaka et al. (2021) were used to validate model performance. The equations and parameters for the growth model are detailed by Onitsuka et al. (2024).

7.2.5 *Projection for the Growth of* U. pinnatifida *Sporophytes in the SIS*

We conducted simulations across the entire coastal area of the SIS for both the 1990s and 2090s. To simulate the growth of *U. pinnatifida* sporophytes, we utilized environmental datasets comprising daily average water temperature, DIN concentration, and current speed at the surface generated using a coupled hydrodynamic-biogeochemical model with a horizontal resolution of 1 km (Higashi et al. 2020; Higashi 2022), because the bathymetry of the SIS is more complex than that of the SCA. Additionally, the M_2 tidal current amplitude, calculated using the two-dimensional tidal model (Guo et al. 2013), was incorporated into the daily average current speed in each grid.

In the cultivation process, juvenile sporophytes attached to seed ropes are transferred to the ambient sea area for nursery cultivation once the water temperature decreases below 23 °C. After 20 days of nursery cultivation, full cultivation begins at farming area. Therefore, the growth model simulations were initiated 20 days after the water temperature decreased below 23 °C in each grid. The initial sporophyte size was set at 1 cm. All simulations for both the 1990s and 2090s in climate change scenarios (RCP2.6, 4.5, 6.0, and 8.5) were conducted until April 30 each year.

In recent years, new cultivars with high temperature tolerance have been developed through crossbreeding as part of climate change adaptation strategies (Murase et al. 2021). We conducted another simulation focusing on climate change adaptation in the 2090s using a new cultivar. Assuming that the new cultivar could thrive upon early start timing (i.e., high water temperature conditions) for nursery and subsequent full cultivation, the adaptation simulation commenced 20 days after the water temperature decreased below 25 °C, while maintaining the other conditions same.

7.3 Results

7.3.1 *Projection of Impact of Climate Change on* U. pinnatifida *Growth in the SCA*

Projection results in the SCA revealed that the start and end of cultivation was regulated by the nutrient concentration (20 μg L^{-1}) in all cases and there was no case in which temperatures >20 °C limited these times. The *TL* at the end of the cultivation period in the 2090s and the 2010s at six GPs (GP 1–6) are shown in Fig. 7.3 and Table 7.1. Results obtained using MIROC5 in RCP2.6 showed that the *TL* in the 2090s was significantly longer than that in the 2010s at GP 1–4. *TL* in the 2090s was significantly shorter than that in the 2010s at GP 4–6 as seen from results obtained using MIROC5 in RCP8.5. The final *TL* as projected for the 2090s in results of MRI-CGCM3 in RCP2.6 was longer than that in the 2010s for all GPs, although the differences were not significant. In contrast, the final *TL* in the 2090s in MRI-CGCM3 in RCP8.5 was similar to or shorter than that in the 2010s, and the difference was significant at GP 5 and 6. Thus, it was projected that a latitudinal difference in the *TL* would occur, with a boundary at approximately 39 °N. For cultivation duration, a significant reduction was mainly found in RCP8.5 regardless of the

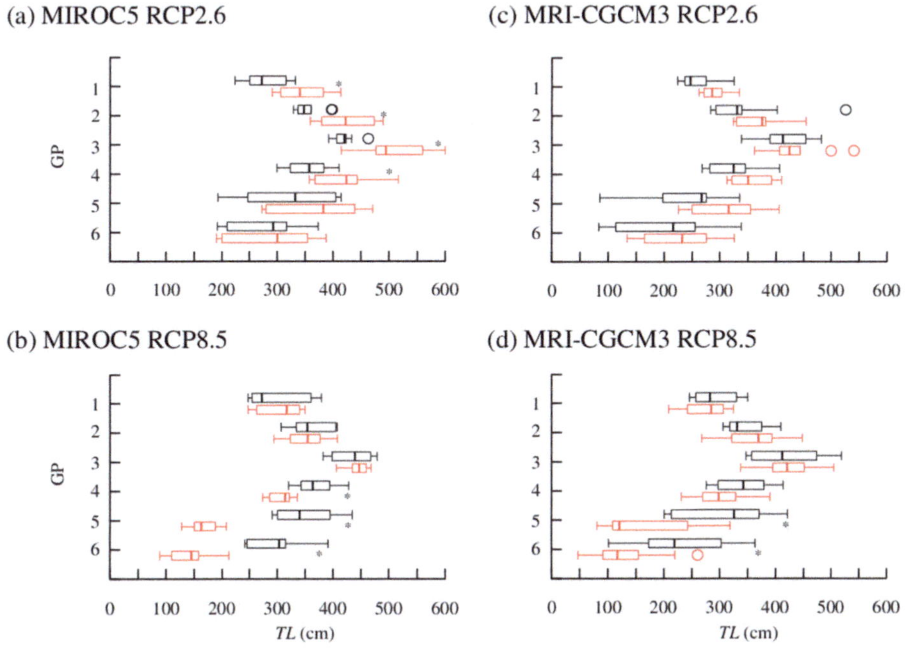

Fig. 7.3 Box plot showing total length of cultured *Undaria* sporophytes in the 2010s (black) and 2090s (red) in the SCA. Asterisks indicate significant difference ($p < 0.01$) based on Welch's t test. (Modified from Kakehi et al. (2024) with permission from Springer Nature)

Table 7.1 Averaged difference between the 2090s and 2010s in total length (*TL*) of cultured *Undaria* sporophytes, cultivation duration, and averaged temperatures during cultivation (experienced temperatures)

Model	RCP	GP	Latitude (°)	*TL* (cm)	Duration (day)	Ex. temp. (°C)
MIROC5	2.6	1	40.12	65.3*	8.2	1.4*
		2	39.46	69.7*	5.3	1.5*
		3	39.25	83.6*	7.5	1.6*
		4	38.96	59.6*	4.8	1.5*
		5	38.83	43.9	3.3	1.5*
		6	38.70	3.0	−12.7	1.2*
	8.5	1		7.7	−42.6*	4.3*
		2		−9.5	−53.9*	4.1*
		3		9.0	−50.3*	4.8*
		4		−59.7*	−67.7*	3.0*
		5		−183.3*	−104.1*	2.0*
		6		−156.9*	−101.2*	3.2*
MRI-CGCM3	2.6	1		32.7	10.3	0.4
		2		29.3	7.3	0.3
		3		18.9	−4.5	0.1
		4		33.0	7.4	0.5
		5		71.2	26.0	0.8
		6		27.9	17.1	0.2
	8.5	1		−17.4	−28.2*	2.1*
		2		12.6	−31.5*	2.6*
		3		0.7	−29.7*	3.1*
		4		−38.5	−50.3*	2.3*
		5		−141.5*	−82.7*	1.6*
		6		−98.4*	−78.0*	2.5*

Modified from Kakehi et al. (2024) with permission from Springer Nature
Results from the projections with two models and two scenarios at grid points (GP) 1–6 in the Sanriku coastal area (SCA). Asterisks indicate significant difference ($p < 0.01$) based on Welch's *t*-test

projection model used (Table 7.1). A significant increase in the experienced temperatures was projected in all cases except for results obtained using MRI-CGCM3 in RCP2.6, regardless of the GP.

7.3.2 *Projection of Impact of Climate Change on* U. pinnatifida *Growth in the SIS*

We compared the growth dynamics of sporophytes between the 1990s and the 2090s using an environmental dataset from the primary cultivation region around the Naruto Strait in the SIS (Fig. 7.4). During the 1990s, the start dates of full

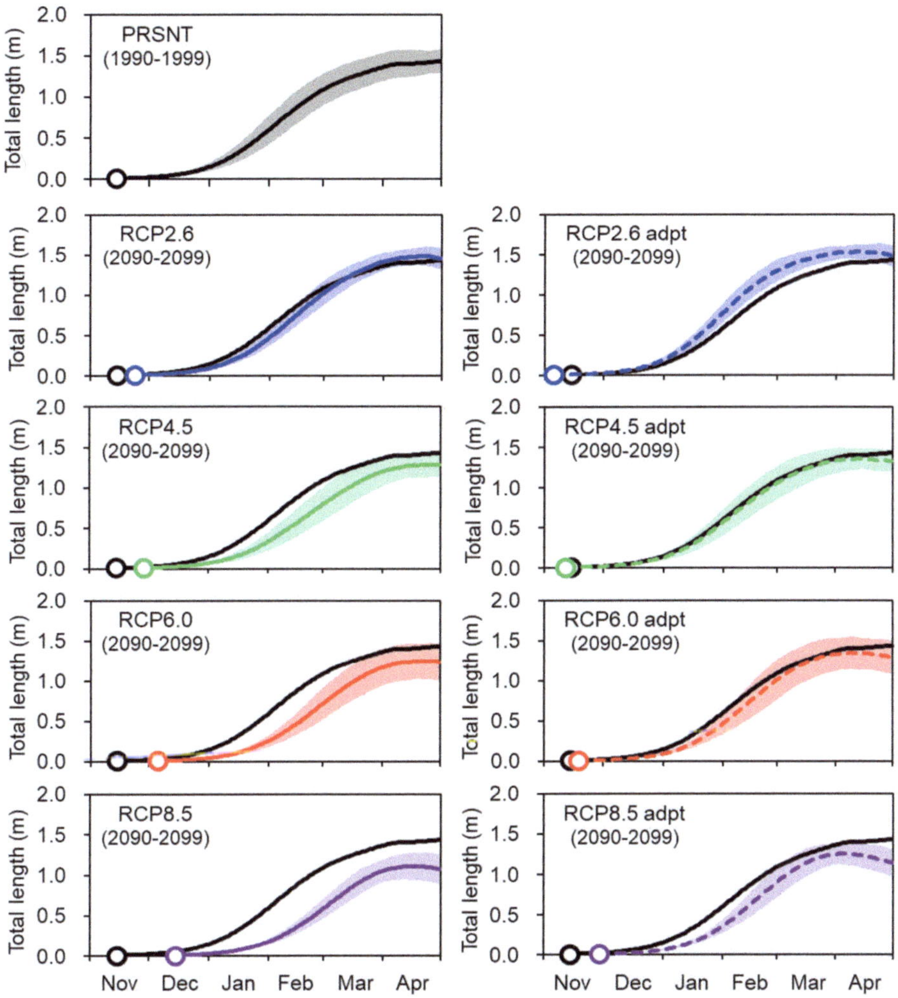

Fig. 7.4 Temporal changes in the total length of *Undaria pinnatifida* sporophytes in the primary cultivation region around the Naruto Strait in the SIS during the 1990s in present climatic conditions (PRSNT) and that projected for the 2090s based on the RCP2.6, 4.5, 6.0, and 8.5 scenarios. The cases for climate change adaptation using a high-temperature-tolerant cultivar are depicted in the right four figures (RCP2.6, 4.5, 6.0, and 8.5 adpt). Open circles indicate the median start dates of cultivation. Solid and dashed lines, and colored areas represent the means and standard deviation for 10 years, respectively. The start date of cultivation and the mean total length of sporophytes in the 1990s are presented overlaying those projected to be in the 2090s. (Reproduced from Onitsuka et al. (2024) with permission from Springer Nature)

cultivation, averaged over the area for a decade, were mid-November. The *TL* of sporophytes was approximately 1 m in late February, surpassing 1.4 m in early April. However, in the 2090s, the initiation of cultivation was consistently delayed across all RCP scenarios owing to ocean warming. By the end of March, the mean

TL ratios to the 1990s were 104%, 88%, 86%, and 77% in RCP2.6, 4.5, 6.0, and 8.5, respectively.

In simulations conducted in the 2090s for climate change adaptation using a high-temperature-tolerant cultivar (RCP2.6, 4.5, 6.0, and 8.5 adpt), the mean start dates of cultivation were advanced compared with those of the conventional cultivar. At the end of March, the mean *TL* ratios to the 1990s were 112%, 98%, 97%, and 92% in RCP2.6, 4.5, 6.0, and 8.5, respectively.

Notable geographical variations in sporophyte growth were observed during the 1990s (Fig. 7.5a).While *TL*s projected for 2090s in the RCP2.6 scenario exhibited a partial increase of up to 10% in the eastern SIS compared with those observed in the 1990s, *TL*s in the RCP8.5 scenario decreased in most regions (Fig. 7.5b, c). Particularly, a decrease rate of >40% was observed in the western SIS areas in the RCP8.5 scenario. In simulations conducted for the 2090s for climate change adaptation using a high-temperature-tolerant cultivar, *TL*s were projected to increase across all areas compared to those without adaptation in the 2090s (Fig. 7.5d, e). In the RCP2.6 scenarios with adaptation, *TL*s projected for the 2090s surpassed those of the 1990s in nearly all areas of the SIS, with the highest increase in the eastern SIS. Even in the RCP8.5 scenario, *TL*s in the eastern SIS were greater than those observed in the 1990s.

7.4 Discussion

Geographical variations in *TL* of *U. pinnatifida* sporophyte were projected to occur in the future: in SCA, the difference between the north and south of the location, and in the SIS, the difference between the east and west of the location. In particular, in RCP8.5, a decrease in *TL* (−50%) is projected south of 39° N in the SCA and a greater decrease in *TL* (−40%) is projected in the western SIS, indicating a large-scale reduction in *U. pinnatifida* yield in the future.

Decreased duration of cultivation, the delay of nutrient supply in autumn, and the forward shift of nutrient depletion in spring were projected for the SCA. The delay and forward shift of nutrient dynamics was induced by the delay of the occurrence of vertical mixing and forward shift of the onset of stratification, respectively. This suggests that the cultivation schedule will change significantly from the current schedule. As the capacity for harvesting and processing sporophytes and shipping and distributing the products is limited, efficient management of harvest operations is necessary. Monitoring environmental conditions and conducting cultivation operations at appropriate times are important.

In the SIS, majority of areas exhibited decreased sporophyte growth in the 2090s compared to that seen in the 1990s, except for the eastern area in the RCP2.6 scenario. This decline is attributed to delayed cultivation start times associated with ocean warming and reduced DIN concentrations. Simulations of climate change adaptation using a high-temperature-tolerant cultivar indicated that yield losses could be mitigated, even in the RCP 8.5 scenario.

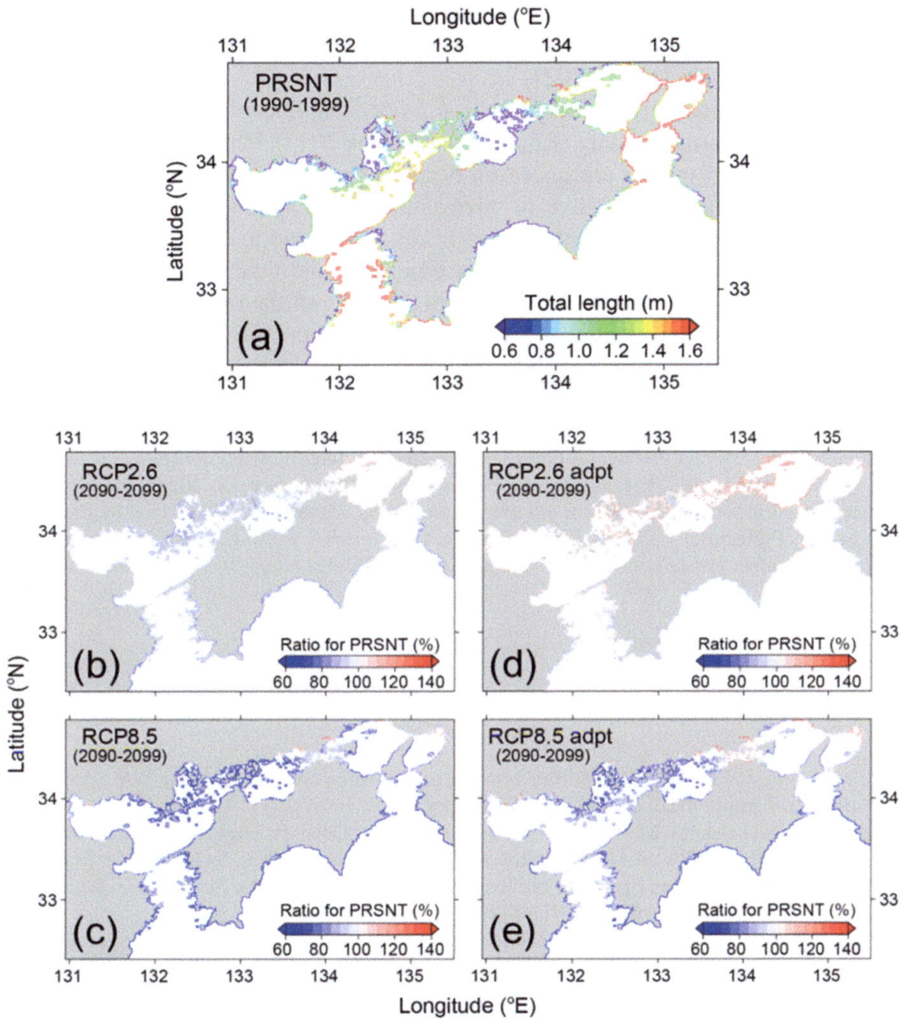

Fig. 7.5 (**a**) Distribution of the total length of *Undaria pinnatifida* sporophytes on 31 March, averaged over that in the 1990s in the SIS. (**b**) and (**c**) Distribution of the ratios of the total length of *Undaria pinnatifida* sporophytes in 2090s in RCP2.6 and 8.5 scenarios, compared to that in the 1990s. (**d**) and (**e**) The same as (**b**) and (**c**), however, for climate change adaptation. (Modified from Onitsuka et al. (2024) with permission from Springer Nature)

Thus, negative effect (decrease in *TL*) of climate change was projected in many locations in both the areas. However, projection results showed that there were some locations where positive effect of climate change (increase in *TL*) would occur depending on models and scenarios (e.g., MIROC5 in RCP2.6 in the SCA and eastern area in RCP2.6 in the SIS). In these cases, warmer temperature accelerated growth rates under adequate nutrient conditions in the low temperature area.

Nevertheless, temperature increase may negatively affect cultivated *Undaria* sporophytes because of factors that were not incorporated into our growth models, such as herbivory risk (Endo et al. 2021).

Uncertainties included in the projection models significantly affected the accuracy of the results of this study (Knutti and Sedláček 2013). Particularly, projections of nutrient concentrations are uncertain because they are calculated using ecosystem processes that are sensitive to environmental changes (Doney et al. 2012). Nutrient inputs from rivers may change in the future due to changes in river discharge and human activity with high uncertainties, and their effects are likely to vary geographically owing to differences in the scale of the rivers. Therefore, it is essential to monitor nutrient concentrations and conduct cultivation under suitable nutrient conditions for *U. pinnatifida* growth.

Acknowledgments Daily temperature data were measured using a fixed-station buoy deployed by the Iwate and Miyagi Prefectural Fisheries Research Institute (https://www.suigi.pref.iwate.jp/teichi; https://suisan-navi.pref.miyagi.jp/suion_top; last accessed July 12, 2024). We also used nitrate concentrations obtained from monthly hydrographic observations conducted by the Iwate and Miyagi Prefectural Fisheries Research Institute (https://www2.suigi.pref.iwate.jp/research_log/nutrient and https://www.pref.miyagi.jp/life/1/63/12/index.html; last accessed on July 12, 2024).

References

Broch OJ, Slagstad D (2012) Modelling seasonal growth and composition of the kelp Saccharina latissima. J Appl Phycol 24:759–776. https://doi.org/10.1007/s10811-011-9695-y

Doney SC, Ruckelshaus M, Emmett Duffy J, Barry JP, Chan F, English CA, Galindo HM, Grebmeier JM, Hollowed AB, Knowlton N, Polovina J, Rabalais NN, Sydeman WJ, Talley LD (2012) Climate change impacts on marine ecosystems. Annu Rev Mar Sci 4:11–37. https://doi.org/10.1146/annurev-marine-041911-111611

Duarte CM, Wu J, Xiao X, Bruhn A, Krause-Jensen D (2017) Can seaweed farming play a role in climate change mitigation and adaptation? Front Mar Sci 4:1–8. https://doi.org/10.3389/fmars.2017.00100

Endo H, Okumura Y, Sato Y, Agatsuma Y (2017) Interactive effects of nutrient availability, temperature, and irradiance on photosynthetic pigments and color of the brown alga *Undaria pinnatifida*. J Appl Phycol 29:1683–1693. https://doi.org/10.1007/s10811-016-1036-8

Endo H, Sato Y, Kaneko K, Takahashi D, Nagasawa K, Okumura Y, Agatsuma Y (2021) Ocean warming combined with nutrient enrichment increases the risk of herbivory during cultivation of the marine macroalga *Undaria Pinnatifida*. ICES J Mar Sci 78:402–409. https://doi.org/10.1093/icesjms/fsaa069

Epstein G, Smale DA (2017) *Undaria pinnatifida*: a case study to highlight challenges in marine invasion ecology and management. Ecol Evol 7:8624–8642. https://doi.org/10.1002/ece3.3430

FAO (2022) The state of world fisheries and aquaculture 2022. Towards blue transformation. FAO, Rome

Gao X, Endo H, Taniguchi K, Agatsuma Y (2013) Combined effects of seawater temperature and nutrient condition on growth and survival of juvenile sporophytes of the kelp *Undaria pinnatifida* (Laminariales; Phaeophyta) cultivated in northern Honshu, Japan. J Appl Phycol 25:269–275. https://doi.org/10.1007/s10811-012-9861-x

Guo X, Harai K, Kaneda A, Takeoka H (2013) Simulation of tidal currents and nonlinear tidal interactions in the Seto Inland Sea, Japan. Report of Research Institute Applied Mechanics Kyushu University 145:43–52. https://doi.org/10.15017/1526096

Harley CD, Anderson KM, Demes KW, Jorve JP, Kordas RL, Coyle TA, Graham MH (2012) Effects of climate change on global seaweed communities. J Phycol 48:1064–1078. https://doi.org/10.1111/j.1529-8817.2012.01224.x

Higashi H (2022) Prediction dataset of climate change impact on water environment in the Seto Inland Sea, ver2021. NIES

Higashi H, Yokoyama A, Nakada S, Yoshinari H, Koshikawa H (2020) Climate change impacts on primary production and water quality in the Seto Inland Sea under RCP8.5 scenario. Journal of Japan society. Civ Eng B2 76:I-1147–I-1152. https://doi.org/10.2208/kaigan.76.2_I_1147. (in Japanese)

Kakehi S, Naiki K, Kodama T, Wagawa T, Kuroda H, Ito SI (2018) Projections of nutrient supply to a wakame (*Undaria pinnatifida*) seaweed farm on the Sanriku Coast of Japan. Fish Oceanogr 27:323–335. https://doi.org/10.1111/fog.12255

Kakehi S, Narimatsu Y, Okamura Y, Yagura A, Ito SI (2021) Bottom temperature warming and its impact on demersal fish off the Pacific coast of northeastern Japan. Mar Ecol Prog Ser 677:177–196. https://doi.org/10.3354/meps13852

Kakehi S, Kidokoro H, Setoh T, Onitsuka G (2024) Assessment of climate change impacts on large brown seaweed (*Undaria pinnatifida*) growth in the Sanriku coastal area, Japan. J Appl Phycol. https://doi.org/10.1007/s10811-024-03382-z

Kim H, Ra CH, Kim SK (2013) Ethanol production from seaweed (*Undaria pinnatifida*) using yeast acclimated to specific sugars. Biotechnol Bioprocess Eng 18:533–537. https://doi.org/10.1007/s12257-013-0051-8

Kitadai Y, Kadowaki S (2004) Sennkai yougyojou ni okeru saibai wakame, the growth process and N, P uptake rates of *Undaria pinnatifida* cultured in coastal fish farms. Aquaculture 52:365–374. https://doi.org/10.11233/aquaculturesci1953.52.365. (in Japanese)

Knutti R, Sedlacek J (2013) Robustness and uncertainties in the new CMIP5 climate model projections. Nat Clim Chang 3:369–373. https://doi.org/10.1038/nclimate1716

Mimura N (2025) Chapter 1 climate change impacts and adaptation strategies: Japan's challenges to integrated research. In: Mimura N, Takewaka S (eds) Climate change impacts and adaptation strategies in Japan. Springer, Tokyo, pp 3–16

Ministry of Agriculture, Forestry and Fisheries (2023) Fisheries and aquaculture statistics in 2021. https://www.e-stat.go.jp/stat-search/files?page=1&layout=datalist&toukei=00500216&tstat=000001015174&cycle=7&year=20220&month=0&tclass1=000001015175&tclass2=000001214760 (in Japanese)

Murase N, Tanada N, Tada A, Shimabukuro H, Yoshida G, Abe M, Noda M (2021) The growth characteristic under high water temperature of young sporophyte of intraspecific crossbreeding of *Undaria pinnatifida* (Laminariales, Phaeophyta) between Naruto cultivar and Tsubaki natural strain, Tokushima prefecture, Japan. J Natl Fisher Univ 69(4):81–88. (in Japanese)

Murata M, Sano Y, Ishihara K, Uchida M (2002) Dietary fish oil and *Undaria pinnatifida* (wakame) synergistically decrease rat serum and liver triacylglycerol. J Nutr 132:742–747. https://doi.org/10.1093/jn/132.4.742

Nishikawa S, Wakamatsu T, Ishizaki H, Sakamoto K, Tanaka Y, Tsujino H, Yamanaka G, Kamachi M, Ishikawa Y (2021) Development of high-resolution future ocean regional projection datasets for coastal applications in Japan. Prog Earth Planet Sci 8:1–22. https://doi.org/10.1186/s40645-020-00399-z

Onitsuka G, Yoshida G, Shimabukuro H, Takenaka S, Tamura T, Kakehi S, Setou T, Guo X, Higashi H (2024) Modeling the growth of cultivated seaweed *Undaria pinnatifida* under climate change scenarios in the Seto Inland Sea, Japan. J Appl Phycol 36:3077. https://doi.org/10.1007/s10811-024-03291-1

Sato Y, Nagoe H, Nishihara GN, Terada R (2021) The photosynthetic response of cultivated juvenile and mature *Undaria pinnatifida* (Laminariales) sporophytes to light and temperature. J Appl Phycol 33:3437–3448. https://doi.org/10.1007/s10811-021-02535-8
Takenaka S, Kohno Y, Tamura T, Sakaguchi H, Takechi A, Shimabukuro H, Yoshida G (2021) Growth of *Sargassum fusiforme*, *Undaria pinnatifida* and *Meristotheca papulosa* when cultivated in different water temperature conditions from the Seto Inland Sea to Bungo Channel. Nippon Suisan Gakkishi 87:375–385. https://doi.org/10.2331/suisan.20-00045. (in Japanese)
Yamanaka R, Akiyama K (1993) Cultivation and utilization of *Undaria pinnatifida* (wakame) as food. J Appl Phycol 5:249–253. https://doi.org/10.1007/BF00004026
Yara Y, Oshima K, Fujii M, Yamano H, Yamanaka Y, Okada N (2011) Projection and uncertainty of the poleward range expansion of coral habitats in response to sea surface temperature warming: a multiple climate model study. Journal of Coral Reef Study 13:11–20. https://doi.org/10.3755/galaxea.13.11

Part III
Natural Ecosystems

Chapter 8
The Advancement of Climate Change Impact Assessment Methodology Project at National Institute for Environmental Studies

Naota Hanasaki, Masashi Okada, Noriko Ishizaki, Yuji Masutomi, Tomomi Inoue, Qinxue Wang, Dai Koide, Ayato Kohzu, Naoki Kumagai, Hironori Higashi, Kazutaka Oka, Fumiko Ishihama, and Seiji Hayashi

Abstract Various impacts of climate change are becoming apparent in Japan. It is important to predict and adapt to the impacts of climate change on various scales and in various fields. The National Institute for Environmental Studies' fifth five-year plan (2021–2025) conducts the Climate Change Adaptation Research Program "the Advancement of Climate Change Impact Assessment Methodology Project (PJ2)." PJ2 targets four fields worldwide, three in the Asia-Pacific region, six in Japan, and one in the Abukuma River Basin in Fukushima Prefecture. It analyzes adaptation measures by predicting the effects of climate change under common climate and socioeconomic scenarios. The main results of the predictions are published on the Climate Change Adaptation Information Platform (A-PLAT) and the Asia-Pacific Climate Change Adaptation Information Platform (AP-PLAT). This chapter introduces the structure and activities of PJ2 and discusses future prospects.

Keywords Water resources · Food production · Terrestrial ecosystems · Aquatic ecosystems · Human health · Renewable energy · Natural hazards · Codesign and coproduction

---

N. Hanasaki (✉) · M. Okada · N. Ishizaki · Y. Masutomi · T. Inoue · Q. Wang · D. Koide
A. Kohzu · N. Kumagai · H. Higashi · K. Oka · F. Ishihama · S. Hayashi
National Institute for Environmental Studies, Tsukuba, Japan
e-mail: hanasaki@nies.go.jp

N. Mimura, S. Takewaka (eds.), *Climate Change Impacts and Adaptation Strategies in Japan*, https://doi.org/10.1007/978-981-96-2436-2_8

8.1 Introduction

The global mean temperature has risen by about 1.1 degrees Celsius compared to the pre-industrial level and is expected to continue to rise (MEXT and JMA 2020). With this climate change, abnormal weather conditions, such as heat waves and heavy rains, are observed frequently, and various damages are reported. It is important to predict the impacts of climate change that may occur in the future and implement appropriate adaptation measures in all fields.

Many predictions of the impacts of climate change have been made so far. In the Assessment Report (AR) of the Working Group 2 (WG2) of the Intergovernmental Panel on Climate Change (IPCC), which is published every 5–8 years, the latest findings on the studies of climate change impacts in various fields are summarized based on more than 30,000 research papers (IPCC 2022). Also, in the Climate Change Impact Assessment Report (MOE 2020) issued in 2020, which is expected to be published every five years, the findings of Japan's impact assessment are summarized.

Thus, regular and comprehensive climate change impact assessments have been conducted for a long time. The basis of such evaluations is the implementation of cutting-edge and multifaceted climate change impact studies and publishing results. The subjects, regions, and periods for conducting climate change impact assessments vary widely, and a complete picture of impacts and adaptation measures cannot be drawn unless many researchers cooperate in performing them. Therefore, it is important for many researchers to standardize future prediction assumptions in advance and to aggregate and publish the results of future predictions and the knowledge gained. In Japan, five-year research projects called S-4 (2005–2009), S-8 (2010–2014), and S-18 (2020–2024), which received research funding called the Environmental Research Promotion Fund, were conducted, and comprehensive climate change impact assessments were performed nationwide. These projects were compiled into an extensive report (for example, Climate Change Impact Prediction Project Team 2009), and they contributed significantly to Japan's climate policy.

In December 2018, the Climate Change Adaptation Act was enacted in Japan. As a part of this, the Center for Climate Change Adaptation (CCCA) was established at the National Institute for Environmental Studies (NIES). The most important role of the CCCA is to aggregate information on Japan's climate change impact assessment and adaptation measures for government and private sector decision makers and publish it on a website called Climate Change Adaptation Information Platform (A-PLAT). In addition, a website called the Asia-Pacific Climate Change Adaptation Information Platform (AP-PLAT) has also been developed and made public to support climate change adaptation in the Asia-Pacific region. Another task of the CCCA is to conduct research for climate change impact assessment and adaptation measures.

As a pillar of CCCA's research activities, the Climate Change Adaptation Research Program was launched in December 2018. In April 2021, it was

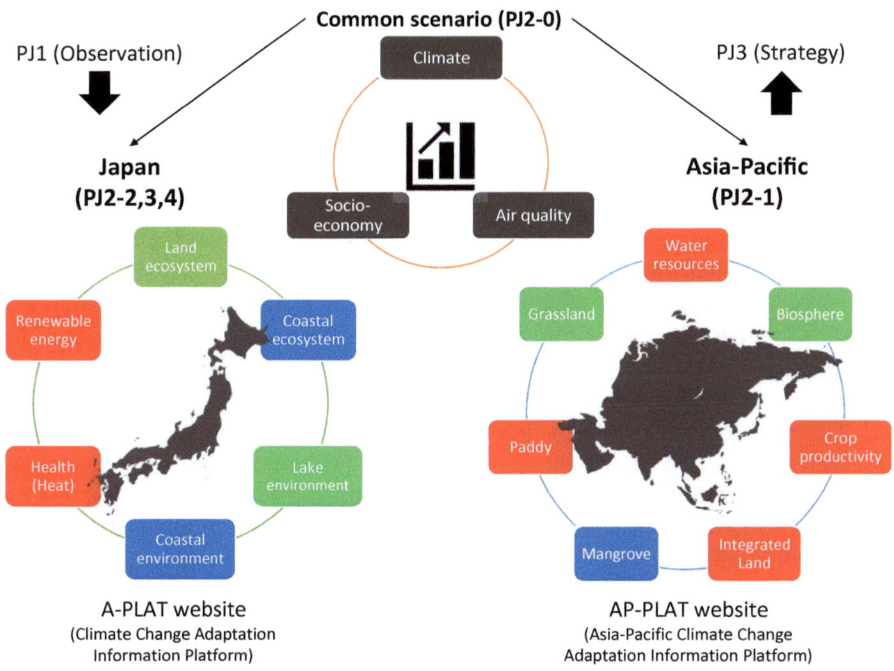

Fig. 8.1 Schematic diagram of PJ2

reorganized to coincide with the start of the NIES's fifth five-year plan (2021–2025). The Climate Change Adaptation Research Program is an activity within NIES. While cooperation and coordination with external organizations are encouraged, research is conducted primarily by researchers within the institute. The Climate Change Adaptation Research Program consists of a research project mainly on the monitoring and detection of climate change impacts, called "Research on evaluation and clarification of effects and mechanisms of climate change," a research project mainly on the modeling and prediction of climate change impacts, called "Research on Advancement of Climate Change Impact Assessment Methodology," and a research project mainly on the formulation and practice of climate change adaptation measures, called "Research to enhance adaptation strategies and practices based on scientific projections," each referred to as PJ1, PJ2, PJ3 (Fig. 8.1). PJ2 is a comprehensive project for predicting the impacts of climate change, similar to S-4, S-8, and S-18. These projects are conducted only by NIES researchers. Still, NIES is a prominent research institute with a wide range of expertise among its researchers, making it possible to perform a wide range of impact assessments. NIES has particular depth and tradition in the field of natural ecosystems and health, stemming from its predecessor, the National Institute for Pollution Research.

This chapter introduces PJ2, which is responsible for research on climate change impact prediction within the Climate Change Adaptation Research Program conducted by CCCA, a permanent institution for climate change established at NIES. It discusses the contribution to impact assessment in Japan and the Asia-Pacific region, as well as the path to providing information and implementing adaptation measures.

8.2 Structure of PJ2

PJ2 is conducting impact assessments and adaptation studies on climate change at four scales: global, Asia-Pacific region, nationwide in Japan, and the Abukuma River basin in Fukushima Prefecture. The reasons for defining these four areas are as follows. First, the nationwide scale in Japan is due to the expectation that the CCCA will primarily provide information on climate change impact assessments and adaptation measures through A-PLAT. In particular, with local climate change adaptation centers established in almost all prefectures, there is a demand for impact assessments covering the whole of Japan. Next, the Asia-Pacific regional scale is due to the expectation that the CCCA will disseminate information to the region through AP-PLAT. The global scale is due to the existence of multiple global impact assessment models at NIES that have received international evaluations that fully cover the Asia-Pacific region. Finally, for the Abukuma River basin in Fukushima Prefecture, it was decided to cooperate with the NIES Fukushima Regional Collaboration Research Base to consider disaster recovery and climate change adaptation.

PJ2 consists of 12 research topics (Fig. 8.1; Table 8.1). The research topics can be divided into five major categories. The first is a topic to consider future assumptions necessary for climate change impact assessments and to provide ready-to-use data (PJ2-0). The second is a topic on the extension and application of global-scale impact assessment models (PJ2-1). Specifically, the extension and application of four models of agricultural productivity, water resources, terrestrial ecosystems, and integrated land are carried out. The third is a topic on the extension and application of Asia-Pacific regional scale impact assessment models (PJ2-2). Based on their expertise, three research groups focus on rice and health, mangroves, and Mongolian grasslands. The fourth is a topic on the extension and application of nationwide scale impact assessment models in Japan (PJ2-3). The main targets are terrestrial ecosystems focusing on plants; lake water quality and ecosystems; coral and seaweed-bed water environment and ecosystems; coastal and closed seawater environment and ecosystems; health and energy; and species distribution mainly focusing on plants. The fifth topic is to examine the mitigation effects of water disaster management through water environment and ecosystem management in the Abukuma River basin in Fukushima Prefecture (PJ2-4).

PJ2 sets two major goals. The first is to conduct cutting-edge research on climate change impact assessments in each scale/field and publish as many high-quality papers as possible, as soon as possible. These papers are expected to be cited from the IPCC AR and the Japanese Climate Change Impact Assessment Report. The latter part is essential in the activities of the CCCA. Busy local government and corporate officials do not read individual research papers but use reports from

Table 8.1 Titles and principal investigators (PIs) of the subprojects

ID	Title	PI
PJ2-0	Development of climate, air quality, and socio-economic scenarios for impact assessment of climate change	Noriko Ishizaki
PJ2-1	Development and application of global scale sector-wise climate change impact assessment models	Masashi Okada
PJ2-2a	Assessment of adaptation measures for the combined impacts of climate change and air pollution on rice production and health in Asia	Yuji Masutomi
PJ2-2b	Mangrove functions for climate change adaptation: Projections and hazard assessments on storm surge and sea level rise	Tomomi Inoue
PJ2-2c	Climate change impact monitoring and adaptation assessment in grasslands	Qinxue Wang
PJ2-3a	Projection of climate change impacts on biodiversity, ecosystem functions, and ecosystem service levels in terrestrial ecosystems	Dai Koide
PJ2-3b	Projection of climate change impacts on lake environments	Ayato Kohzu
PJ2-3c	Improvement of prediction of changes in the distribution of coral and algal communities due to climate change	Naoki Kumagai
PJ2-3d	Climate change impact prediction and adaptation for water environment in coastal and semi-enclosed seas	Hironori Higashi
PJ2-3e	Climate change impact assessment in health and energy sectors	Kazutaka Oka
PJ2-3f	Advancement of methods for predicting and assessing changes in the distribution of biological species due to climate change	Fumiko Ishihama
PJ2-4	Assessment of climate change adaptation effectiveness through ecosystem-based management at a watershed scale	Seiji Hayashi

public institutions as their main source of information. The second is to conduct climate change impact predictions under common assumptions and disseminate the results through A-PLAT and AP-PLAT. The items for impact prediction are pre-adjusted so as not to overlap with the S-18 project carried out simultaneously, and the future assumptions are almost the same as for S-18. Combining the results of S-18 and PJ2 makes it possible to realize impact predictions in a wide range of fields in Japan.

8.3 Activities of PJ2

8.3.1 Common Future Assumptions (PJ2-0)

It is important to set common future assumptions to conduct comprehensive climate change impact predictions across multiple fields. This requires the creation of a document called a protocol that outlines the specifications of the prediction and the development of numerical data called scenarios that quantitatively represent future assumptions. The protocol includes information such as what period will be considered the past reference period, what period will be considered the future period, and what assumptions will be made about changes in climate and socioeconomics. Scenarios are time series and gridded data of weather, land use, and others. This is because many impact assessment models are grid-based (dividing the earth's surface into equal grid cells as the smallest unit) and require daily and monthly weather conditions.

PJ2 has set two protocols and scenarios for the globe and Japan. First, for the globe, it was made identical to the international project Inter-Sectoral Impact Model Intercomparison Project (ISIMIP; https://www.isimip.org/). ISIMIP is an international research project launched in 2012 under the leadership of the Potsdam Institute for Climate Impact Research (PIK) in Germany to conduct climate change impact assessments on a global scale across multiple fields. ISIMIP has gained the support of many researchers and has successfully evaluated the impacts of climate change using various models and fields under common future assumptions, which had been difficult until then (Warszawski et al. 2014). This has greatly contributed to the IPCC's fourth and subsequent assessment reports. ISIMIP has been developing a protocol that aggregated opinions from various fields, and distributing scenarios based on the latest climate science (Frieler et al. 2024). Many research groups at NIES are participating in ISIMIP, so PJ2 decided to use ISIMIP's protocols and scenarios as they are. Next, for Japan, PJ2 and S-18 participants developed their own after extensive discussions. The protocol determined the climate change impact prediction items to be implemented with prioritization, taking into account the availability of data that can cover the whole of Japan with uniform quality, the needs of decision-makers, and the feasibility of individual impact assessment models. Details about PJ2's protocols and scenarios can be found in Ishizaki et al. (2023).

8.3.2 Climate Change Impact Prediction (PJ2-1, PJ2-2, PJ2-3, PJ2-4)

PJ2-1, PJ2-2, PJ2-3, and PJ2-4 conduct impact predictions at their respective scales. PJ2-1 is conducting research using biological and physical process models related to global-scale agricultural productivity, water resources, terrestrial ecosystems, and integrated land. A biological and physical process model is a type of model that calculates various processes in time steps of not greater than a day and integrates the results. In addition to their own model development and simulation analysis, all models are participating in the ISIMIP project—they are also conducting intercomparison with other impact assessment models in the world and model ensemble impact prediction (Jägermeyr et al. 2021; Heinecke et al. 2024). Global-scale research is mainly conducted by simulation, and individual researcher-based field observation is seldom carried out.

PJ2-2 is conducting calculations using a biological and physical process model of rice in Southeast Asia, a statistical model of health, a physical process model related to coastal wave and mangrove's buffering effects in the Asia-Pacific region, and a physical and social model related to grasslands in Mongolia (Wang et al. 2024). They are also actively conducting local surveys.

PJ2-3 focuses on domestic ecosystems and issues related to health and energy, with many research topics where observation and modeling are proceeding simultaneously. The prediction methods vary, including process, statistical, and machine learning models. Among many research examples, the ones related to terrestrial plant species distribution and coral reefs and algae fields will be detailed in the following chapters by Koide et al. (2025) and Abe and Yamano (2025). Another significant research result is a study that estimated the nationwide distribution of pollinator's functional diversity based on the distribution of 14 species of bees by collaborating citizen science and modeling (Suzuki-Ohno et al. 2024).

PJ2-4 is conducting research using a physical process model on flood disasters in the Abukuma River basin in Fukushima Prefecture. Through detailed modeling, such as quantitative estimation of the effects of paddy dams that enhance the water storage function of paddy fields and inland flooding at the point where tributaries merge with the main river, they are conducting research to propose specific adaptation measures for the region.

8.3.3 Aggregation and Publication of Results

All research results of PJ2 are published as papers. Among them, the primary impact prediction results calculated based on the common future assumptions will be published from the Future Prediction of Climate Change WebGIS (hereinafter referred to as WebGIS; https://adaptation-platform.nies.go.jp/webgis/) for Japan and the Climate Impact Viewer (CIV; https://a-plat.nies.go.jp/ap-plat/asia_pacific/index.html) of AP-PLAT for the Asia-Pacific region (those calculated globally are extracted only for this region).

The main results of the climate change impact prediction in the climate change adaptation research program (2018–2020) implemented during the fourth five-year plan period (2016–2020) of NIES are being released from WebGIS and CIV. The former can be viewed by prefecture, and the latter can be viewed by a 30-minute latitude and longitude grid (about 55 km near the equator) for the entire Asia-Pacific region. Detailed information on how the predictions were made, how to interpret the results, and what to be careful of when using them is described in Hanasaki et al. (2023). WebGIS and CIV are Japan's unique information infrastructure elements that provide impact predictions in multiple fields for Japan and the Asia-Pacific region.

8.4 Conclusions and Future Prospects

PJ2 is a research project on climate change impact predictions and the effectiveness of adaptation measures in a wide range of fields led by a group of NIES researchers. By publishing many papers with the latest findings, we aim to contribute to creating climate change information. In the future, we hope that these papers will be cited in the IPCC WG2 AR and Japan's Climate Change Impact Assessment Report and that many decision-makers on climate change adaptation in Japan and the Asia-Pacific region will read these reports. In addition, we are providing detailed spatio-temporal information that cannot be fully conveyed in the above reports through the publication of common assumptions (protocols and scenarios) on A-PLAT and AP-PLAT and the posting of prediction results (numerical data) on WebGIS and CIV.

In terms of implementing and publicizing the results of cutting-edge climate change impact predictions, ISIMIP is leading the way and is worth introducing. At ISIMIP, individual climate change impact assessments are conducted voluntarily (i.e., for free) by researchers around the world who are developing global models. The representative institution, PIK, skillfully coordinates to encourage researchers worldwide to participate enthusiastically and continuously. PIK has a management group and a cross-sectoral science team. The management group supports the ever-expanding project by formulating strategies such as understanding the needs of the IPCC and research funders, securing funds, holding meetings, and managing data. The cross-sectoral science team supports the implementation of numerous cutting-edge studies published in high-profile journals, including Nature and its sister journals, by proposing timely protocols and implementing ambitious interdisciplinary collaborations in line with the needs of the IPCC's special reports. Also, the ISIMIP Repository, a site for providing input and output data (https://data.isimip.org/), allows all the input data used and the output data submitted by researchers free of charge, including for commercial purposes. This site has now become an information platform for climate change and environmental impact assessments worldwide, supporting not only researchers but also private sector and NGO activities.

By comparing with ISIMIP, it becomes clear what PJ2 and CCCA, Japan's unique permanent institution for climate change impact assessment, should aim for in the future. They have already managed to immediately and freely publish protocols and scenarios, so the next challenges to be realized are to propose attractive protocols (research themes) that make people want to participate voluntarily from outside, further align protocols with the needs of influential institutions (stakeholders), coordinate to realize ambitious cross-disciplinary research, and widely publish prediction result data after obtaining the consent of researchers.

References

Abe H, Yamano H (2025) Chapter 9 description of coastal ecosystems (coral, macroalgae, and seagrass) in local climate change adaptation plans of Japanese prefectures. In: Mimura N, Takewaka S (eds) Climate change impacts and adaptation strategies in Japan. Springer-Nature, Tokyo:119–135

Climate Change Impact Prediction Project Team (2009) Climate change impacts on Japan—long-term climate stabilization levels and impact risk assessment—pp 38. https://www.env.go.jp/press/files/jp/13617.pdf

Frieler K, Volkholz J, Lange S, Schewe J, Mengel M, del Rocío Rivas López M, Otto C, Reyer CPO, Karger DN, Malle JT, Treu S, Menz C, Blanchard JL, Harrison CS, Petrik CM, Eddy TD, Ortega-Cisneros K, Novaglio C, Rousseau Y, Watson RA, Stock C, Liu X, Heneghan R, Tittensor D, Maury O, Büchner M, Vogt T, Wang T, Sun F, Sauer IJ, Koch J, Vanderkelen I, Jägermeyr J, Müller C, Rabin S, Klar J, Vega del Valle ID, Lasslop G, Chadburn S, Burke E, Gallego-Sala A, Smith N, Chang J, Hantson S, Burton C, Gädeke A, Li F, Gosling SN, Müller Schmied H, Hattermann F, Wang J, Yao F, Hickler T, Marcé R, Pierson D, Thiery W, Mercado-Bettín D, Ladwig R, Ayala-Zamora AI, Forrest M, Bechtold M (2024) Scenario setup and forcing data for impact model evaluation and impact attribution within the third round of the Inter-Sectoral Impact Model Intercomparison Project (ISIMIP3a). Geosci Model Dev 17:1–51. https://doi.org/10.5194/gmd-17-1-2024

Hanasaki N, Nagashima T, Koide D, Ashina S, Oka K, Ito A, Okada M, Honda Y, Ikegami M, Ishizaki N, Masago Y, Masutomi Y (2023) Projection of climate change impacts under the Climate Change Adaptation Research Program at the National Institute for Environmental Studies and publication of the results. Glob Environ Res 23(1):45–58

Heinicke S, Volkholz J, Schewe J, Gosling SN, Müller Schmied H, Zimmermann S, Mengel M, Sauer IJ, Burek P, Chang J, Kou-Giesbrecht S, Grillakis M, Guillaumot L, Hanasaki N, Koutroulis A, Otta K, Qi W, Satoh Y, Stacke T, Yokohata T, Frieler K (2024) Global hydrological models continue to overestimate river discharge. Environ Res Lett 19(7):074005. https://doi.org/10.1088/1748-9326/ad52b0

IPCC (2022) In: Pörtner H-O, Roberts DC, Tignor M, Poloczanska ES, Mintenbeck K, Alegría A, Craig M, Langsdorf S, Löschke S, Möller V, Okem A, Rama B (eds) Climate change 2022: impacts, adaptation, and vulnerability. Contribution of Working Group II to the Sixth Assessment Report of the Intergovernmental Panel on Climate Change. Cambridge University Press, Cambridge, New York, p 3056. https://doi.org/10.1017/9781009325844

Ishizaki N, Hanasaki N, Shiogama H, Takahashi K (2023) Common climate and socio-economic scenarios for climate change impact assessment. Glob Environ Res 28(1):35–44

Jägermeyr J, Müller C, Ruane AC et al (2021) Climate impacts on global agriculture emerge earlier in new generation of climate and crop models. Nat Food 2:873–885. https://doi.org/10.1038/s43016-021-00400-y

Koide D, Yoshikawa T, Ishihama F (2025) Detection, assessment, and projection of climate change effects on terrestrial ecosystems. In: Mimura N, Takewaka S (eds) Climate change impacts and adaptation strategies in Japan. Springer-Nature, Tokyo:137–150

Ministry of Education, Culture, Sports, Science and Technology (MEXT) and Japan Meteorological Agency (JMA) (2020) Climate change in Japan 2020—observations and predictions on atmosphere, land and ocean assessment report—(full report). Ministry of Education, Culture, Sports, Science and Technology and Japan Meteorological Agency, p 263. https://www.data.jma.go.jp/cpdinfo/ccj/2020/pdf/cc2020_shousai.pdf

Ministry of Environment (MOE) (2020) Assessment report on climate change impacts in Japan. Ministry of Environment, p 107. https://www.env.go.jp/content/000047546.pdf

Suzuki-Ohno Y, Ishihama F, Yokoyama J, Inoue MN, Nakashizuka T, Kawata M (2024) Estimating bee distributions and their functional range to map important areas for protecting bee species and their functions. Sci Rep 14:12842. https://doi.org/10.1038/s41598-024-61848-z

Wang Q-X, Okadera T, Nakayama T, Batkhishig O (2024) Estimation of the carrying capacity and relative stocking density of Mongolian grasslands under various adaptation scenarios. Sci Total Environ 913:169772. https://doi.org/10.1016/j.scitotenv.2023.169772
Warszawski L, Frieler K, Huber V, Piontek F, Serdeczny O, Schewe J (2014) The Inter-Sectoral Impact Model Intercomparison Project (ISI–MIP): project framework. Proc Natl Acad Sci 111(9):3228–3232. https://doi.org/10.1073/pnas.1312330110

Chapter 9
Description of Coastal Ecosystems (Coral, Macroalgae, and Seagrass) in Local Climate Change Adaptation Plans of Japanese Prefectures

Hiroya Abe and Hiroya Yamano

Abstract Coastal ecosystems, composed of corals, macroalgae, and seagrasses, provide numerous benefits to human life. These ecosystems are found along Japan's coasts, but their distribution and health have been impacted by global and regional environmental stresses. With rising sea temperatures expected to bring significant changes to coastal ecosystems, it is essential to base the use and conservation of these species on their specific regional conditions. This study examined how these biological communities are addressed in the climate change adaptation plans of Japanese prefectures. The current distribution of corals, macroalgae, and seagrasses varies by prefecture, and there have been observable changes in the past, with potential shifts anticipated in the future. The extent to which these species are addressed in administrative documents also varies widely by species and region. Corals, in particular, were often found to lack specific conservation and adaptation measures, compared to macroalgae and seagrasses. As scientific understanding of future projections and adaptation strategies continues to grow, it is critical to incorporate this knowledge into administrative documents to support the sustainable use and conservation of these vital ecosystems.

Keywords Coastal ecosystem · Text analysis · Administrative document · Distribution · Conservation · Coral · Macroalgae · Seagrass

H. Abe (✉)
National Institute for Environmental Studies, Tsukuba, Japan
e-mail: abe.hiroya@nies.go.jp

H. Yamano
National Institute for Environmental Studies, Tsukuba, Japan

The University of Tokyo, Tokyo, Japan

N. Mimura, S. Takewaka (eds.), *Climate Change Impacts and Adaptation Strategies in Japan*, https://doi.org/10.1007/978-981-96-2436-2_9

9.1 Introduction

The reef-building corals (hereafter corals), macroalgae, and seagrasses that constitute these ecosystems are dominant in shallow waters and often function as foundation species by providing habitat for other organisms and building landscapes. These communities are more easily monitored than migratory organisms because they are more persistent on the seafloor in shallow waters. Changes in coastal ecosystems and impacts on human livelihoods associated with climate change are a phenomenon of concern, but the response to rising water temperatures varies among biological communities (IPCC 2019). These coastal communities are subject to global and/or regional-scale environmental stresses. The bleaching and mortality of coral communities due to elevated water temperatures and the decline in macroalgal beds and seagrass meadows due to high water temperatures and increased feeding pressure are well-known (adverse) effects of environmental changes on coastal ecosystems. Many cases of tropicalization of ecosystems in temperate zones have been reported. As mentioned earlier, coastal ecosystems have outstanding value for human life. Therefore, to sustainably receive ecosystem services, it is essential to understand the impacts of environmental variations such as climate change, conduct future projections, and consider and promote adaptation measures.

The sea area around Japan is subtropical to subarctic, spanning from 24°N to 45°N in the north–south direction (Fig. 9.1). Owing to the wide latitudinal zone and the influence of ocean currents such as the Kuroshio and Oyashio currents, the

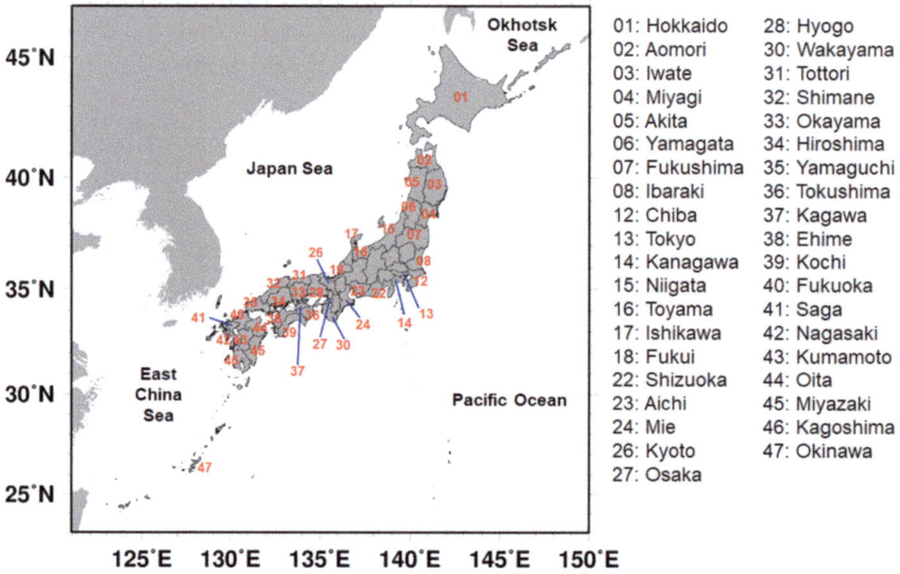

Fig. 9.1 The study area and prefectures in Japan. Prefectures are numbered and named only for those facing the sea ($n = 39$)

waters surrounding Japan have high biodiversity. In addition, a north–south gradient of coastal ecosystems according to latitude (sea surface temperature) is a prominent feature (Sugihara et al. 2009; Abe et al. 2021b; Sato et al. 2021). As described below, corals are distributed in subtropical to temperate zones around Japan (Kitano et al. 2020), and macroalgae and seagrass are distributed in almost all areas (Kumagai et al. 2016; Ohba and Miyata 2020; Sudo et al. 2020). Because of the significant regional differences in the marine environment and ecosystems in Japan, each region (e.g., prefecture) must individually assess the impacts of climate change and other factors and consider and promote the necessary conservation and adaptation measures.

Japan has 47 prefectures, 39 of which face the sea (Fig. 9.1). These 39 prefectures were included in the analysis of this study. This study aimed to summarize the present status of coral reefs (including temperate coral communities), macroalgal beds, and seagrass meadows in Japan and the challenges associated with climate change adaptation by reviewing administrative documents and scientific papers. Specifically, the objectives of this study were (1) to summarize the status of corals, macroalgae, and seagrass in the local climate change adaptation plans of Japanese prefectures; (2) to compile previous studies on future projections and adaptation measures for corals, macroalgae, and seagrasses in Japan; and (3) to discuss issues related to the consideration and promotion of adaptation measures in each region.

9.2 Materials and Methods

9.2.1 Characteristics of the Study Area

The sea surface temperature during the coldest and/or warmest months is an environmental factor that strongly controls the broad-scale distributions of corals, macroalgae, and seagrasses (Yara et al. 2011; Takao et al. 2015a, b; Abe et al. 2022d). Therefore, to identify the sea surface temperature (SST) around Japan in February and August, we used the CoralTemp (Skirving et al. 2020) dataset to plot the 10-year average horizontal distribution of the SST from 2014 to 2023. The north–south gradient and seasonal variation in SST around Japan are considerable (Fig. 9.2). The February SST is above 20 °C around Okinawa but below 5 °C in Hokkaido. On the other hand, in August, the temperature increases to approximately 25 °C to 30 °C from Okinawa to the central part of Honshu, and around Hokkaido, it is also approximately 15 °C to 20 °C. Focusing on long-term changes in the SST, the average annual rate of increase in the SST in the seas around Japan over the last 100 years through 2023 was +1.28 °C/100 years. In the waters around Japan, the trend of increasing water temperature differs depending on the sea area, and an exceptionally high increasing trend was observed in the southern to central parts of the Sea of Japan and southeastern Hokkaido (JMA 2024).

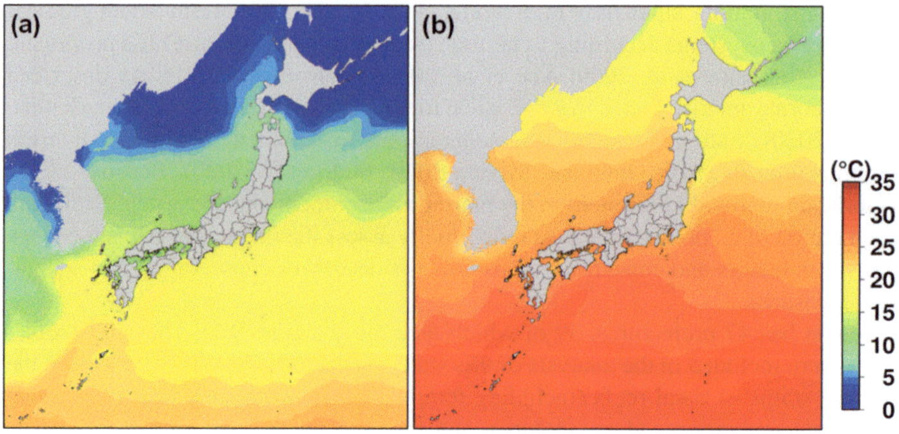

Fig. 9.2 Mean sea surface temperatures (SSTs) in (**a**) February and (**b**) August from 2014–2023. (The data were obtained from CoralTemp)

Fig. 9.3 An example of an underwater seascape composed of reef-building corals, macroalgae, and seagrasses

Significant geographic differences exist in the distributions of corals, macroalgae, and seagrasses (the underwater landscape composed of these communities) along the coast of Japan (Fig. 9.3). However, many reported decline events have been reported in coral, macroalgae, and seagrass communities in Japan. Typical phenomena include mass coral bleaching and a decrease in coral cover (Kawagoe 2017) and the decline or disappearance of macroalgal beds and seagrass meadows (Kiyomoto et al. 2021; Terada et al. 2021; Shimboku and Terada 2023). The factors involved in these declines vary by region and community, but exposure to high water temperatures (Yatsuya et al. 2014; Kayanne 2017) and excessive feeding damage (Ishikawa et al. 2017; Kitamura et al. 2022) are often contributing drivers. Long-term changes in the distribution range and species composition have also been observed in coral communities (Yamano et al. 2011; Kumagai et al. 2018) and macroalgal beds (Tanaka et al. 2012; Kumagai et al. 2018).

9.2.2 Collection and Text Analysis of Prefectural Administrative Documents

Analyzing the contents of administrative documents issued by each prefecture makes it possible to understand the phenomena observed or predicted and the effects of the concerns or expectations in each region (Baba and Tanaka 2015; Abe et al. 2022a). The administrative documents used in this study were the plans for climate change adaptation. Regional climate change adaptation plans issued by sea-facing prefectures were collected from the internet in June 2024. However, in prefectures where climate change adaptation plans have yet to be established, documents of global warming countermeasures plans or environmental plans were collected and used in the analysis.

First, search terms were set, and the frequency of occurrence of each keyword in the documents in each prefecture was determined. Search keywords were defined with respect to ecosystems, phenomena, and functions. The specific keywords used were coral (*sango* in Japanese), macroalgae or seaweed (*kaisou*), seagrass (*umikusa*), macroalgal bed or seagrass meadow (*moba*), coral bleaching (*sango no hakka*), *isoyake* (*isoyake*), and blue carbon. *Isoyake* refers to the decline or disappearance of seaweed beds, which is a problem in many Japanese waters. Coastal vegetation is expected as blue carbon.

If the frequency of occurrence of a keyword is high, it is assumed that the biological community or phenomenon is essential in the area. Conversely, there may be cases where the target biological community is distributed but not mentioned in administrative documents, and the detection of such gaps can also be evaluated in this analysis. However, it is also possible that the keyword is used in a context that has nothing to do with the situation or policies of the prefecture. Therefore, from the sentences before and after the keywords, we extracted descriptions related to past or future changes and impacts in each prefecture and conservation and adaptation measures.

9.2.3 *Review of Previous Studies on Future Distribution Projections of Biological Communities and Adaptation Measures*

During our review of scientific papers, we collected previous studies on future projections of Japan's coral, macroalgae, and seagrass distributions. We compiled information on the spatial resolution of future projections, the areas covered, and an overview of the studies.

As one of the adaptation measures in this study, we collected case studies on selecting protected areas, refuge areas, and priority conservation areas. The Ministry of the Environment, Japan, has published a guide to the concept of climate change adaptation in the field of biodiversity (https://www.env.go.jp/content/900523948.pdf) and a guide to adaptation measures to consider for climate change in national parks and other protected areas (https://www.biodic.go.jp/biodiversity/about/library/files/tekiou_tebiki.pdf). The measures to be taken (conservation and adaptation measures) differ depending on the degree of impact of climate change. Conventional conservation and use can continue if the potential for ecosystem degradation is minimal. On the other hand, in cases where there are refuge areas within the protected area, it is necessary to identify priority areas to be preserved, maintain or limit their use, expand the protected area, and alleviate environmental stresses simultaneously. This study summarized previous studies in Japan on selecting protected areas and priority conservation areas under climate change scenarios. If these findings (methods and concepts) can be utilized, they are expected to promote efficient conservation activities in addition to climate change impact assessment in each region.

9.3 Results and Discussion

9.3.1 *Frequency of Occurrence and Description of Keywords in the Regional Climate Change Adaptation Plan*

Regional climate change adaptation plans were not available for all of the targeted prefectures. Six of the 39 prefectures (Hokkaido, Aomori, Tokyo, Niigata, Shizuoka, and Aichi) had regional climate change adaptation plans available in June 2024. Twenty-five prefectures had adaptation plans included in their global warming countermeasures plans, and eight prefectures (Yamagata, Ishikawa, Fukui, Wakayama, Shimane, Saga, Kumamoto, and Miyazaki) had adaptation plans included in their environmental plans.

There were significant differences in the frequency of coral, macroalgae, and seagrass-related keywords between prefectures and among keywords (Fig. 9.4). In Hokkaido, Ibaraki, Ishikawa, Aichi, Mie, Kyoto, Shimane, and Kagawa, the total

Code	01	02	03	04	05	06	07	08	12	13
Keyword \ Prefecture	Hokkaido	Aomori	Iwate	Miyagi	Akita	Yamagata	Fukushima	Ibaraki	Chiba	Tokyo
Coral	0	1	1	0	1	0	0	2	6	2
Seaweed	1	6	2	2	3	3	0	0	4	0
Seagrass	0	0	0	0	0	1	0	0	4	0
Seaweed bed / Seagrass meadow	0	2	13	4	1	1	4	0	4	1
Coral bleaching	0	0	0	0	1	0	0	0	0	1
Isoyake	0	13	0	1	0	0	1	0	0	0
Blue carbon	0	0	12	4	0	2	0	0	4	0

Code	14	15	16	17	18	22	23	24	26	27
Keyword \ Prefecture	Kanagawa	Niigata	Toyama	Ishikawa	Fukui	Shizuoka	Aichi	Mie	Kyoto	Osaka
Coral	3	3	1	0	0	5	0	0	0	3
Seaweed	12	0	3	1	1	6	0	0	1	0
Seagrass	0	0	0	0	0	0	0	0	0	0
Seaweed bed / Seagrass meadow	22	10	22	0	6	9	1	0	1	7
Coral bleaching	1	0	0	0	0	2	0	0	0	1
Isoyake	5	0	0	0	0	5	0	0	0	0
Blue carbon	6	1	6	0	0	0	0	0	0	7

Code	28	30	31	32	33	34	35	36	37	38
Keyword \ Prefecture	Hyogo	Wakayama	Tottori	Shimane	Okayama	Hiroshima	Yamaguchi	Tokushima	Kagawa	Ehime
Coral	1	3	0	0	2	0	2	7	0	5
Seaweed	2	1	0	0	1	2	2	2	0	1
Seagrass	1	0	0	0	0	2	1	1	0	1
Seaweed bed / Seagrass meadow	5	2	5	2	6	2	11	7	0	6
Coral bleaching	0	1	0	0	0	0	1	0	0	0
Isoyake	0	1	0	0	0	0	0	0	0	0
Blue carbon	8	0	1	0	0	9	5	1	0	3

Code	39	40	41	42	43	44	45	46	47
Keyword \ Prefecture	Kochi	Fukuoka	Saga	Nagasaki	Kumamoto	Oita	Miyazaki	Kagoshima	Okinawa
Coral	23	2	0	1	0	0	7	0	28
Seaweed	3	0	2	6	2	1	1	1	3
Seagrass	1	0	1	0	1	0	1	1	2
Seaweed bed / Seagrass meadow	19	6	5	13	19	10	13	6	5
Coral bleaching	14	0	0	0	0	0	0	0	10
Isoyake	4	0	3	3	0	0	0	0	1
Blue carbon	5	2	1	0	0	2	5	0	11

Fig. 9.4 Heatmap showing the frequency of occurrence of keywords (coral, seaweed, seagrass, seaweed bed or seagrass meadow, coral bleaching, *isoyake*, blue carbon) in each prefecture's regional climate change adaptation plan or global warming countermeasures plan and environmental plan. Document searches were conducted in Japanese. If each keyword appears more frequently than 5, 10, or 20 times, it is shown in green, yellow, or red, respectively

number of hits for the set of keywords was two or fewer. On the other hand, the total number of hits was more than 40 in Kanagawa, Kochi, and Okinawa. The frequency of mentioning corals was highest in Okinawa, followed by Kochi. Descriptions of seaweed and seagrass tended to be more common in Kanagawa, Toyama, and Kochi and less common in Hokkaido/Tohoku and Kanto, except in the Kanagawa, Kinki, and Tokai areas. There were few descriptions of coral bleaching except in Kochi and Okinawa. Although *isoyake* was mentioned the greatest number of times in Aomori, differences between regions were unclear. The blue carbon keyword was mentioned in more than half of the prefectures, especially in Iwate and Okinawa, where there were more than ten occurrences.

Focusing on gaps, there were no descriptions of corals in Kagoshima, although coral reefs are formed there (Nansei Islands), as they are in Okinawa. In Hokkaido, kelp forests form extensively and are essential fishery resources (Sudo et al. 2020), but these vegetations were not mentioned in the adaptation plan. The challenge is that the actual conditions of the local ecosystem need to be adequately reflected in the adaptation plan. Blue carbon is a feature that has received much attention in recent years, and it was described in many prefectures. It is considered necessary as a mitigation measure and carbon credit policy.

Tables 9.1 and 9.2 provide a brief description of corals and macroalgae/seagrasses in the administrative documents (climate change adaptation plans) of each prefecture, respectively. There were more macroalgae and seagrass descriptions than corals (Tables 9.1 and 9.2). Among the 39 prefectures, 13 had specific descriptions of coral communities. Only six prefectures described coral conservation or adaptation measures (Table 9.1). On the other hand, 34 prefectures presented descriptions of seaweed, seagrass, seaweed beds, and seagrass meadows. In particular, the detailed description of the conservation and adaptation measures revealed notable differences between the seaweed and seagrass samples and the coral samples (Table 9.2). Most of the statements about past or future changes and impacts were made from the perspective of biodiversity, fisheries, and CO_2 absorption. Although there were few mentions from the perspective of the tourism industry, there was a note on the possible impact on tourism in Chiba Prefecture due to the northward expansion of the coral distribution. In addition, many prefectures described the creation and functional evaluation of blue carbon in seaweed beds and seagrass meadows. Climate change is expected to cause significant changes in the distributions of corals and macroalgae, necessitating changes in traditional conservation and utilization practices (Abe et al. 2021b, 2022d). However, local stakeholders' perceptions of the changes in and value of coastal ecosystems may differ among regions and industries (Abe et al. 2022c). Therefore, while changes in ecosystems in coastal areas due to climate change are expected to significantly impact fisheries operations, it is also necessary to consider and promote necessary measures from the tourism perspective.

9.3.2 Case Studies on Future Projections and Concepts of Climate Change Adaptation

Many studies have investigated the future distribution of corals and macroalgae in Japan, but few studies have focused on seagrasses. As shown in Tables 9.1 and 9.2, there were few specific mentions of seagrasses in administrative documents, which may be because the impact of distribution changes due to climate change on seagrasses is smaller than that on corals and macroalgae.

With respect to methods for projecting future changes in distribution, most studies were based on the idea that sea surface temperatures strongly control the distribution areas of organisms. Therefore, the spatial resolution of distribution change projections often depends on the spatial resolution of the dataset for future projections of sea surface temperature. Studies in the early 2010s often used global climate model projections with a coarse spatial resolution of tens to hundreds of kilometers (Yara et al. 2011, 2012; Komatsu et al. 2014; Takao et al. 2015a). Later,

Table 9.1 Description of observed or projected changes/impacts and conservation and adaptation measures for coral reefs (including corals and coral communities) in each prefecture's regional climate change adaptation plan, global warming countermeasures plan, or environmental plan

Code	Prefecture	Changes/Impacts	Measures
12	Chiba	A northward shift in coral distribution has been observed. Due to rising water temperatures and ocean acidification, suitable areas for growth are projected to disappear in the future.	To understand the impact of the northward expansion of coral distribution on leisure activities and to promote the tourism industry.
13	Tokyo	Coral bleaching has been observed. Due to rising water temperatures and ocean acidification, suitable areas for coral growth are projected to disappear in the future.	NA
15	Niigata	Corals are appearing more frequently.	NA
22	Shizuoka	Expansion of the coral distribution has been observed.	NA
28	Hyogo	Future increases in water temperatures may cause seaweed beds to shift to coral communities.	NA
30	Wakayama	Coral bleaching and coral predation are expected.	Continue coral monitoring and population control of predators.
35	Yamaguchi	Coral bleaching is occurring due to rising water temperatures. The coral-growing area may disappear in the future.	NA
36	Tokushima	Coral predation damage occurred as a result of rising water temperatures. Coral declined due to the deterioration of the growing environment.	Coral rehabilitation projects are being implemented.
38	Ehime	Increases in water temperature have been observed to expand the distribution area of corals and increase the number of coral species. Due to rising water temperatures, the suitable distribution area is projected to move northward or disappear in the future.	NA
39	Kochi	Bleaching events have occurred frequently. Declines in coral communities have occurred due to rising water temperatures.	Conduct surveys to determine coral communities' distribution, disturbance, and regenerative capacity.

(continued)

Table 9.1 (continued)

Code	Prefecture	Changes/Impacts	Measures
40	Fukuoka	A transition from seaweed beds to coral communities is predicted.	NA
45	Miyazaki	Degradation of coral communities due to predation has been observed.	To work with stakeholders and municipalities to conserve coral. Exterminate coral-eating predators and survey coral habitats.
47	Okinawa	There is concern about the impact of coral bleaching on coral reef ecosystems due to high water temperatures and sediment inflows due to heavy rainfall.	Support monitoring surveys, genetic analysis of coral tolerant to high water temperatures, and coral reef conservation activities. Elucidate environmental conditions that prevent bleaching and develop technologies to mitigate bleaching.

NA means that there was no statement in that prefecture that specifically related to corals

downscaling led to high-resolution simulations, and Nishikawa et al. (2021) published a 2 km-resolution simulation of future hydrodynamics around Japan. This dataset allows for future projections and impact assessments at a regional scale (Abe et al. 2021b, 2022d; Yuan et al. 2023).

As shown in Tables 9.1 and 9.2, many prefectures have specific examples of conservation and adaptation measures for corals, macroalgae, and seagrasses. However, there was no mention of where climate change impacts are likely (or unlikely) to occur or where measures should be taken except for macroalgal conservation in Tottori Prefecture (Table 9.2). Because the extent to which human intervention can be used in coastal areas is minimal, location-specific measures are essential. Since impact assessments at high spatial resolutions and the selection of refuge areas, protected areas, and priority conservation areas are important in such cases, we reviewed this information. Although the spatial resolution is approximately 100 km when the output results of global climate models are used (Makino et al. 2015), it is possible to evaluate the data at a finer resolution than that in these prefectures through the use of available high-resolution downscaling datasets (Abe et al. 2022b; Abe and Yamano 2024) and the application of local hydrodynamic models (Abe et al. 2021a, 2024). While various assessment methods are used in each study, some examples consider use and conservation perspectives in addition to basic information such as the ocean environment and species distribution (Abe et al. 2021a; Abe and Yamano 2024). In addition, because corals and coral-eating starfish have high dispersal capacities (Nakabayashi et al. 2019; Horoiwa et al. 2022), connectivity is also an essential consideration in protected area selection concerning coral communities (Abe et al. 2022b; Abe and Yamano 2024).

Table 9.2 Description of observed or projected changes/impacts and conservation and adaptation measures for macroalgal beds and seagrass meadows in each prefecture's regional climate change adaptation plan, global warming countermeasures plan, or environmental plan

Code	Prefecture	Changes/Impacts	Measures
02	Aomori	Changes in the marine environment and predation have resulted in *isoyake*. Further progression of *isoyake* is a concern	NA
03	Iwata	Decrease of *moba* due to high water temperature has been observed. There is concern about changes in seaweed species, abundance of *moba*, and the decrease of *moba*	Work to build momentum for using blue carbon, examine its effectiveness, and rehabilitate and create *moba*
04	Miyagi	NA	Promote creation and maintenance of *moba*, promotion of *isoyake* countermeasures, monitoring *moba* distribution, and disseminating information to raise awareness of blue carbon
05	Akita	The emergence of warm-water seaweeds has been observed. There is concern about the impact on the ecosystem due to changes in seaweed species composition caused by rising water temperatures	Develop technologies related to the maintenance and increment of *moba*
06	Yamagata	NA	Promote blue carbon through *moba* creation
07	Fukushima	NA	Support ecosystem conservation activities by fishermen to preserve the function of *moba*
12	Chiba	Loss of *moba* has been confirmed. There is concern that the loss of *moba* will continue to expand in the future, affecting rocky shore resources	Support efforts for conservation and rehabilitation of *moba*. Support efforts to prevent loss and restore *moba*, including monitoring of *moba*
13	Tokyo	The loss of *moba* has become apparent, resulting in a decrease in rocky shore resources	NA

(continued)

Table 9.2 (continued)

Code	Prefecture	Changes/Impacts	Measures
14	Kanagawa	*Isoyake* has been observed and is also a concern for the future. Rising water temperatures are projected to degrade the *moba* ecosystem	Developing *moba* creation, restoration, and conservation techniques to utilize blue carbon
15	Niigata	A decrease in the area of *moba* with increasing seawater temperature has been observed	Develop technology to restore *moba* and consider fisheries that respond to changes in *moba* ecosystems. Examine the development of blue carbon assessment methods and *moba* formation technology
16	Toyama	The decline of *moba* has been reported. It is projected that *moba* will almost completely disappear in the future	Develop effective *moba* creation methods and evaluate carbon sequestration by *moba*. Engage in research to understand *moba*'s status on a broad scale and to properly assess its distribution area and changes in it
18	Fukui	Declines in *moba* are occurring due to rising sea temperatures. It is predicted that the decrease in *moba* will continue to progress further in the future	To support conservation activities of *moba* with the participation of residents, as well as to carry out the creation of *moba*
22	Shizuoka	Changes in the constituent species of *moba* and the occurrence of *isoyake* have been observed	Support mother-algae inputs and the removal of herbivorous fish species to promote *moba* recovery
23	Aichi	Degradation of the *moba* ecosystem is projected to occur due to rising water temperatures and the northward expansion of the distribution of herbivorous fish species	NA
26	Kyoto	The degradation of the *moba* ecosystem is projected to occur as a result of rising water temperatures and the northward expansion of the distribution of herbivorous fish species	NA
27	Osaka	NA	Create *moba* that produces blue carbon
28	Hyogo	Possible transition from seaweed beds to coral communities due to increased water temperatures	Promote efforts to restore seagrass meadows. Create *moba* to increase blue carbon
30	Wakayama	Decline of *moba* is expected	The conservation of *moba* will increase the carbon sinks

(continued)

Table 9.2 (continued)

Code	Prefecture	Changes/Impacts	Measures
31	Tottori	There is concern about seaweed die-off and a decrease in *moba* due to high water temperatures	Promote blue carbon through conservation activities by creating *moba* and exterminating sea urchins. Consideration will be given to creating *moba* in areas with high water temperatures and adequate seawater exchange and to creating *moba* that are resistant to predation damage
32	Shimane	The decline in *moba* is a concern	Conserve *moba* based on scientific findings
33	Okayama	NA	Conduct research for the restoration of *moba* and support restoration activities
34	Hiroshima	NA	Consider effective approaches to blue carbon
35	Yamaguchi	A massive decline phenomenon in seaweed is occurring	Investigate the impact of high summer water temperatures on *moba*. Promote conservation and functional recovery of *moba*
36	Tokushima	Declines in *moba* and rocky shore resources have been observed and are expected to continue	Work on measures to prevent predation on *moba*, as well as promote conservation and creation
38	Ehime	Decline of *moba* has been reported as a result of rising sea temperatures	Work on restoration and conservation of *moba*, and promote measures for sinks using blue carbon
39	Kochi	The area of *moba* was significantly reduced, and *isoyake* progressed. The decline of temperate species and the expansion of the distribution range of subtropical species have been observed	Promote measures for sinks by marine ecosystems through the creation and preservation of *moba*
40	Fukuoka	Rising water temperatures and the northward expansion of the distribution of herbivorous fish species are projected to degrade the *moba* ecosystem	Assist in the implementation of population control of predators. Work on conservation and restoration of *moba*
41	Saga	NA	Promote efforts related to *moba* conservation measures. To maintain healthy *moba*, efforts will be made to raise awareness and provide guidance on restoration methods for *isoyake*
42	Nagasaki	Changes in sea surface temperature have led to changes in the types of *moba*, the timing of their formation, and the seaweed species distributed in the area	To promote the creation of *moba* by understanding the overview of *moba* and producing seedlings of seaweed species that can adapt to the increase in water temperature

(continued)

Table 9.2 (continued)

Code	Prefecture	Changes/Impacts	Measures
43	Kumamoto	The area of *moba* is decreasing due to the development of coastal areas and environmental changes	Promote conservation and rehabilitation through the creation of *moba*
44	Oita	Rising sea water temperatures may be causing poor seaweed growth	Promote *moba* formation and prevent *moba* decline
45	Miyazaki	NA	Restore and maintain *moba* for sustainable use of *moba*. Maintain and improve the public use that *moba* provides and promote understanding among the citizens of the prefecture
46	Kagoshima	NA	Developing technology to create *moba* and promote activities to maintain and conserve *moba*, which contribute to the prevention of global warming
47	Okinawa	There is concern about the impact on fisheries resources associated with the decline of seagrass meadow	Monitoring surveys will be conducted primarily in seagrass meadows that are likely to be affected, and data will be accumulated to assess the presence or absence of climate change impacts

NA means that there was no statement in that prefecture that specifically related to macroalgae, macroalgal beds, seagrasses, or seagrass meadows. *moba* refers to macroalgal beds and/or seagrass meadows

9.3.3 Summary

In the analysis of the administrative documents, the adaptation plans of the prefectures were targeted. However, descriptions of biological communities may differ depending on the document type, such as environmental plans, fisheries plans, tourism plans, and biodiversity strategies (Abe et al. 2022a). The coastal environment and ecosystem conditions can also vary widely within each prefecture. Therefore, a more detailed understanding could be achieved by increasing the number of types of administrative documents targeted and by conducting analyses at the municipal scale. Furthermore, by utilizing questionnaires and other surveys, it is possible to identify regional and sectoral differences in climate change perceptions among citizens, government officials, and researchers. Challenges in promoting climate change adaptation vary from region to region (Fujita et al. 2023). Many challenges remain in reflecting on the scientific findings in administrative documents and returning them to local stakeholders. Therefore, implementing a social science approach in parallel with environmental and ecological monitoring and climate change impact assessment is essential for promoting climate change adaptation.

Acknowledgements This study was partially supported by the Climate Change Adaptation Research Program by the National Institute for Environmental Studies (NIES), Japan, JST, CREST Grant Number JPMJCR23J2, Japan, JSPS KAKENHI Grant Number JP23K17069, and the Sumitomo Foundation (2230150).

References

Abe H, Yamano H (2024) Simulated connectivity of crown-of-thorns starfish around Ashizuri-Uwakai National Park (western Japan) based on a high-resolution hydrodynamic modeling. Coral Reefs 43:371–390. https://doi.org/10.1007/s00338-024-02471-2

Abe H, Kumagai NH, Yamano H, Kuramoto Y (2021a) Coupling high-resolution coral bleaching modeling with management practices to identify areas for conservation in a warming climate: Keramashoto National Park (Okinawa prefecture, Japan). Sci Total Environ 790:148094. https://doi.org/10.1016/j.scitotenv.2021.148094

Abe H, Suzuki H, Kitano YF, Kumagai NH, Mitsui S, Yamano H (2021b) Climate-induced species range shift and local adaptation strategies in a temperate marine protected area, Ashizuri-Uwakai National Park, Shikoku Island, western Japan. Ocean Coast Manag 210:105744. https://doi.org/10.1016/j.ocecoaman.2021.105744

Abe H, Kitano YF, Fujita T, Yamano H (2022a) Distribution, use, management, regulation, and future concerns of reef-building corals based on administrative documents in Japan. Mar Policy 141:105090. https://doi.org/10.1016/j.marpol.2022.105090

Abe H, Kumagai NH, Yamano H (2022b) Priority coral conservation areas under global warming in the Amami Islands, Southern Japan. Coral Reefs 41:1637–1650. https://doi.org/10.1007/s00338-022-02309-9

Abe H, Mitsui S, Yamano H (2022c) Conservation of the coral community and local stakeholders' perceptions of climate change impacts: examples and gap analysis in three Japanese national parks. Ocean Coast Manag 218:106042. https://doi.org/10.1016/j.ocecoaman.2022.106042

Abe H, Mitsui S, Yamano H (2022d) Management of reef-building corals, macroalgae and seagrasses in Japanese national parks: issues and suggestions. Japan J Conserv Ecol 27:1–19. https://doi.org/10.18960/hozen.2120. (in Japanese with English abstract)

Abe H, Hayashi S, Sakuma A, Yamano H (2024) Priority sites for coral aquaculture in Kume Island based on numerical simulation. Estuar Coast Shelf Sci 303:108797. https://doi.org/10.1016/j.ecss.2024.108797

Baba K, Tanaka M (2015) Challenges of implementing climate change adaptation policy for disaster risk reduction—implications from framing gap among stakeholders and the general public. J Disaster Res 10(3):404–419. https://doi.org/10.20965/jdr.2015.p0404

Fujita T, Mameno K, Kubo T, Masago Y, Hijioka Y (2023) Unraveling the challenges of Japanese local climate change adaptation centers: a discussion and analysis. Clim Risk Manag 39:100489. https://doi.org/10.1016/j.crm.2023.100489

Horoiwa M, Nakamura T, Yuasa H, Kajitani R, Ameda Y, Sasaki T, Taninaka H, Kikuchi T, Yamakita T, Toyoda A, Itoh T, Yasuda N (2022) Integrated population genomic analysis and numerical simulation to estimate larval dispersal of *Acanthaster* cf. *solaris* between Ogasawara and other Japanese regions. Front Marine Sci 8:688139. https://doi.org/10.3389/fmars.2021.688139

Intergovernmental Panel on Climate Change (IPCC) (2019) Special Report on the Ocean and Cryosphere in a Changing Climate. https://www.ipcc.ch/srocc/. Accessed on 12 Aug 2024

Ishikawa T, Tose T, Abe M, Iwao T, Morita T, Maegawa M, Kurashima A (2017) Changes in algal flora by removing Diadema in Haidaura Bay, Mie prefecture. Nippon Suisan Gakkaishi 83(4):599–606. (in Japanese with English abstract). https://doi.org/10.2331/suisan.16-00085

Japan Meteorological Agency (JMA) (2024) Long-term trend in sea surface temperature (near Japan) https://www.data.jma.go.jp/kaiyou/data/shindan/a_1/japan_warm/japan_warm.html. Accessed on 3 July 2024

Kawagoe H (2017) Mass coral bleaching in 2016 reported by the monitoring sites 1000 project. J Japan Coral Reef Soc 19(1):21–28. (in Japanese with English abstract). https://doi.org/10.3755/jcrs.19.21

Kayanne H (2017) Validation of degree heating weeks as a coral bleaching index in the northwestern Pacific. Coral Reefs 36:63–70. https://doi.org/10.1007/s00338-016-1524-y

Kitamura T, Iwai T, Shigematsu Y, Miura C, Miura T (2022) Previous outbreaks and current abundance of corallivorous gastropods in southwestern Shikoku. Japan J Conserv Ecol 27(2):247–256. (in Japanese with English abstract). https://doi.org/10.18960/hozen.2124

Kitano YF, Hongo C, Yara Y, Sugihara K, Kumagai NH, Yamano H (2020) Data on coral species occurrences in Japan since 1929. Ecol Res 35:975–985. https://doi.org/10.1111/1440-1703.12136

Kiyomoto S, Yamanaka H, Yoshimura T, Yatsuya K, Shao H, Kadota T, Tamaki A (2021) Long-term change and disappearance of Lessoniaceae marine forests off Waka, Ikishima Island, northwestern Kyushu, Japan. Nippon Suisan Gakkaishi 87(6):642–651. (in Japanese with English abstract). https://doi.org/10.2331/suisan.21-00013

Komatsu T, Fukuda M, Mikami A, Mizuno S, Kantachumpoo A, Tanoue H, Kawamiya M (2014) Possible change in distribution of seaweed, *Sargassum horneri*, in Northeast Asia under A2 scenario of global warming and consequent effect on some fish. Mar Pollut Bull 85:317–324. https://doi.org/10.1016/j.marpolbul.2014.04.032

Kumagai NH, Yamano H, Fujii M, Yamanaka Y (2016) Habitat-forming seaweeds in Japan (fucoids and temperate kelps). Ecol Res 31:759. https://doi.org/10.1007/s11284-016-1404-5

Kumagai NH, García Molinos J, Yamano H, Takao S, Fujii M, Yamanaka Y (2018) Ocean currents and herbivory drive macroalgae-to-coral community shift under climate warming. Proc Natl Acad Sci USA 115:8990–8995. https://doi.org/10.1073/pnas.1716826115

Makino A, Klein CJ, Possingham HP, Yamano H, Yara Y, Ariga T, Matsuhasi K, Beger M (2015) The effect of applying alternate IPCC climate scenarios to marine reserve design for range changing species. Conserv Lett 8:320–328. https://doi.org/10.1111/conl.12147

Nakabayashi A, Yamakita T, Nakamura T, Aizawa H, Kitano YF, Iguchi A, Yamano H, Nagai S, Agostini S, Teshima KM, Yasuda N (2019) The potential role of temperate Japanese regions as refugia for the coral *Acropora hyacinthus* in the face of climate change. Sci Rep 9:1892. https://doi.org/10.1038/s41598-018-38333-5

Nishikawa S, Wakamatsu T, Ishizaki H, Sakamoto K, Tanaka Y, Tsujino H, Yamanaka G, Kamachi M, Ishikawa Y (2021) Development of high-resolution future ocean regional projection datasets for coastal applications in Japan. Prog Earth Planet Sci 8:7. https://doi.org/10.1186/s40645-020-00399-z

Ohba T, Miyata M (2020) Seagrasses of Japan, Revised Edition. xii + 139 pp. Hokkaido University Press, Sapporo

Sato M, Nakamura Y, Hori M (2021) Potential stocks of reef fish-based ecosystem services in the Kuroshio current region: their relationship with latitude and biodiversity. Popul Ecol 63:75–91. https://doi.org/10.1002/1438-390X.12061

Shimboku N, Terada R (2023) Seagrass bed in Kagoshima Bay: the distribution status in 2021 and long-term decline. Nippon Suisan Gakkaishi 89(6):510–520. (in Japanese with English abstract). https://doi.org/10.2331/suisan.23-00020

Skirving W, Marsh B, De La Cour J, Liu G, Harris A, Maturi E, Geiger E, Eakin CM (2020) CoralTemp and the coral reef watch coral bleaching heat stress product suite version 3.1. Remote Sens 12(23):3856. https://doi.org/10.3390/rs12233856

Sudo K, Watanabe K, Yotsukura N, Nakaoka M (2020) Predictions of kelp distribution shifts along the northern coast of Japan. Ecol Res 35:47–60. https://doi.org/10.1111/1440-1703.12053

Sugihara K, Sonoda N, Imafuku T, Nagata S, Ibusuki T, Yamano H (2009) Latitudinal changes in hermatypic coral communities from West Kyushu to Oki Islands in Japan. Journal of the Japanese Coral Reef Society 11:51–67. (in Japanese with English abstract). https://doi.org/10.3755/jcrs.11.51

Takao S, Kumagai NH, Yamano H, Fujii M, Yamanaka Y (2015a) Projecting the impacts of rising seawater temperatures on the distribution of seaweeds around Japan under multiple climate change scenarios. Ecol Evol 5:213–223. https://doi.org/10.1002/ece3.1358

Takao S, Yamano H, Sugihara K, Kumagai NH, Fujii M, Yamanaka Y (2015b) An improved estimation of the poleward expansion of coral habitats based on the inter-annual variation of sea surface temperatures. Coral Reefs 34:1125–1137. https://doi.org/10.1007/s00338-015-1347-2

Tanaka K, Taino S, Haraguchi H, Prendergast G, Hiraoka M (2012) Warming off southwestern Japan linked to distributional shifts of subtidal canopy-forming seaweeds. Ecol Evol 2:2854–2865. https://doi.org/10.1002/ece3.391

Terada R, Shindo A, Tanaka M, Esaki S (2021) Long-term changes in seaweed assemblages in Nagashima Island, Kagoshima prefecture: disappearance of the canopy-forming kelp and fucoid algae from coastal areas facing the East China Sea. Nippon Suisan Gakkaishi 87(6):631–641. (in Japanese with English abstract). https://doi.org/10.2331/suisan.21-00023

Yamano H, Sugihara K, Nomura K (2011) Rapid poleward range expansion of tropical reef corals in response to rising sea surface temperatures. Geophys Res Lett 38:L04601. https://doi.org/10.1029/2010GL046474

Yara Y, Oshima K, Fujii M, Yamano H, Yamanaka Y, Okada N (2011) Projection and uncertainty of the poleward range expansion of coral habitats in response to sea surface temperature warming: a multiple climate model study. Galaxea 13:11–20. https://doi.org/10.3755/galaxea.13.11

Yara Y, Vogt M, Fujii M, Yamano H, Hauri C, Steinacher M, Gruber N, Yamanaka Y (2012) Ocean acidification limits temperature-induced poleward expansion of coral habitats around Japan. Biogeosciences 9:4955–4968. https://doi.org/10.5194/bg-9-4955-2012

Yatsuya K, Kiriyama T, Kiyomoto S, Taneda T, Yoshimura T (2014) On the deterioration process of *Ecklonia* and *Eisenia* beds observed in 2013 at Gounoura, Iki Island, Nagasaki prefecture, Japan. Initiation of the bed degradation due to high water temperature in summer and subsequent cascading effect by the grazing of herbivorous fish in autumn. Algal Res 7(2):79–94. (in Japanese with English abstract). https://doi.org/10.20804/jsap.7.2_79

Yuan FL, Yamakita T, Bonebrake TC, McIlroy SE (2023) Optimal thermal conditions for corals extend poleward with oceanic warming. Divers Distrib 29:1388–1401. https://doi.org/10.1111/ddi.13765

Chapter 10
Detection, Assessment, and Projection of Climate Change Effects on Terrestrial Ecosystems

Dai Koide, Tetsuro Yoshikawa, and Fumiko Ishihama

Abstract This chapter reports research on Japanese terrestrial natural ecosystems from the Climate Change Adaptation Research Program by the National Institute for Environmental Studies. We detected colder juvenile shifts compared to adults of the same species, suggesting past colder shifts of forest tree species. We assessed the vulnerability of forest tree biodiversity against future climate change using national forest inventory data and the Info-gap decision theory. Regional fine-scale vegetation shifts were projected and evaluated in the Mt. Taisetsu region using vegetation maps, species distribution models, and a decision-supportive SecSel model. We then summarized candidate refugia habitats and realistic target intervention sites for the use of ecosystem service and management of vegetation. These methodologies are applicable to other regions and would help approach complex ecological mechanisms behind and create better decisions to adapt to future climate conditions. We hope this report further provides positive stimulations to other research fields, connecting them and approaching generalized climate change adaptation.

Keywords Terrestrial natural ecosystem · Vulnerability · Biodiversity · Vegetation · Distribution shift · Ecosystem service

10.1 Introduction

Contemporary global climate change has been reported as an important driver causing several changes in upland natural ecosystems such as species distribution shifts in both plants and animals (Lenoir et al. 2020), phenological changes in onset and end of terrestrial plants (Piao et al. 2019), ecosystem productivity with difference

D. Koide (✉) · F. Ishihama
National Institute for Environmental Studies, Ibaraki, Japan
e-mail: koide.dai@nies.go.jp

T. Yoshikawa
Osaka Metropolitan University, Osaka, Japan

N. Mimura, S. Takewaka (eds.), *Climate Change Impacts and Adaptation Strategies in Japan*, https://doi.org/10.1007/978-981-96-2436-2_10

along water condition (Hogan et al. 2024), and human well-being through ecosystem services (Pecl et al. 2017; Scheffers et al. 2016). These impacts of climate change on natural ecosystems were also evident in Japan. Some of the cold-adapted species showed a northern shift or regional population shrinks at the warmer edge (Amagai et al. 2018; Mizunaga et al. 2005; Takegawa et al. 2017), whereas the warm-adapted species showed a population increase and spread in their northern distribution areas as reported in evergreen broadleaf tree species (Suzuki et al. 2015) and southern butterflies in the third report of SATOYAMA survey (2013–2017) (https://www.biodic.go.jp/moni1000/findings/reports/pdf/third_term_satoyama.pdf). Phenological changes are reported in several ecosystems from temperate to alpine regions (Doi and Takahashi 2010; Kudo 2020), and a phenological mismatch between herbal flowers and pollinators was reported to reduce the fruition rate for entomophilous species (Kudo 2014; Kudo and Cooper 2019). Complex changes in forest productivity and subsequent changes in carbon flows are reported by the flux tower networks (Murayama et al. 2024; Sha et al. 2021). Impact assessments and climate change adaptation plans are important for proper natural ecosystem management under such dynamic conditions. To approach mechanisms behind and provide efficient information about climate change adaptation, the National Institute for Environmental Studies has conducted the climate change adaptation research program including research for terrestrial natural ecosystems in parallel with the S-18 Project. Here we provide some outcomes from the program in each basic step for the adaptation planning, (1) detection, (2) assessment, and (3) future projection and assessment for specific regions. Each step was reported from each case study hereafter. The detection step was shown from the national survey of tree species using juvenile–adult distribution difference method. The assessment step was reported for forest vulnerability using national forest inventory and the Info-Gap theory. The regional future projection and assessment was reported for alpine vegetations in Mt. Taisetsu region in Hokkaido.

10.2 Detection: Assessing Historical Shifts in Tree Species Distribution

Forest tree species are fundamental species that are basic producers supporting the terrestrial ecosystem. Since tree species are reported to move slowly and are vulnerable to climate change according to their long lifespan and sessile character (Lenoir et al. 2008, 2020), several monitoring programs started for forest tree species (e.g., JaLTER (http://www.jalter.org/en/), Monitoring sites 1000 (https://www.biodic.go.jp/moni1000/), Forest ecosystem diversity basic survey (http://forestbio.jp/)). However, these monitoring programs would take several decades to capture actual changes in climate change. For quick monitoring and adaptation planning, quickly detecting past-time changes is quite useful using past-time records. However, such records are still limited for space, time, and species, preventing pattern analyses for broad spatial and species ranges. Such limited research range also prevented us from approaching overall mechanisms and drivers of the changes in natural ecosystems. To overcome these limitations, we used the tricky idea of analyzing temperature distribution differences between juveniles and adults of the same species. Juveniles regenerated in current warmer conditions and adult trees regenerated in past-time cold conditions, making distribution difference between the size classes along past climate change directions (Fig. 10.1). We focused on this juvenile–adult difference as an index for tall tree species distribution shift in the past (Lenoir et al. 2009; Woodall et al. 2009) and clarified the overall pattern in Japan to approach possible drivers and mechanisms behind the shifts.

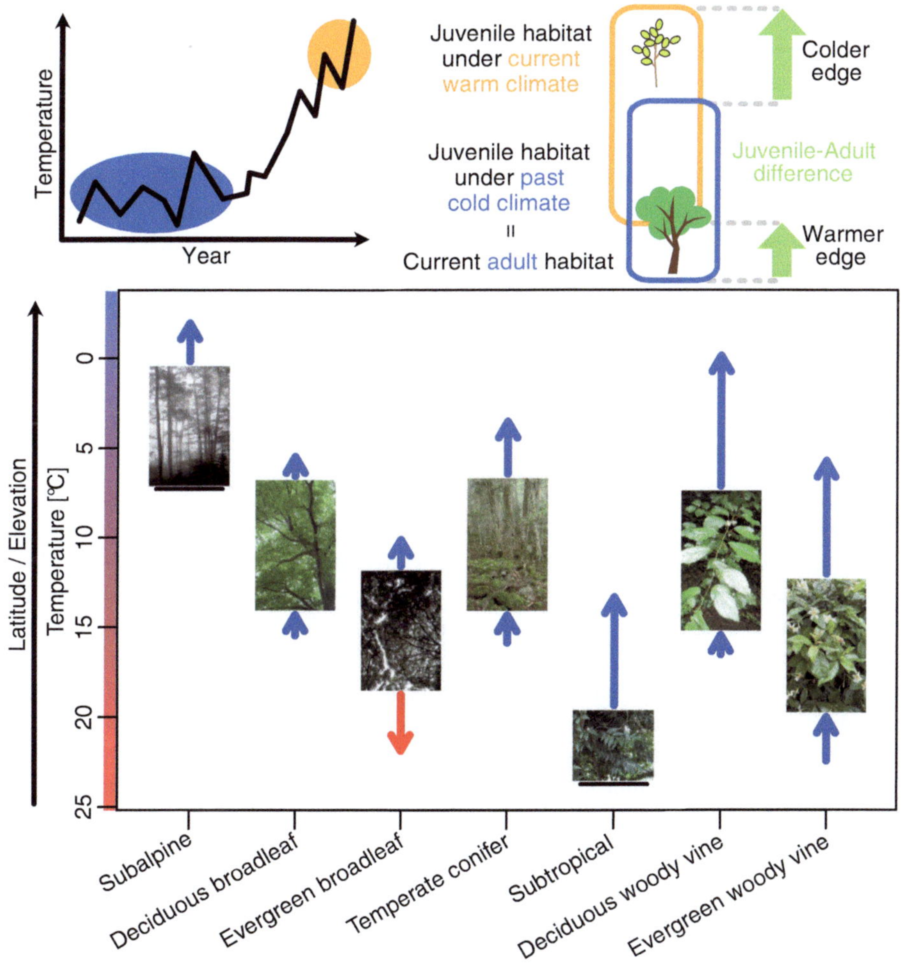

Fig. 10.1 Juvenile–adult difference patterns in the seven forest functional types. (Source: Adopted from Fig. 3 in Koide et al. (2022) with permission of John Wiley and Sons)

10.2.1 Materials and Methods

Using 13,409 forest plot vegetation survey data from the 6th and 7th National Vegetation Survey data (Biodiversity Center, Ministry of the Environment; http://www.biodic.go.jp/trialSystem/EN/info/vg67.html) and the Agro-Meteorological Grid Square Data (https://amu.rd.naro.go.jp/wiki_open/doku.php?id=start), we detected climatic distribution along the annual mean temperature gradient in each species and each size class. According to the vegetation survey methodology counting each species coverage in each forest layer, adult size was detected from the canopy (i.e., canopy and sub-canopy layer) occurrence, and juvenile size was detected from the forest floor (i.e., shrub and herb layer). The target species was selected when its occurrence was more than ten plots in each size class. The juvenile–adult difference was calculated in the three distribution points as the mean temperature, the colder edge (2.5th percentile), and the warmer edge (97.5th percentile) to capture differences between the distribution points. Different distribution points would have different drivers such as dispersal ability and cold resistance for the leading edge and competitive ability for the trailing edge (Ettinger and Hillerislambers 2017). We also used seven functional types (i.e., Subalpine, Deciduous broadleaf, Evergreen broadleaf, Temperate conifer, Subtropical, Deciduous woody vine, and Evergreen woody vine) to wrap up each species and analyzed differences among them. The permutation test was conducted to detect a significant difference in temperature distribution between juveniles and adults by setting a null hypothesis that both size temperature distributions are the same.

10.2.2 Results and Discussion

The results showed an overall colder shift of juveniles compared to adults among 302 tree species (Fig. 10.1). Colder shifts of juveniles were typical at their colder edge (i.e., leading edge under temperature warming situation), showing a significant colder shift of juveniles in the whole functional types. On the other hand, the warmer edge (i.e., trailing edge) showed mostly a colder shift of juveniles but different patterns appeared in several functional types. Typically, the warmer edge of the evergreen broadleaf functional type showed a reversed warmer shift of juveniles. This warmer shift of evergreen broadleaf juveniles was typical in the southwestern island part of Japan. This warmer shift would be affected by increased stochastic conditions in the island meta-populations (i.e., island habitat can sustain less competitive species through limiting strong competitor's dispersal, making open space by chance) and increased typhoon disturbances (i.e., reducing competitive superiority of subtropical species). The relatively less competitive difference between the evergreen broadleaf functional type and subtropical functional type (i.e., both are evergreen broadleaf) may also increase the possibility of regeneration for the evergreen broadleaf functional type species under the subtropical functional type forests.

Further, light seed mass species showed wider juvenile–adult differences both in leading and trailing edges, suggesting the importance of this functional trait for tree species movement. However, the meaning of the seed mass should be different between the edges as dispersal abilities for the leading edge (i.e., light-seeded species can distribute their seeds farther) and shade tolerance for the trailing edge (i.e., light-seeded species tend to show weak shade tolerance (weak competitive ability) creating wider die out range in their trailing edge though adult trees can persist in canopy layer).

Although monitoring and validation of the above mechanisms are still needed, this in situ detection of past-time distribution shifts in wide spatial and species ranges should be crucial to estimating and understanding the mechanisms of climate change effects on natural ecosystems. Broadly common colder shifts of juveniles would suggest a universal driver such as global warming as a main driver. The other area-specific drivers (e.g., precipitation, nitrogen deposition, anthropogenic disturbance) would be relatively less evident because these drivers are spatially more complex and these mosaiced distributions would not directly cause the universal juvenile–adult difference. The effects of species interactions (competitions and facilitations) between functional types and geographical conditions (island geography) were also suggested as ecological background mechanisms for these patterns in several zones. These in situ detections and suggestions are crucial to construct proper climate change adaptation plans and our juvenile–adult difference methodology should be a useful tool to detect these patterns in the early stage of assessments for climate change effects in terrestrial natural ecosystems.

10.3 Assessment: Vulnerability of the Forest Tree Community Under Future Climate Change

The biodiversity is reported to be a fundamental essence of ecosystem functions, altering a wide range of ecosystem services (Hong et al. 2022). Not only the amount of ecosystem service through the complementarity effect (i.e., different species complement each other in their roles, leading to improved ecosystem performance), but biodiversity is also a main source providing stability of the important functions through the redundancy effect (i.e., in a highly diverse ecosystem, multiple species perform similar ecological roles or functions, maintaining ecosystem functions under several species losses) (Gamfeldt et al. 2008). For the sustainable use of ecosystem services and conservation or adaptation planning, the vulnerability of biodiversity under future climate change should be considered. Future vulnerability assessment includes a variety of uncertainties in each step (e.g., climate model, species distribution models (SDMs)). To incorporate such uncertainties, previous works used multi-scenario and multi-SDM future projections creating the range of uncertainty and searching for optimal goals for the decision-making. However, this range of uncertainty tends to be quite large in highly uncertain systems such as natural ecosystems, and this wide variety of projection outputs prevented researchers and stakeholders from smoothly deciding on one plan. On the other hand, there is another reversed directional decision-making method such as the Info-gap decision theory (IGDT) framework (Johnson and Geldner 2019; Yokomizo et al. 2014). This framework starts with setting acceptable goals for target performance of natural or social systems, then searching the range of acceptable uncertainties to satisfy or maintain the desirable condition (Fig. 10.2). The estimated acceptable range directly means the vulnerability of the target systems (i.e., a small acceptable uncertainty range suggests higher vulnerability of the system). We implemented this IGDT framework to assess the vulnerability of Japanese forest tree communities under future climate change conditions.

10.3.1 Materials and Methods

We selected 42 natural forest inventory plot data of the Monitoring site 1000 project (https://www.biodic.go.jp/moni1000/), and created simple temperature envelope models for all canopy tree species that occurred in these plots (176 species in total) using presence data from the 6th and 7th National Vegetation Survey data (Biodiversity Center, Ministry of the Environment; http://www.biodic.go.jp/trial-System/EN/info/vg67.html) and the nationwide 1-km^2 mean annual temperature (MAT) data for 1981–2000 (Ishizaki et al. 2020). The species richness was selected as the target performance and the acceptable goals were set as maintaining 90, 75, and 50% of species richness. The species richness under current and future "best-guess" condition (2031–2050, the average among four GCMs and two RCPs) was predicted in each plot. Further temperature departure (warming) from the future "best-guess" condition was simulated in 0.1 °C intervals and the decline of species richness along the departure was analyzed (Fig. 10.2). Note that we only counted the species reduction phase, and the species increase phase was not considered in this analysis.

10.3.2 Results and Discussion

Our simulation assessment successfully mimicked warming-induced reductions in species richness, and the vulnerability (i.e., the range of acceptable uncertainty to maintain the target species richness level) showed different patterns among different forest functional types and forest plots (Fig. 10.2). Among the four forest functional types (i.e., boreal coniferous forest, deciduous broadleaved forest, temperate coniferous forest, and evergreen broadleaved forest), the evergreen broadleaved forest showed the highest reduction rate of species richness in the "best-guess" warming condition in 2031–2050. In the vulnerability assessment, the temperate coniferous

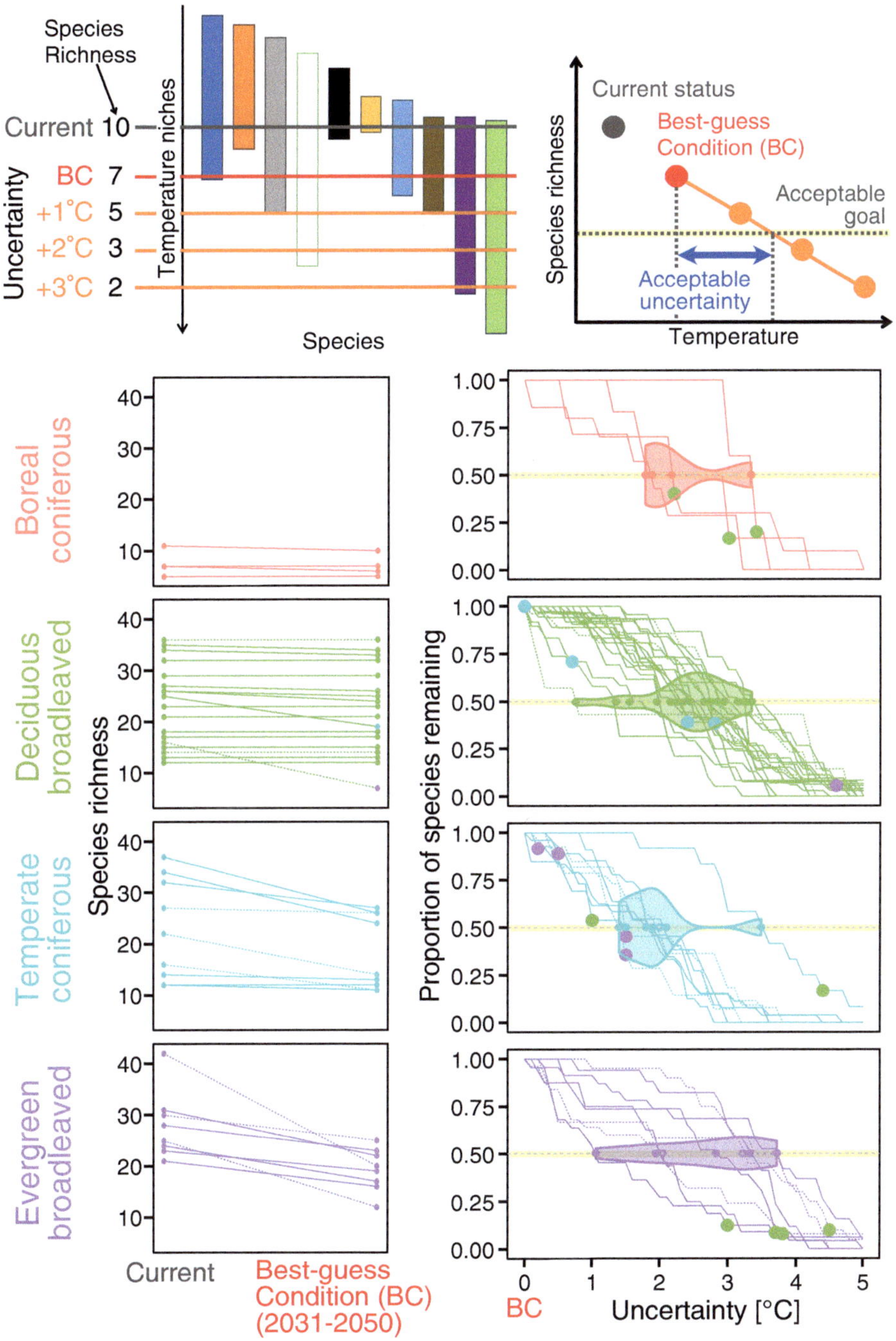

Fig. 10.2 Predicted future change of species richness and its vulnerability assessment in uncertainty along with further temperature warming. The info-gap theory starts by setting the acceptable goal condition (yellow line of right column figures). By analyzing how much further temperature warming is needed to break the acceptable goal condition, the acceptable uncertainty range (i.e., vulnerability) (blue arrow in the top right figure and violin plots in the bottom right figures) was calculated. The violin plots show vulnerabilities under the acceptable goal of maintaining 50% of species richness. Different color symbols in the bottom row show the change of dominant functional type in the research plot. (Source: Reproduced from Yoshikawa et al. 2023)

forest showed relatively higher vulnerability (i.e., smaller acceptable uncertainty) to future temperature warming though this difference was insignificant.

This IGDT framework can assess the vulnerability of the ecosystem based not only on the structure of the species community (i.e., each species' thermal niche breath and its combinations) but also on social and natural factors when deciding acceptable goals. Suppose the local species community contains a large number of warm-adapted species or a low level of acceptable goal (i.e., lower species richness) is accepted. In that case, the local system can get a wider acceptable uncertainty (i.e., lower vulnerability). This vulnerability should also be treatable as an index for adaptive capacity which can be a source of ecosystem resilience. The methodology for how to set the acceptable goal still needs discussion but this framework should be useful to assess natural ecosystem vulnerability and construct a robust strategy to adapt to future climate changes, especially for highly uncertain systems such as natural ecosystems.

10.4 Future Projection and Assessment for Conservation of Alpine Vegetation in Daisetsuzan National Park

In the implementation phase of conservation or adaptation plans, stakeholders or researchers tend to set a target region and search for detailed fine-scale information about the region. This fine-scale assessment (e.g., resolution <1 km^2) is crucially needed for future projections and actual planning of adaptation or conservation under future climate change especially in regional-level assessments for finely mosaiced vegetations. In such detailed fine-scale future projections, popular methodology based on each species' presence/absence data seems to be a bit rough since these binary data cannot capture dominances that can precisely describe conditions between presence and absence. The presence data are also uncertain since it's unclear whether it was created as a dominant stable distribution or an occasional rare distribution such as a sink population. Moreover, species of high conservation value are often rare and there are only a few presence data, creating a difficult situation to construct a reliable species distribution model. On the other hand, the actual implementation phase must consider many optional and region-specific conditional situations such as accessibility, cost, budget, labor, and the spatial distribution of target vegetation. To overcome these limitations for actual implementation, we focused on vegetation area cover from the vegetation map as a response variable instead of species presence/absence data to identify the probability of the presence of rare conservation target species. We also applied our original software tool (SecSel) for conservation prioritization considering accessibility and differences in required conservation measures among vegetation types. We then selected Mt. Taisetsu region in Hokkaido Island as a target for conservation planning of valuable alpine vegetation which was reported to be vulnerable to climate change (Amagai et al. 2018). We conducted regionally detailed future projections of alpine and subalpine vegetation distributions and assessed priority areas for conservation while considering realistic conditions.

10.4.1 Future Projection of Alpine Vegetation

10.4.1.1 Materials and Methods

Using GIS data of a 1:50,000-scale vegetation map by the Biodiversity Center of Japan, Ministry of the Environment (http://gis.biodic.go.jp/webgis/sc-025.html?kind=vg), we collected vegetation covers of five target vegetations (snowbeds, alpine heathland/fellfield, *Sasa* spp., subalpine shrubs, and subarctic forest). We analyzed the cover in each cell of approximately 1 km (30″ latitude) × 1 km (45″ longitude). We collected five climatic variables and five topographic variables as explanatory variables. We conducted an ensemble analysis using five habitat-suitability models (generalized linear models, generalized additive models, generalized linear model with tobit distribution, generalized additive model with tobit distribution, random forest, and boosted regression tree). Future distribution changes were projected based on three GCMs (MIROC5, MRI-CGCM3, and IPSL-CM5A-LR), two RCPs (2.6 and 8.5), and two time slices (2046–2050 and 2096–2100).

10.4.1.2 Results and Discussion

The habitat suitability models showed that summer temperature and snow-cover period were the main drivers creating spatial differences in coverages in each vegetation type. Typical alpine vegetation (snowbeds and alpine heathland/fellfield) drastically decreased under the RCP2.6 scenario with a limited number of refuges, but it will be totally replaced by subarctic forests under RCP8.5 (Fig. 10.3). The projected distribution area for suitable habitats did not differ markedly among GCMs and habitat suitability models.

As in other future projection studies, this study also has limitations such as low spatial resolution to capture spatially fine mosaiced alpine vegetation distributions. However, it also remains a possibility to include further microrefugia in small spots which have desirable geological, topographic, and environmental (e.g., high snow accumulation) characteristics for alpine vegetation. The invasion of lower elevational vegetation also has limitations due to the lack of ecological mechanisms such as dispersal ability and species interactions (i.e., competition and facilitation) in the model. Although further monitorings and validations are crucial, our future projections are fundamental revealing potentially vulnerable vegetation against future climate changes and important to prioritize actual conservation planning.

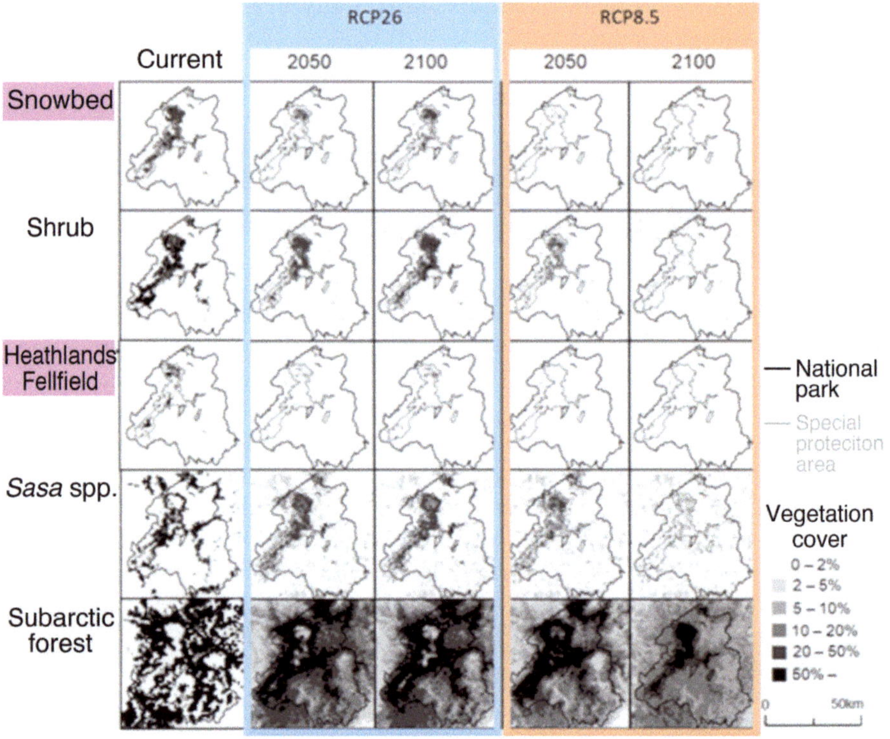

Fig. 10.3 Future projection of five vegetation covers in Mt. Taisetsu. (Source: Reproduced from https://www.nies.go.jp/whatsnew/20221213-1/20221213-1.html originally from Amagai et al. (2022))

10.4.2 Conservation Prioritization

Other than the vulnerable local populations as in above, spatial planning of conservation areas should prioritize sites that can efficiently satisfy multiple conservation target features. Bigger and more stable local populations (e.g., many individuals and wide vegetation cover) should also be prioritized as conservation targets to maintain the species or biodiversity certainly. However, sites that have a large amount of conservation value tend to invite human attractiveness in eco-tourism. This spatial conflict between biodiversity conservation and human use of ecosystem services frequently occurs in the real world. This conflict occurs even between biodiversity features if a conservation treatment for one species is harmful to the other. Conservation for grasslands often includes annual mowing to increase the light condition but this treatment is harmful when conserving the place as forests. The prioritization tools were intended to solve these conflicts among conservation target features, creating well-balanced site prioritization, by defining incompatible features and quantitatively analyzing which feature should be prioritized in each site

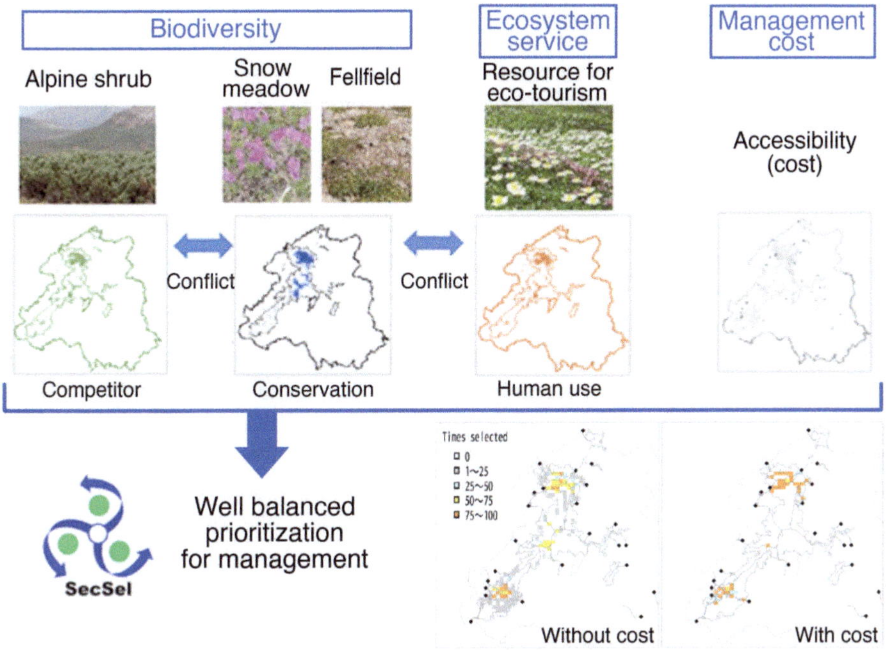

Fig. 10.4 Concept of the SecSel model and its outcomes in Mt. Taisetsu. (Source: Adapted from https://www.nies.go.jp/biology/research/publish/202112.html and Takenaka et al. (2021))

(Fig. 10.4). The SecSel model has a characteristic to start selecting from the best cell for all target features, then the second and third cells. As with the other prioritization tools, this site selection continues until fulfilling all the conservation targets (inputted as the number of cells or some unit amounts) for all target features included in the conservation areas. The model also considers pre-reserved sites and site cost data to support more realistic decision-making. The model outputs a set of selected sites with the frequency to be selected in each local unit using a range created by multi-run to cover random selection-based uncertainties.

10.4.2.1 Materials and Methods

We applied the SecSel model to search prioritized conservation/implementation cells (1 × 1 km grid) for alpine vegetation in Mt. Taisetsu region. The target features include four alpine vegetation covers (snow meadow, fellfield, wilderness, and shrubs) and one ecosystem service (the area sum of snow meadow and fellfield, representing an attractive flower field as a tourism resource). Since tourism can be a threat to the conservation of alpine vegetation through trampling or illegal digging and shrub vegetation can out-perform other vegetation types through competition, we considered these conflicts of features in the model. The costs of management (e.g., patrolling, monitoring, cutting back sasa-bamboo and shrubs) were also uniformly considered using the accessibility (i.e., the time required to reach each cell). We set a conservation target of eight cells for each biodiversity feature and ecosystem service and conducted 100 iterative calculations.

10.4.2.2 Results and Discussion

Results reported a high conservation/utilization priority for the northern and southern parts of the mountain range and a bit less priority for the central part (Fig. 10.4). But when considering the cost of accessibility, the importance of the central part diminished, and dominance of the northern and southern parts increases because the main trail entrances were concentrated in these parts. The SecSel model has wide flexibility to consider several actual and necessary conditions for the target region, providing a high potential to be applied to many regions. The model should include stakeholders from a priori timing to consider their needs and interests in the model, interactive timing to manipulate model settings, and to a posteriori timing to discuss the interpretation of the results. With such dense communications, the task of this model as a support tool for decision-making will be fulfilled.

10.5 Conclusion

Here we summarized some results in the terrestrial natural ecosystem field of the climate change adaptation research program by the National Institute for Environmental Studies. These results provide important insights into climate change effects and adaptations in the natural ecosystems in this country. Current model-based future projections in this research field still have a certain range of uncertainty based on the lack of in situ mechanistic processes and limited methodological tips and frameworks for treating such uncertainties. Our achievements here would contribute to filling the gap in such ecological mechanisms and help treat such unstable features for creating better decisions to adapt to future climatic conditions. We hope this report further provides positive stimulations to the other places and research fields, connecting to the next step for generalized climate change adaptation works in near future.

References

Amagai Y, Kudo G, Sato K (2018) Changes in alpine plant communities under climate change: dynamics of snow-meadow vegetation in northern Japan over the last 40 years. Appl Veg Sci 21(4):561–571. https://doi.org/10.1111/avsc.12387

Amagai Y, Oguma H, Ishihama F (2022) Predicted scarcity of suitable habitat for alpine plant communities in northern Japan under climate change. Appl Veg Sci 25(4):e12694. https://doi.org/10.1111/avsc.12694

Doi H, Takahashi M (2010) Macro-scale investigation of climate change effect on plant phenological timings using the phenological dataset of Japan metrological agency. Jpn J Ecol 60:241–247. https://doi.org/10.18960/seitai.60.2_241

Ettinger A, Hillerislambers J (2017) Competition and facilitation may lead to asymmetric range shift dynamics with climate change. Glob Chang Biol 23(9):3921–3933. https://doi.org/10.1111/gcb.13649

Gamfeldt L, Hillebrand H, Jonsson PR (2008) Multiple functions increase the importance of biodiversity for overall ecosystem functioning. Ecology 89(5):1223–1231. https://doi.org/10.1890/06-2091.1

Hogan A, Domke GM, Zhu K, Johnson DJ, Lichstein JW (2024) Climate change determines the sign of productivity trends in US forests. Proc Natl Acad Sci USA 121(4):e2311132121. https://doi.org/10.1073/pnas.2311132121

Hong PB, Schmid B, De Laender F, Eisenhauer N, Zhang XW, Chen HZ, Wang SP (2022) Biodiversity promotes ecosystem functioning despite environmental change. Ecol Lett 25(2):555–569. https://doi.org/10.1111/ele.13936

Ishizaki NN, Nishimori M, Iizumi T, Shiogama H, Hanasaki N, Takahashi K (2020) Evaluation of two bias-correction methods for gridded climate scenarios over Japan. SOLA 16:80–85. https://doi.org/10.2151/sola.2020-014

Johnson DR, Geldner NB (2019) Contemporary decision methods for agricultural, environmental, and resource management and policy. Annu Rev Resour Econ 11:19–41. https://doi.org/10.1146/annurev-resource-100518-094020

Koide D, Yoshikawa T, Ishihama F, Kadoya T (2022) Complex range shifts among forest functional types under the contemporary warming. Glob Chang Biol 28(4):1477–1492. https://doi.org/10.1111/gcb.16001

Kudo G (2014) Vulnerability of phenological synchrony between plants and pollinators in an alpine ecosystem. Ecol Res 29(4):571–581. https://doi.org/10.1007/s11284-013-1108-z

Kudo G (2020) Dynamics of flowering phenology of alpine plant communities in response to temperature and snowmelt time: analysis of a nine-year phenological record collected by citizen volunteers. Environ Exp Bot 170:103843. https://doi.org/10.1016/j.envexpbot.2019.103843

Kudo G, Cooper EJ (2019) When spring ephemerals fail to meet pollinators: mechanism of phenological mismatch and its impact on plant reproduction. Proc R Soc B 286(1904):20190573. https://doi.org/10.1098/rspb.2019.0573

Lenoir J, Gégout JC, Marquet PA, De Ruffray P, Brisse H (2008) A significant upward shift in plant species optimum elevation during the 20th century. Science 320:1768–1771

Lenoir J, Gégout JC, Pierrat JC, Bontemps JD, Dhôte JF (2009) Differences between tree species seedling and adult altitudinal distribution in mountain forests during the recent warm period (1986–2006). Ecography 32:765–777. https://doi.org/10.1111/j.1600-0587.2009.05791.x

Lenoir J, Bertrand R, Comte L, Bourgeaud L, Hattab T, Murienne J, Grenouillet G (2020) Species better track climate warming in the oceans than on land. Nat Ecol Evol 4(8):1044–1059. https://doi.org/10.1038/s41559-020-1198-2

Mizunaga H, Sako S, Nakao Y, Shimono Y (2005) Factors affecting the dynamics of the population of *Fagus crenata* in the Takakuma Mountains, the southern limit of its distribution area. J For Res 10:481–486. https://doi.org/10.1007/s10310-005-0165-8

Murayama S, Kondo H, Ishidoya S, Maeda T, Saigusa N, Yamamoto S et al (2024) Interannual variation and trend of carbon budget observed for more than two decades at Takayama in a cool-temperate deciduous forest in central Japan. J Geophys Res Biogeosci 129(6):e2023JG007769. https://doi.org/10.1029/2023JG007769

Pecl GT, Araujo MB, Bell JD, Blanchard J, Bonebrake TC, Chen IC et al (2017) Biodiversity redistribution under climate change: impacts on ecosystems and human well-being. Science 355(6332):eaai9214. https://doi.org/10.1126/science.aai9214

Piao SL, Liu Q, Chen AP, Janssens IA, Fu YS, Dai JH et al (2019) Plant phenology and global climate change: current progresses and challenges. Glob Chang Biol 25(6):1922–1940. https://doi.org/10.1111/gcb.14619

Scheffers BR, De Meester L, Bridge TCL, Hoffmann AA, Pandolfi JM, Corlett RT et al (2016) The broad footprint of climate change from genes to biomes to people. Science 354(6313):aaf7671. https://doi.org/10.1126/science.aaf7671

Sha LQ, Teramoto M, Noh NJ, Hashimoto S, Yang M, Sanwangsri M, Liang NS (2021) Soil carbon flux research in the Asian region: review and future perspectives. J Agric Meteorol 77(1):24–51. https://doi.org/10.2480/agrmet.D-20-00013

Suzuki SN, Ishihara MI, Hidaka A (2015) Regional-scale directional changes in abundance of tree species along a temperature gradient in Japan. Glob Chang Biol 21(9):3436–3444. https://doi.org/10.1111/gcb.12911

Takegawa Y, Kawaguchi Y, Mitsuhashi H, Taniguchi Y (2017) Implementing species distribution models to predict the impact of global warming on current and future potential habitats of white-spotted char (*Salvelinus leucomaenis*) for effective conservation planning in Japan. Jpn J Conserv Ecol 22:121–134. https://doi.org/10.18960/hozen.22.1_121

Takenaka A, Oguma H, Amagai Y, Ishihama F (2021) SecSel, a new software tool for conservation prioritization that is applicable to ordinal-scale data for multiple biodiversity features. PLoS One 16(7):e0247737. https://doi.org/10.1371/journal.pone.0247737

Woodall CW, Oswalt CM, Westfall JA, Perry CH, Nelson MD, Finley AO (2009) An indicator of tree migration in forests of the eastern United States. For Ecol Manag 257:1434–1444. https://doi.org/10.1016/j.foreco.2008.12.013

Yokomizo H, Coutts SR, Possingham HP (2014) Decision science for effective management of populations subject to stochasticity and imperfect knowledge. Popul Ecol 56(1):41–53. https://doi.org/10.1007/s10144-013-0421-2

Yoshikawa T, Koide D, Yokomizo H, Kim JY, Kadoya T (2023) Assessing ecosystem vulnerability under severe uncertainty of global climate change. Sci Rep 13(1):5932. https://doi.org/10.1038/s41598-023-31597-6

Part IV
Coastal Zone, Natural Disaster and Water Resources

Chapter 11
Assessing Costs of Adaptations to Sea Level Rise in Japanese Coastal Areas

Kohei Imamura, Makoto Tamura, and Hiromune Yokoki

Abstract Sea level rise poses significant inundation risks to coastal areas worldwide. Chap. 11 evaluates the impacts of inundation, the effectiveness of adaptation measures, and the associated costs in Japanese coastal areas. First, we estimate the potentially inundated areas, affected populations, and economic damages under two greenhouse gas emission scenarios: RCP8.5-SSP5 (high emissions) and RCP2.6-SSP1 (low emissions). We then examine two primary adaptation measures—protection and relocation—by estimating their effectiveness and costs, and we identify the most suitable approaches for climate change adaptation in Japanese coastal regions. The results show that RCP2.6-SSP1 resulted in smaller inundated areas, fewer affected populations, and lower economic damages compared to RCP8.5-SSP5, highlighting the importance of mitigation efforts. Additionally, the findings indicate that relocation costs are higher overall than protection costs in Japan. However, the cost relationship between these two strategies varies depending on population distribution and socioeconomic conditions across different regions.

Keywords Sea level rise · Inundation area · Affected population · Economic damage · Protection cost · Relocation cost

11.1 Introduction

According to the sixth Assessment Report of the IPCC (IPCC 2022), future sea level rise (SLR) is inevitable, even with climate change mitigation efforts. From 1902 to 2010, the global mean sea level (GMSL) rose by 16 [12–21] cm (hereafter, values in brackets indicate a 95% confidence interval). Since the 1980s, SLR has been

K. Imamura (✉) · M. Tamura
Global and Local Environment Co-creation Institute (GLEC), Ibaraki University, Mito, Japan
e-mail: kohei.imamura.st87@vc.ibaraki.ac.jp

H. Yokoki
Graduate School of Science and Engineering, Ibaraki University, Ibaraki, Japan

N. Mimura, S. Takewaka (eds.), *Climate Change Impacts and Adaptation Strategies in Japan*, https://doi.org/10.1007/978-981-96-2436-2_11

observed in the seas around Japan, with an annual average increase of 4.1 [0.1–8.2] mm observed between 2006 and 2015 (MEXT, JMA 2020). The GMSL is expected to continue rising beyond 2100, depending on future greenhouse gas emissions (IPCC 2021).

SLR poses significant risks to coastal areas worldwide (IPCC 2022). As an island nation surrounded by the sea, Japan is particularly vulnerable to the effects of SLR. Therefore, in addition to efforts aimed at mitigating global warming, it is crucial to implement adaptation measures to minimize its impact.

Adaptation options to sea level rise are generally categorized into three types: protection, accommodation, and relocation (or retreat) (Fig. 11.1; IPCC 2019). Protection includes the use of gray infrastructure, such as reinforcing and elevating dikes, as well as green infrastructure or Ecosystem-based Adaptation (EbA), such as beach nourishment and mangroves. Relocation involves moving residences to inland or higher ground, while accommodation includes raising the floor level of houses and reassessing land use patterns. However, as climate change progresses, adaptation options for low-lying coastal areas are becoming more limited (Haasnoot et al. 2021), making it essential to adopt measures tailored to the specific conditions of each region.

The projection of impacts and evaluation of adaptation strategies for SLR due to climate change have long been discussed but are still under development. Tamura et al. (2019) and Kumano et al. (2021) assessed the impacts of inundation and the costs of protection using gray-green infrastructure on a global scale, respectively, demonstrating the validity of early adaptation, as the cost of protection is lower than the economic damage caused by inundation. Diaz (2016) developed the Cost-Effective Adaptation Model to address global SLR and storm surges, concluding that in many countries, protection is more expensive than relocation. In contrast, most studies on relocation in Japan have focused on relocations due to tsunami disasters or floods in specific cities (e.g. Takeda and Tsuda 2015; Higashino and Murao 2021), with no analyses related to relocation caused by SLR. These previous studies evaluated adaptation measures independently, making it difficult to compare them under a unified standard. Therefore, it is necessary to integrate existing

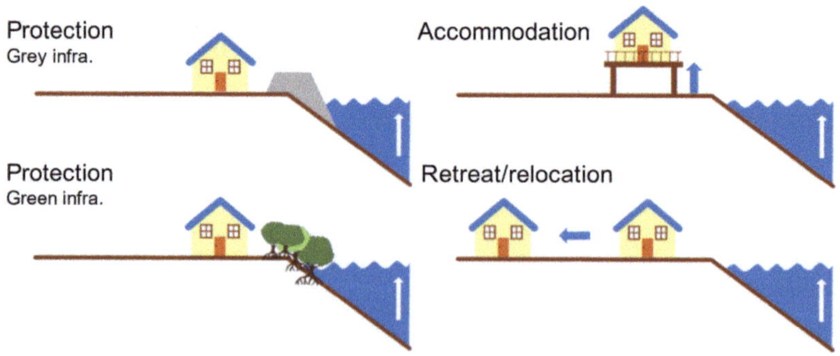

Fig. 11.1 Coastal adaptations

knowledge and conduct consistent research on impact assessments and adaptation evaluations.

This study evaluates the impacts of inundation, the effectiveness of adaptation measures, and the costs associated with sea level rise in Japanese coastal areas. First, we compare the inundation impacts under high and low greenhouse gas emission scenarios, specifically RCP8.5-SSP5 (fossil fuel-dependent society) and RCP2.6-SSP1 (sustainable society). Next, we focus on protection and relocation as adaptation measures, estimating their effectiveness and costs. By comparing these strategies, we discuss the most appropriate approaches for climate change adaptation in Japanese coastal areas.

11.2 Methodology

11.2.1 Assessing Impacts of SLR

Figure 11.2 shows the outline of the research method. In the impact assessment, inundation calculations were performed every 10 years from 2020 to 2100, and the potentially inundated area (hereinafter, inundated area) and potentially affected population (hereinafter, affected population) were estimated. In calculating inundation, each grid of the Digital National Land Information tertiary mesh was analyzed by comparing the sea level, which accounts for SLR and tides, with the average elevation. Grids where the sea level exceeded the average elevation were classified as inundated. The MIROC-ESM-CHEM values were used for SLR, the TPXO9.0 values for tides, and the Digital National Land Information elevation/slope tertiary mesh for average elevation. The area of inundated grid was totaled to determine the inundated area at each time. The population for each grid at each time point was obtained from the SSP population scenario (NIES 2021). The population of the inundated grids was then summed to determine the affected population at each time point. By 2100, the future population of Japan is expected to decrease in both scenarios, with about 79.4 million people in SSP5 and about 72.9 million people in SSP1.

It is important to note that this impact assessment does not account for inundation prevention by existing coastal structures, such as dikes and river barriers, which may result in an overestimation of the impact as "potentially inundated impact." This is because data on the height of such structures is not available on a nationwide scale in Japan. Several studies have evaluated inundation impacts using detailed structural data, but these are limited to specific areas, such as Tokyo Bay. However, to the best of our knowledge, no comprehensive data on coastal structures is available nationwide, and there are very few studies that assess inundation impacts on a national scale. Currently, inundation is being prevented by coastal and river structures, so the 2020 calculations reflect the potential impact if these structures did not exist. The projections for 2030 and beyond assume a hypothetical scenario in which

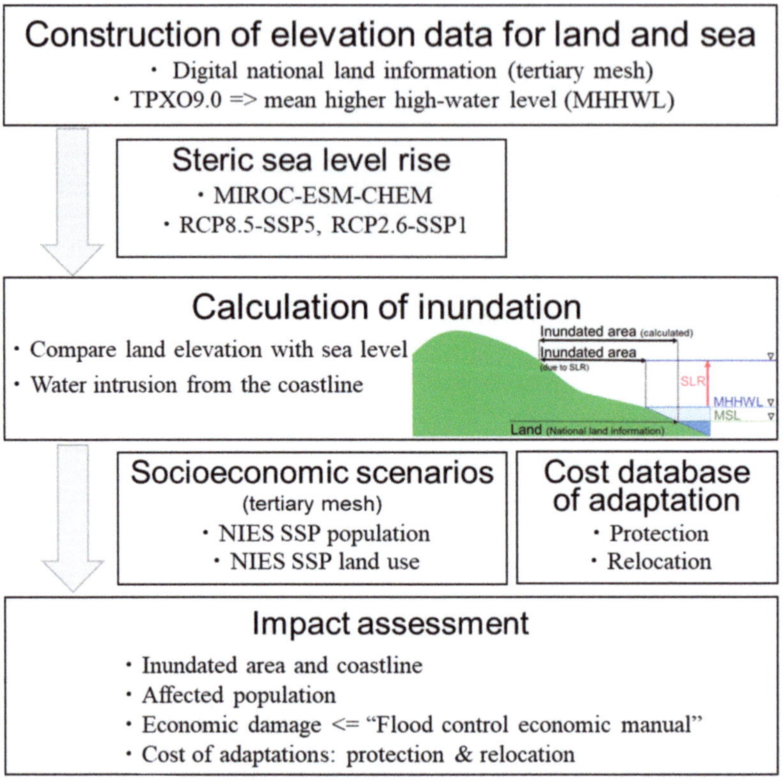

Fig. 11.2 Outline of the method. (Adapted from Tamura et al. (2023))

sea levels surpass the current height of all such structures across the country, as their exact heights are unknown.

11.2.2 Assessing Economic Damage

Economic damage from inundation was estimated based on the Ministry of Land, Infrastructure, Transport and Tourism's Manual for Economic Evaluation of Flood Control Investment (Draft) (hereinafter referred to as the Flood Control Economic Manual: MLIT 2020). The Flood Control Economic Manual is designed to assess the cost-effectiveness of flood control infrastructure, such as dikes and dams, and outlines methods for calculating the direct economic damage to various assets that have been temporarily inundated by river flooding, among other events. Inundation due to SLR is ongoing, which differs slightly from the temporary flooding assumed in the Flood Control Economic Manual. However, since a method for estimating the economic damage from continuous inundation has not yet been established, this manual was used in the study. Additionally, the authors assessed economic damage using the disaster database (EM-DAT) and land prices. For further details and comparisons, please refer to Oba et al. (2022), Kodama et al. (2022), and Ujike et al. (2023).

In terms of inundation-related economic damage, we considered houses, household goods, businesses, public facilities, production losses of rice and vegetables, farmland and agricultural equipment, and golf courses. The economic damage was estimated by multiplying the unit price of each item, as listed in the Flood Control Economic Manual, by the inundated area by land use, affected population, and damage rate based on inundation depth. For future economic damages, the values were converted to nominal figures for each time period by applying a deflator calculated for the RCP8.5-SSP5 and RCP2.6-SSP1 scenarios, taking into account future price increases.

11.2.3 Assessing Cost of Adaptation

11.2.3.1 Protection

To estimate protection costs, we calculated the cost of raising the height of dikes from their current levels to match sea level at each time point. Additionally, we estimated replacement costs, assuming a service life of 30 years for the dikes, and summed both costs. Estimating the cost of raising the dikes requires knowledge of their current height; however, no nationwide data on dike heights is available (cf. Sect. 11.2.1). Therefore, this study assumed that the current height of dikes corresponds to the sea level in 2020 and that the dikes would be raised in accordance with future sea level changes.

The cost of raising and replacing dikes was estimated by multiplying the unit cost of dikes by the length of the coastline within inundated grids. Two methods were used for the unit cost estimation: the first (hereafter, Type 1) employed a regression formula based on dike height and GDP per capita, using a database of 455 dikes across 20 countries, similar to the approach of Kumano et al. (2021). The second method (hereafter, Type 2) used a unit price proposed in the report by the Technical Committee on Resilience Assurance of the Japan Society of Civil Engineers (JSCE 2018). All monetary values were converted to nominal values at each time point.

11.2.3.2 Relocation

Relocation costs were estimated as the administrative cost of relocating the entire affected population at each time point. Previous studies have faced challenges, such as a lack of data to justify the estimated amounts and the arbitrary treatment of cost items that do not reflect existing systems. Since there have been no cases of relocation due to sea level rise in Japan, this study empirically estimated relocation costs by following the framework of the Disaster Prevention Group Relocation Promotion Project, which has been used in reconstruction projects such as the Great East Japan Earthquake, and by utilizing data on civil engineering costs.

The affected population at each time point is determined by overlaying inundation calculations with population scenarios. These population scenarios do not account for population movements caused by SLR (cf. Sect. 11.2). Therefore, the affected population at each time represents the number of people impacted if inundation occurs for the first time when sea levels exceed protective structures. Similarly, the relocation cost at each time reflects the cost in this scenario.

The framework of the Disaster Prevention Group Relocation Promotion Project is based on the number of households, so the affected population was converted into the number of households (i.e. affected households). We conducted a simulation to allocate the number of affected households in order of proximity from the inundated grid to candidate relocation grids and determined the relocation sites (Imamura et al. 2023). The candidate relocation grids were assumed to be areas that would not be inundated for 30 years from the evaluation point. Since the inundation calculations extend only until 2100 (cf. Sect. 11.2), the estimation of relocation costs was limited to 2070.

The cost items for relocation include acquisition and development costs of the relocation site, infrastructure construction costs at the relocation site, purchase costs of farmland and residential land at the original location (i.e. inundated grids), subsidies for land and housing acquisition at the relocation site, and subsidies for miscellaneous expenses such as moving costs. Based on the simulation results for the relocation site, these cost items were estimated individually using actual data from past disaster prevention group relocation projects and civil engineering construction costs. All costs were converted to nominal values for each time point.

11.3 Results

11.3.1 Impacts of SLR

Figure 11.3 shows the potentially inundated areas in 2020 and 2100. Figure 11.4 illustrates the temporal changes in potentially inundated areas under RCP2.6 and RCP8.5 across eight regions, compiled from 47 coastal prefectures. The affected population is shown in Fig. 11.5, and the economic damage in Fig. 11.6. The "Tide" in Fig. 11.3 represents the potential inundation impact assuming no protective structures in 2020. The inundated grids are primarily concentrated around Tokyo Bay, Ise Bay, Osaka Bay, and the Ariake Sea. Regionally, the areas affected are ordered as Kyushu-Okinawa, Chubu, Kinki, and Kanto. The total inundated area in Japan is projected to be approximately 2111–2127 km^2 by 2050 and 2261–2598 km^2 by 2100. The total affected population is estimated to be about 4.45–4.7 million people

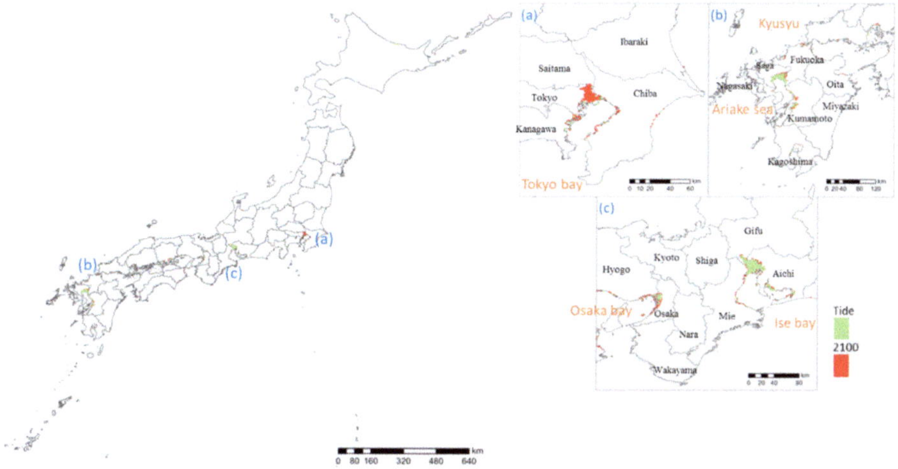

Fig. 11.3 Potentially inundated area in 2020 and 2100 (RCP8.5; Adapted from Tamura et al. (2023))

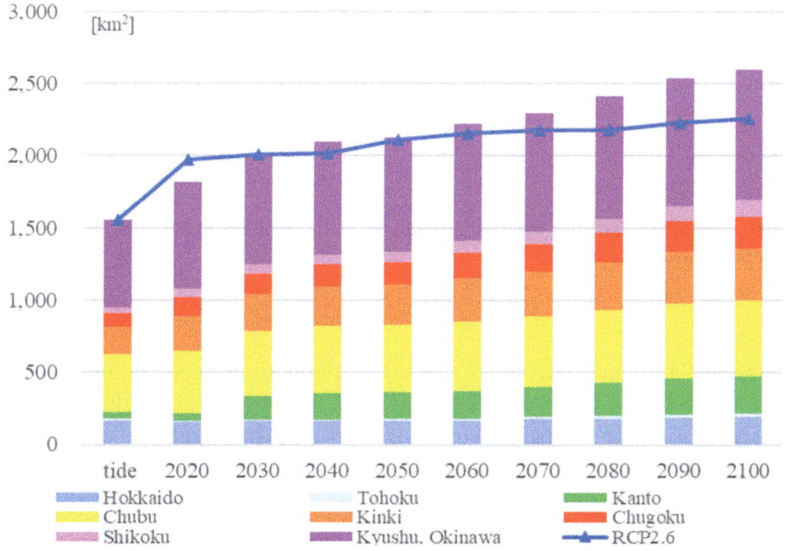

Fig. 11.4 Temporal change of potentially inundated areas in eight regions (bar: RCP8.5; line: RCP2.6; Reproduced from Tamura et al. (2023))

in 2050 and 3.76–4.92 million people in 2100. The total economic damage in Japan is projected to range from 151 to 181 trillion yen in 2050 and from 243 to 455 trillion yen in 2100 (adjusted for future inflation). Under the SSP5 scenario, GDP is expected to reach 764 trillion yen in 2050 and 1354 trillion yen in 2100, while under SSP1, GDP is estimated to be 641 trillion yen in 2050 and 852 trillion yen in 2100 (converted at $1 = 121 yen at 2015 prices) (NIES 2021).

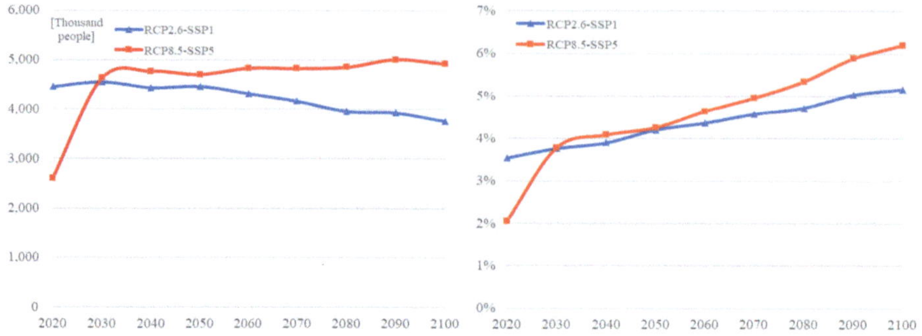

Fig. 11.5 Affected population without adaptation. (Reproduced from Tamura et al. (2023)) (left: total population; right: % of total population in each SSP)

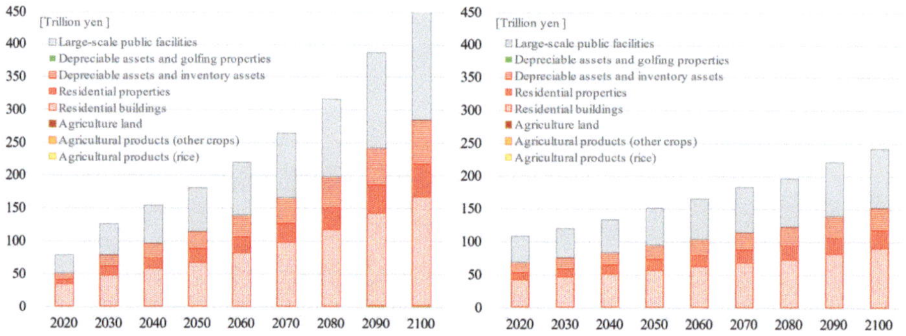

Fig. 11.6 Economic damage without adaptation (left: RCP8.5-SSP5, right: RCP2.6-SSP1; Reproduced from Tamura et al. (2023))

Comparing the inundated area, affected population, and economic damage between RCP8.5-SSP5 and RCP2.6-SSP1, all metrics were smaller for RCP2.6-SSP1, highlighting the importance of mitigation efforts. Regionally, economic damage was more pronounced around the three major bays (Tokyo Bay, Ise Bay, and Osaka Bay) due to the concentration of building sites and affected populations.

11.3.2 Cost of Adaptations

11.3.2.1 Protection

Figure 11.7 shows the protection costs for all of Japan, calculated using both the Type 1 and Type 2 methods (cf. Section 11.2). For Type 1, the cost was estimated to be around 39.7–54.4 trillion yen by 2100, while for Type 2, the estimate was

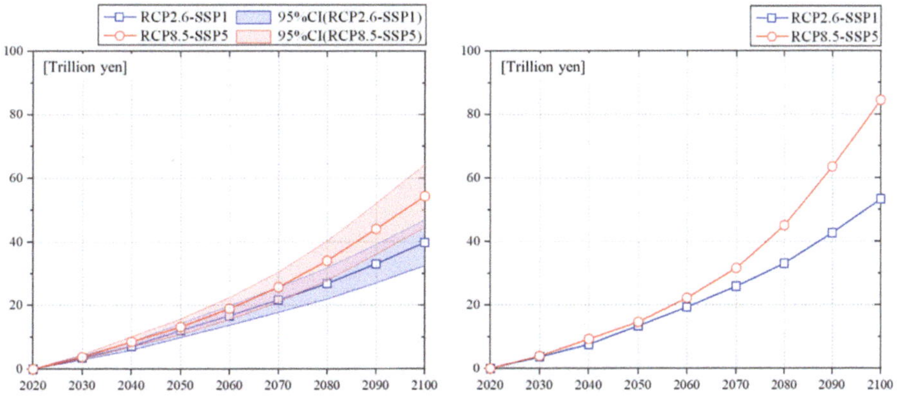

Fig. 11.7 Protection cost (left: Type 1, right: Type 2; Reproduced from Tamura et al. (2023))

approximately 53.4–84.5 trillion yen by 2100. Protection costs under RCP8.5-SSP5 were higher than those under RCP2.6-SSP1. This is because the length and height of the dikes required for protection in RCP2.6-SSP1 are smaller than those in RCP8.5-SSP5.

11.3.2.2 Relocation

Figure 11.8 shows the simulation results for relocation destinations. Taking the area around Tokyo Bay as an example, the majority of households affected by inundation are expected to relocate to Chiba, Saitama, Ibaraki, and Kanagawa prefectures, where ample land is available. Comparing the distribution of relocation destinations in 2030 and 2070, some grids that were not chosen as relocation destinations in 2030 were selected in 2070. This result suggests that land that was previously inhabited and unavailable in 2030 became usable due to population decline and was subsequently chosen as a relocation destination.

Figure 11.9 shows the total relocation cost and its breakdown for the entire country. The estimated cost was approximately 90–94 trillion yen in 2030, about 101–117 trillion yen in 2050, and around 109–150 trillion yen in 2070. The relocation cost under RCP2.6-SSP1 is lower than that under RCP8.5-SSP5, due to differences in the affected population (≈ relocating households) and the deflator scenario used to convert into nominal values. Regarding the breakdown of relocation costs, the largest proportion was attributed to the purchase of farmland and residential land at the original location, followed by land development costs and infrastructure construction costs at the relocation destination, which together accounted for the majority of the total.

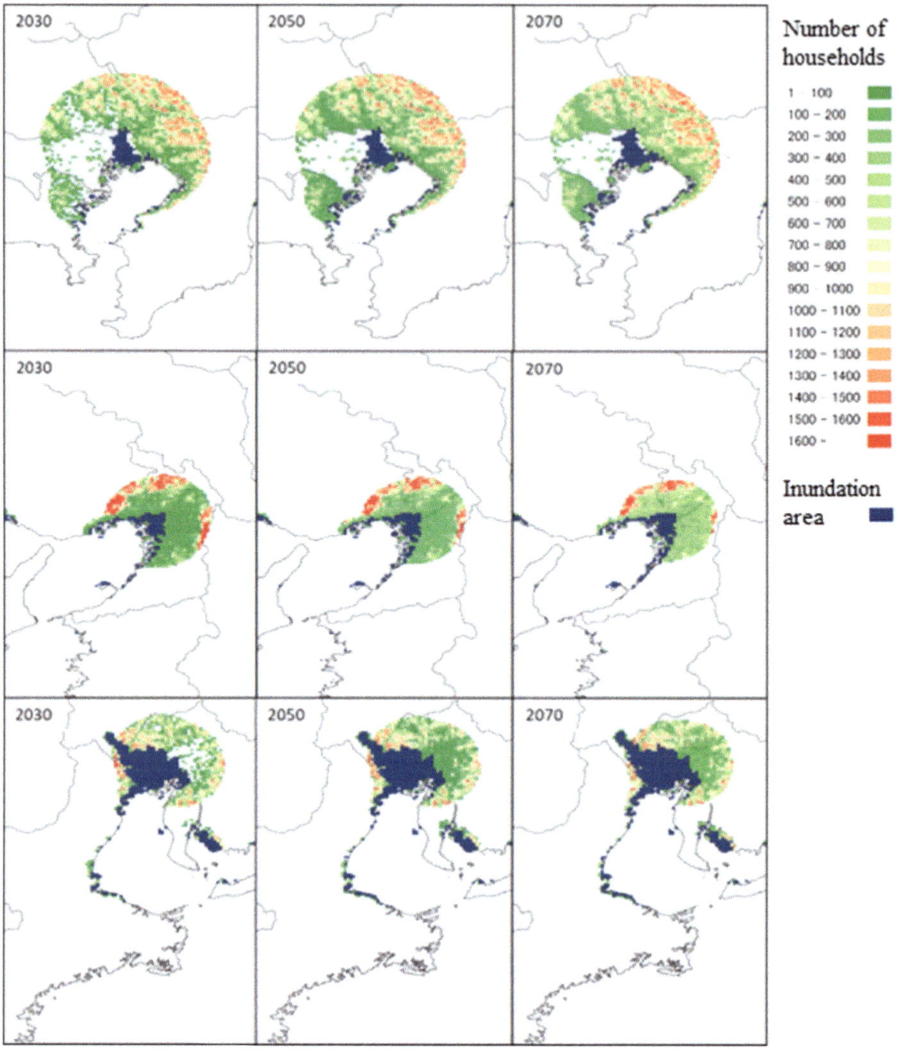

Fig. 11.8 Distribution of relocated households in RCP8.5-SSP5 (top: Tokyo Bay, middle: Osaka Bay, bottom: Ise Bay; Reproduced from Imamura et al. (2023))

11.4 Conclusion

This study initially estimated the potentially inundated areas, affected populations, and economic damages along the Japanese coast due to sea level rise and tides. We compared high and low greenhouse gas emission pathways (RCP8.5-SSP5 and RCP2.6-SSP1) and evaluated the uncertainties of socioeconomic scenarios. The results showed that RCP2.6-SSP1 led to smaller potentially inundated areas, affected populations, and economic damages compared to RCP8.5-SSP5, underscoring the importance of mitigation efforts.

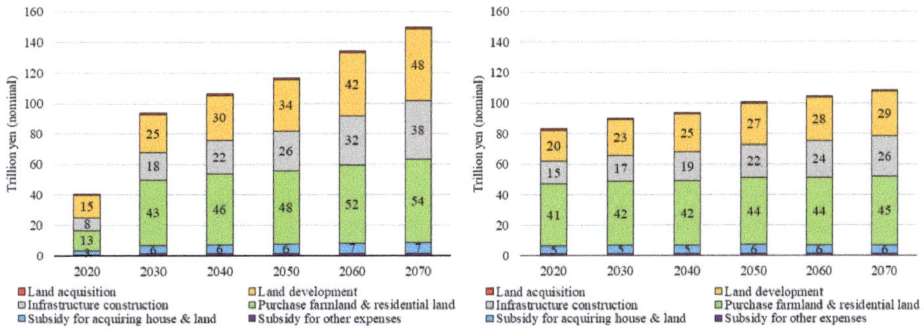

Fig. 11.9 Details of relocation cost (left: RCP8.5-SSP5, right: RCP2.6-SSP1; Adapted from Imamura et al. (2023))

Next, we calculated the effects and total costs of protection-only and relocation-only strategies across Japan. The cost of protection was estimated using a database of past protection measures, such as dikes. Since there have been no cases of relocation due to sea level rise in Japan, relocation costs were estimated based on the Disaster Prevention Group Relocation Promotion Projects, which have been used in reconstruction efforts after the Great East Japan Earthquake. The findings indicated that, in general, relocation costs are higher than protection costs in Japan. However, it was also suggested that the cost relationship between the two strategies varies depending on population distribution and socioeconomic conditions in different regions, such as urban and suburban areas.

Although not considered in this study, if vacant houses and land can be utilized for relocation, the cost of relocation could be significantly reduced. However, if such land is not sufficiently close to the inundated area to accommodate all affected households, the applicable areas are limited. Even in cases where relatively inexpensive relocation is possible, there is still the cost of purchasing the land at the original location. Therefore, if all affected households are relocated, a nationwide cost of at least 40–60 trillion yen is unavoidable. Consequently, it is considered most cost-effective to protect densely populated areas, such as major metropolitan areas around the three major bays, with coastal structures, and to relocate in less densely populated areas. Determining the optimal combination of these adaptation strategies is a subject for future research.

In this study, we primarily discussed the impact assessment and adaptation measures for sea level rise from a cost perspective; however, it goes without saying that cost is just one factor to consider when evaluating adaptation measures. Relocation significantly alters the living environment of residents, so there may be cases where some prefer protection measures regardless of the cost. On the other hand, in recent years, the compactification of urban areas has been promoted as part of town planning in response to population decline and aging. If relocation is implemented as part of such town planning, it could serve as an effective adaptation option, provided that areas with low inundation risk are chosen as relocation destinations.

References

Diaz DB (2016) Estimating global damages from sea level rise with the Coastal Impact and Adaptation Model (CIAM). Clim Chang 137(1):143–156

Haasnoot M, Lawrence J, Magnan AK (2021) Pathways to coastal retreat. Science 372(6548):1287–1290

Higashino M, Murao O (2021) A study on the benefits of relocation to higher ground in advance in the central district of Rikuzentakata city based on the great east Japan earthquake reconstruction projects. J Soc Safe Sci 39:81–90. (in Japanese)

Imamura K, Tamura M, Yokoki H (2023) Simulation for residential relocation to adapt to sea level rise in Japan coasts. J Jpn Soc Civil Eng G 79(10):12p. (in Japanese)

IPCC. (2019). Special report on the ocean and cryosphere in a changing climate

IPCC (2021) Climate change 2021: the physical science basis. Contribution of Working Group I to the Sixth Assessment Report of the IPCC

IPCC (2022) Climate change 2022: impacts. Adaptation and Vulnerability, Contribution of Working Group II to the Sixth Assessment Report of the IPCC

JSCE (2018) Technical study report on countermeasures for mega disasters that cause national disaster. In: Appendix 2: subcommittee on coasts and ports. Technical Study Committee on Ensuring Resilience in Japan Society of Civil Engineers. (in Japanese)

Kodama K, Yokoki H, Tamura M (2022) Assessing socioeconomic impacts of sea level rise on Japanese coasts via scenarios of population and land use. J Jpn Soc Civil Eng G (Environ) 78(5):I_349–I_357. (in Japanese)

Kumano N, Tamura M, Inoue T, Yokoki H (2021) Estimating the cost of coastal adaptation using mangrove forests against sea level rise. Coast Eng J 63(3):263–274

MEXT, JMA (2020) 2020 climate change in Japan: report of atmospheric/oceanic observation and projection. Ministry of Education, Culture, Sports, Science and Technology (MEXT) and the Japan Meteorological Agency (JMA) (in Japanese)

MLIT (2020) Manual for economic evaluation of flood control investment (draft). River bureau, Ministry of Land, Infrastructure, Transport and Tourism (in Japanese)

NIES (2021) Tertiary mesh SSP population scenarios (2nd edition): Product of environmental research and technology development fund 2–1805, National Institute for Environmental Studies. https://adaptation-platform.nies.go.jp/socioeconomic/population.html

Oba M, Yokoki H, Tamura M (2022) Impact of sea level rise on Japanese coastal areas and economic assessment via shared socioeconomic pathways. J Jpn Soc Civil Eng G (Environ) 77(5):I_221–I_231. (in Japanese)

Takeda H, Tsuda T (2015) Feasibility study of predisaster relocation planning of urban functions in Nankai trough earthquake area. J City Plan Inst Jpn 50(3):594–601. (in Japanese)

Tamura M, Kumano N, Yotsukuri M, Yokoki H (2019) Global assessment of the effectiveness of adaptation in coastal areas based on RCP/SSP scenarios. Clim Chang 152(3–4):363–377

Tamura M, Imamura K, Kumano N, Yokoki H (2023) Assessing the effectiveness of adaptation against sea level rise in Japanese coastal areas: protection or relocation? vol 26. Environment, Development and Sustainability, pp 1–17

Ujike K, Yokoki H, Tamura M, Imamura K (2023) Assessing economic damage of sea level rise on Japanese coasts using land prices. J Jpn Soc Civil Eng 79(27):8. (in English)

Chapter 12
Projection of Coastal Impacts: Beach Erosion and Inundation Risk of Urban Areas

Nobuhito Mori and Takuya Miyashita

Abstract This chapter describes the impact of climate change on natural beaches and the inundation of mega-delta cities in Japan. First, the nationwide retreat of beaches is discussed based on the national beach geometry database under several SSP scenarios. Under the SSP1-2.6 and SSP5-8.5 scenarios, it is projected that 30-60% of Japan's major natural beaches could be lost by 2100. Second, the combined risks of sea-level rise and storm surges are shown to significantly increase inundation areas and affected populations, with impacts potentially tripling or more than the effects of sea-level rise alone. The number of residents affected by the combined influence of sea-level rise and storm surge is projected to peak around 2050 due to future population decline in Japan.

Keywords Sea-level rise · Storm surge · Natural beach · Inundation risk · Population change · Mega-delta city

12.1 Introduction

Human activities in coastal areas of Japan are greatly affected by sea-level rise (SLR) and storm surges associated with typhoons. In particular, coastal areas are at increased risk of inundation due to sea-level rise and increased storm surge heights. The IPCC Sixth Assessment Report (AR6) developed the Shared Socioeconomic Pathways (SSPs), scenarios that combine different radiative forcing for global warming with future socioeconomic development pathways to assume future conditions. According to IPCC AR6 WGI Chap. 9 (Fox-Kemper et al. 2021), even under the least warming SSP1-2.6 scenario, the global mean sea level is projected to increase by 0.28 to 0.55 m by 2100 and under the warmest SSP5-8.5 scenario,

N. Mori (✉) · T. Miyashita
Disaster Prevention Research Institute, Kyoto University, Uji, Kyoto, Japan
e-mail: mori@oceanwave.jp

N. Mimura, S. Takewaka (eds.), *Climate Change Impacts and Adaptation Strategies in Japan*, https://doi.org/10.1007/978-981-96-2436-2_12

global mean sea level is projected to increase by 0.63 to 1.01 m for SSP5-8.5 by 2100 by 2100.

Tropical cyclones (TCs) cause severe coastal hazards in the middle latitude. Typhoon Jebi caused coastal flooding in Kansai Airport, Japan, in 2018. TC intensity will be expected to increase due to climate change by global warming (IPCC AR6 WGI Chap. 11; Seneviratne et al. 2021). TC intensity and the other TC characteristics are linked to future changes in extreme storm surges and extreme ocean wave climates in middle latitudes. In addition, the population of Japan is concentrated in low-lying coastal areas, and storm surge inundation is likely to cause extensive damage by coastal flooding. Therefore, considering climate change, it is important to consider the rise of sea levels and increased storm surge heights in the Japanese coastal areas. However, the long-term effect of increased coastal hazards and future population changes in the coastal inundation area is not well understood. Furthermore, a long-term assessment of these risks is also important information for timing adaptation measures (e.g., Mori and Shimura 2023).

In this chapter, we summarize the latest results of quantifying the impact of sea-level rise (SLR) on sandy beaches covering the whole of Japan's coastlines. Furthermore, in addition to the coastal inundation area, the impact on society, such as the affected population, is also important in storm surge risk assessment. This chapter shows the change in the potentially affected population by SLR and storm surge until 2100 according to different SSP scenarios.

12.2 Impact of Sea-Level Rise on Japanese Beaches

For natural beaches, the National Sandy Beach Database has been developed by Kyoto University and the National Institute of Advanced Industrial Science and Technology to project future changes in sandy beaches. In this database, the elevation data of the fifth mesh of the National Land Information (ver. 1.0; GSI, 2011) and Google Earth data were combined to create a nationwide sandy beach database with 806 beaches across Japan with a length of 1 km or more as evaluation targets (Mori et al. 2018). For the decrease of beach area due to sea-level rise, we used two different types of beach gradients based on the beach database to estimate the amount of static area decrease. Details are provided below.

We focus on the estimation of the beach gradient around the shoreline. We use the fifth mesh dataset (ver. 1.0; GSI, 2011; denotes GSI-fifth dataset) produced by the Geospatial Information Authority of Japan (GSI). This official Japan topographic map data includes maximum, average, and minimum altitudes with 10-m resolution covering the whole country. The GSI-fifth dataset is the digital surface model (GSM) dataset that combines laser and aerial photogrammetric surveying. Two different beach slopes were estimated based on the GSI-fifth dataset. The first beach slope $< \theta >$ is calculated by averaging over the local slopes of each beach. The second beach slope θ_{mean} is calculated from the average height of the beach divided by the beach width. See the details of the definition and uncertainty of estimation in Mori et al. (2018).

Sea-level rise scenarios around Japan based on IPCC AR6 were used, and both scenarios and model uncertainties were considered. The scenarios were compared for SSP1-2.6 and SSP5-8.5, and the model uncertainty was evaluated using two types of SLR projections: the mean and the top 5% tiled SLR projections. Figure 12.1 shows future decreases in beach areas in Japan estimated by θ_{mean} and $< \theta >$ based on the SSP1-2.6 and SSP5-8.5 scenarios, respectively. The mean values by IPCC AR6 for sea-level rise were used for projection. The warmer the color, the larger the percentage area change is for the beaches, but there are no regional differences. The SSP5-8.5 projection indicates that more beaches will lose 80% or more of their area. Table 12.1 shows the national average of future changes in the beach area for all 806 beaches. SSP5-8.5 predicts a loss of 39 or 66% of the beach area, indicating that the impact of sea-level rise on sandy beaches is significant. The fraction of the current coastline that undergoes shoreline retreat depends on the slope estimation and the SLR scenario. Therefore, a database of accurate measurements of nearshore topography and bathymetry is necessary for assessing the impact of climate change in terms of the decreases in sandy beaches. Regardless of the accepted level of accuracy of decreases in beach area estimates, there are clear danger areas for severe beach retreat in Japan. These are along the Pacific Ocean and the Seto Inland Sea, with long, gently sloping beaches. The hot spot of beach decrease varies depending on whether the target of evaluation is the rate of change or the area. When evaluated by percentage, loose sandy beaches with slopes have more severe beach recession than steep sandy beaches. On the other hand, long sandy beaches are affected more greatly than short sandy beaches when evaluated by area change.

It is insufficient to project future decreases in beach area changes using the proposed model or existing methodology (e.g., the Bruun rule). We need to develop models and create a database in order to make a quantitative impact assessment of climate change on decreases in the beach areas.

12.3 Impact of Storm Surges on Japanese Major Coastal Cities

As shown in the previous section, the sea-level rise will change the beach area size, and the impact of sea-level rise on urban area protection will also be significant. However, a storm surge is an important factor for coastal protection along the Pacific side of Japan. Therefore, sea-level and extreme sea-level need to be considered for the impact assessment on coastal urban areas. Based on future projections of sea-level rise and storm surge, high-resolution topographic and population data for all of Japan were analyzed to evaluate the distribution of impacted areas and populations at each elevation. Both the influenced area and population of Tokyo were

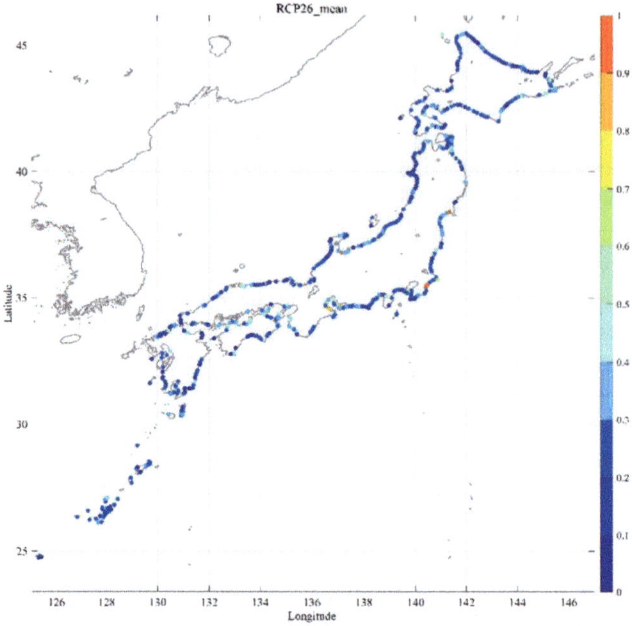

(a) SSP1-2.6 with mean slope θ_{mean}

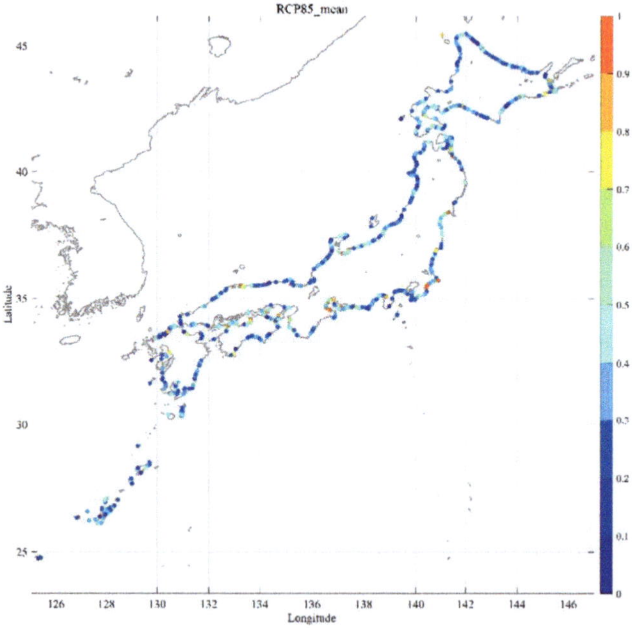

(b) SSP5-8.5 with mean slope θ_{mean}

Fig. 12.1 Changes in beach area at 2100 normalized by 2011

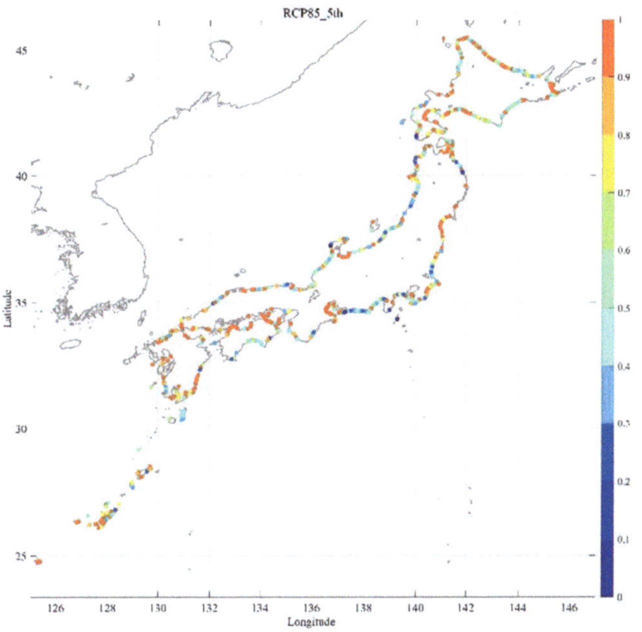

(c)SSP5-8.5 with 5th mesh slope <θ>

Fig. 12.1 (continued)

Table 12.1 National average of future changes in beach area (upper: θ_{mean}, lower: <θ>)

	2100	2150
ssp126	−27%	−35%
	−51%	−69%
ssp585	−39%	−87%
	−66%	−96%

found to have more than doubled with and without storm surges, in addition to a sea-level rise.

Future changes in the impacted area and impacted population were analyzed using elevation data from the fifth order mesh (10 m resolution) of the National Land Information System and NIES population projection data with ZENRIN population mesh data 2019 (100 m resolution) as exposure data. For storm surge, the impacted area and impacted population were evaluated in terms of sea-level rise plus storm surge anomaly for Osaka Bay. We used the same IPCC AR6 data for the sea-level rise in Japan, as shown in Sect. 12.2.

Storm surges are an extreme phenomenon compared to rising sea levels, and it is difficult to predict future changes. In particular, the difficulty of annual assessment is very high. Therefore, based on the theory of Maximum Potential Intensity (MPI) of typhoons, a method is used to estimate the surge height of the Maximum Potential

Storm (MPS), assuming the worst-case scenario from monthly mean climate values calculated by a climate model. The maximum potential of storm surge in the MPS framework assumes the worst-case scenario of the typhoon's track and moving speed (Mori et al. 2021a). The MPI and MPS can be calculated from monthly mean climate values calculated by climate models (Mori et al. 2021b). Therefore, individual typhoon information is not necessary, and no need to correct biases of typhoons of GCMs.

The input climate values for the future change analysis are the historical experiment run (namely HPD) for the period 1950–2014 conducted by the Japan Meteorological Agency Meteorological Research Institute using the atmospheric meteorological model MRI-AGCM3.2H (Mizuta et al. 2012) with a horizontal resolution of 60 km developed by the same institute, the future experiment for the period 2015–2099 under the warming scenario condition, and the future experiment for the period 2099–2050 using the MRI-AGCM3.2H model developed by the same institute. The 150-year run consists of a historical experiment run (HPD) from 1950 to 2014 and a future experiment run (namely HFD) from 2015 to 2099 under a warming scenario (hereafter, a 150-year run). The climate scenarios are RCP2.6 and RCP8.5. In the following, the results of RCP2.6 and RCP8.5 are considered equivalent to those of SSP1-2.6 and SSP5-8.5 and are discussed with the sea-level rise projections.

Figure 12.2 shows an example of the spatial land elevation of Tokyo area. The different colored areas indicate different elevation intervals, and the red area means below the mean sea level. There is a large low-lying area below the mean sea level, so-called zero-meter area. Such zero-meter areas are located in several big cities, such as Osaka and Nagoya, Japan, and other Asian mega-cities (e.g., Shanghai). In the case of Tokyo Bay, the most affected area is the Tokyo Metropolis, and its impacted area will increase significantly with the rise of sea level, from about 32 km^2 at present to about 66 km^2, even with SSP1-2.6.

The area affected by sea-level rise plus storm surge anomaly is approximately 160 km^2 in the present climate, a fivefold increase in the impact on the area below 0 m elevation. Furthermore, the area affected by the water table, which is sea-level rise plus storm surge anomaly, is about 200 km^2 for SSP5-8.5, a 40 km^2 increase. Figure 12.3 shows the results of a similar evaluation of SLR for the affected population of Tokyo using NIES population projections. The solid line shows the case where future population changes are considered according to the SSP scenario, while the dashed line shows the case where the current population is maintained. The effect of the future population change is also large, and the results do not show a monotonic increase. The difference between scenarios increases after 2060. Depending on the SSP scenario, the impact on the influenced population is compatible or larger for population change than that for sea-level rise. This result suggests that impact assessments on the population (or society broadly) need to consider societal changes in parallel with changes in hazards.

Figure 12.4 shows the results of a similar study for the Tokyo population, assuming future changes in storm surge in addition to sea-level rise, and shows that the impacted population is several times larger when storm surge is considered,

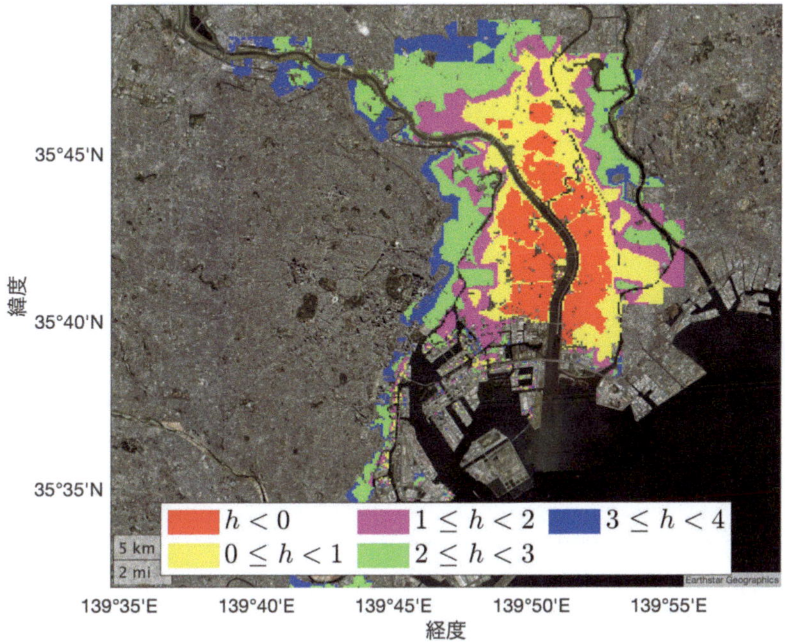

Fig. 12.2 Spatial land elevation of Tokyo area. (Different colors indicate different elevation interval. The red color area indicates below the mean sea level; unit: m) (attribution of background map: Esri in MATLAB R2024a)

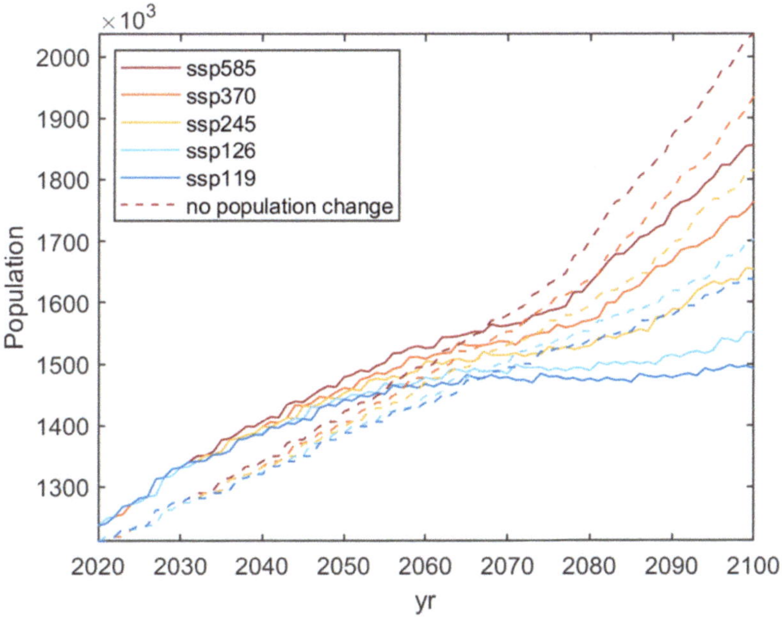

Fig. 12.3 Annual change in population affected by sea-level rise in Tokyo (colored lines: SSP scenario, solid line; considering population change, dashed line: without population change)

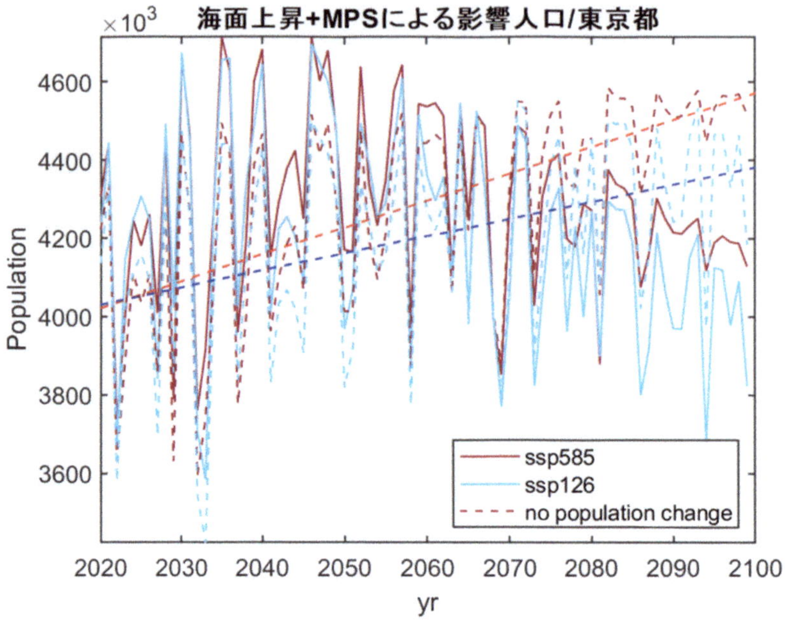

Fig. 12.4 Annual change in population affected by sea-level rise and maximum possible storm surge in Tokyo (colored line: SSP scenario, solid line; considering population change, dashed line: without population change)

compared to the impacted population for sea-level rise shown in Fig. 12.2. A peak-out is observed due to population change, with the greatest impact in the period from 2030 to 2040 and a decrease in the impacted population after 2050. Similar results were obtained for other metropolitan areas such as Osaka and Nagoya.

12.4 Summary

Based on the SSP1-2.6 or SSP5-8.5 scenarios, it was shown that 30–60% of Japan's major beaches could be lost by 2100. The risk assessment of sea-level rise and storm surge to society is the extent of inundation and the impact on society, such as population. Therefore, considering the effects of climate change, we estimated the secular change of each SSP scenario, the area inundated by the storm surge, and the population affected by the storm surge. The results show that the impacted population in Tokyo increases significantly by a factor of three or more when storm surge is taken into account and that the impact peaks around 2050.

Sea-level rise and extreme sea-level rise, which includes sea-level rise and storm surge, have a significant impact not only on natural coasts but also on disaster risk in urban areas. The impact assessment of storm surges is particularly important for disaster risk in coastal areas of Japan. The long-term changes in population and society must be considered at the same time.

References

Fox-Kemper B, Hewitt H, Xiao C, Aðalgeirsdóttir G, Drijfhout S, Edwards T, Golledge NR, Hemer M, Kopp RE, Krinner G, Mix A, Notz D, Nowicki S, Nurhati IS, Ruiz L, Sallée J-B, Slangen ABA, Yu Y (2021) Chapter 9: ocean, cryosphere and sea level change, in climate change 2021: the physical science basis. Contribution of Working Group I to the Sixth Assessment Report of the IPCC

Geospatial Information Authority of Japan (2011) GSI 5th mesh DEM data, https://nlftp.mlit.go.jp/ksj/

Mizuta R, Yoshimura H, Murakami H, Matsueda M, Endo H, Ose T, Kamiguchi K, Kitoh A (2012) Climate simulations using MRI-AGCM3. 2 with 20-km grid. J Meteorol Soc Jpn Ser II 90:233–258

Mori N, Shimura T (2023) Tropical cyclone-induced coastal sea level projection and the adaptation to a changing climate, vol 1. Cambridge Prisms: Coastal Futures, p e4. https://doi.org/10.1017/cft.2022.6

Mori N, Nakajo S, Iwamura S, Shibutani Y (2018) Projection of decrease in Japanese beaches due to climate change using a geographic database. Coast Eng J 60(2):239–246

Mori N, Ariyoshi N, Shimura T, Miyashita T, Ninomiya J (2021a) Future projection of maximum potential storm surge height at three major bays in Japan using the maximum potential intensity of a tropical cyclone. Clim Chang 164:25

Mori S, Mori N, Shimura T, Miyashita T (2021b) Long-term changes in maximum potential storm surge height in Japan's major bays due to climate change. J Jpn Soc Civil Eng., Ser. B2 (Coastal Engineering), (Released on J-STAGE November 04, 2021, Online ISSN 1883-8944) 77:I_937–I_942

Seneviratne SI, Zhang X, Adnan M, Badi W, Dereczynski C, Di Luca A, Ghosh S, Iskandar I, Kossin J, Lewis S, Otto F, Pinto I, Satoh M, Vicente-Serrano SM, Wehner M, Zhou B (2021) Chapter 11: weather and climate extreme events in a changing climate. In: Climate change 2021: the physical science Basis. Contribution of Working Group I to the Sixth Assessment Report of the IPCC

Chapter 13
Assessment of Climate Change Adaptation Measures for Pluvial Flooding

Hayata Yanagihara and So Kazama

Abstract This chapter quantitatively assesses the effectiveness of four adaptation measures in reducing pluvial flood damage: improving the maintenance level of inland water drainage facilities, converting buildings to piloti buildings, installing flood prevention plates, and implementing paddy field dams, both individually and in combination. Under the baseline climate, the amount of pluvial flood damage across Japan will reduce by 52.8%–53.2% through the improvement in the maintenance level of inland water drainage facilities, by 47.9%–55.5% through the conversion of buildings to piloti buildings, by 8.9%–16.2% with the installation of flood prevention plates, and by 3.0%–5.7% through the implementation of paddy field dams. However, not all prefectures will benefit equally under future climate scenarios, as represented by Representative Concentration Pathway (RCP) 2.6 and RCP 8.5. Even with all four adaptation measures adopted, 5 out of 47 prefectures under RCP 2.6 and 17 out of 47 under RCP 8.5 are projected to experience pluvial flood damage levels higher than those under the baseline climate.

Keywords Pluvial flood damage · Climate change · Adaptation measures · Inland water drainage facilities · Piloti buildings · Flood prevention plates · Paddy field dams

H. Yanagihara (✉) · S. Kazama
Tohoku University, Sendai, Japan
e-mail: yanagihara.hayata.r1@dc.tohoku.ac.jp; so.kazama.d3@tohoku.ac.jp

N. Mimura, S. Takewaka (eds.), *Climate Change Impacts and Adaptation Strategies in Japan*, https://doi.org/10.1007/978-981-96-2436-2_13

13.1 Introduction

13.1.1 Overview of this Chapter

This chapter introduces the evaluation results of climate change adaptation measures for pluvial flooding, a type of natural disaster. Pluvial flooding refers to inundation caused by rainfall exceeding the drainage capacity of inland water drainage facilities and inundation caused by the inability to drain rainwater into rivers due to high river levels. We selected the improvement in the maintenance level of inland water drainage facilities, the conversion of buildings to piloti buildings (houses on stilts), the installation of flood prevention plates, and the implementation of paddy field dams (paddy field storage) as adaptation measures and evaluated the ability of these four adaptation measures to mitigate pluvial flood damage in the context of climate change. Some of the results of this chapter have already been published in Yanagihara et al. (2022a).

13.1.2 Research Background

In recent years, not only flood damage from rivers but also inundation damage from pluvial flooding has been a problem in Japan. The heavy rain event of July 2018 and Typhoon Hagibis each caused housing damage due to pluvial flooding to approximately 15,000 and 30,000 households, respectively (Study Group on Countermeasures for Urban Inundation Considering Climate Change 2022). Additionally, 64% of the total number of flooded buildings in Japan from 2009 to 2018 was due to pluvial flooding (Study Group on Countermeasures for Urban Inundation Considering Climate Change 2022). Such pluvial flood damage is feared to expand due to an increase in precipitation associated with global warming. With respect to changes in precipitation due to global warming, Fujita et al. (2019) reported that even if the 2 °C target of the Paris Agreement is achieved, the maximum daily precipitation in almost all regions of the Japanese archipelago will increase by more than 10%. Given these circumstances, it is necessary to implement adaptation measures for pluvial flooding while considering the impact of climate change. To discuss specific adaptation measures, it is necessary to quantitatively evaluate the ability of adaptation measures to mitigate pluvial flood damage considering climate change.

13.1.3 Issues in Previous Research and the Purpose of this Study

In previous studies, the evaluation of adaptation measures for pluvial flooding has been conducted on an urban scale (for example, Löwe et al. 2017; Zhou et al. 2018; Qiu et al. 2021). Therefore, there is a lack of knowledge about the effects of adaptation measures for pluvial flooding at the national scale, including in Japan. By evaluating adaptation measures on a national scale, it is possible to discuss the necessary adaptation measures to mitigate the impact of climate change by region. Therefore, this study aims to evaluate the mitigation effects of different adaptation measures for pluvial flood damage across Japan.

13.2 Dataset and Methodology

This study used the same dataset as Yanagihara et al. (2022a). In addition, for the landform classification data, we used the 1:500,000 scale landform classification maps from the Digital National Land Information (Ministry of Land, Infrastructure, Transport and Tourism 1968). We used landform classification data when setting the ridge height of paddy field dams (refer to Sect. 13.2.4). Following the methodology of Yanagihara et al. (2022a), we evaluated the mitigating effect of adaptation measures on pluvial flood damage. The methodological strategy is as follows: (1) calculate the inundation depth via pluvial flood analysis, (2) calculate the damage amount on the basis of the inundation depth, and (3) calculate the damage reduction amount via adaptation measures. The details of the methodology are described below.

13.2.1 Pluvial Flood Analysis

For flood analysis, we applied a two-dimensional unsteady flow model (Tezuka et al. 2014) without distinguishing between rivers and floodplains and analyzed all of Japan at the same time. The grid size of the flood analysis is approximately 250 m. In the analysis of the baseline climate (1981–2000), we gave the extreme rainfall data of Kawagoe et al. (2010) for 24 h at a constant intensity. In the analysis of the early twenty-first-century climate (2006–2025), near-future climate (2031–2050), and late twenty-first-century climate (2081–2100), we used future climate extreme rainfall data obtained via the method of Sect. 13.2.3 for 24 h at a constant intensity. The parameters used in the flood analysis are the same as those of Tezuka et al. (2014). Yanagihara et al. (2022a) conducted a flood analysis assuming the worst-case scenario where the river water level is high and rainwater is not drained into the river. In contrast, following Yanagihara et al. (2022b), this study focused on the presence or absence of drainage to the river and set two inundation conditions: drainage and poor drainage. Drainage is the condition with the minimum risk of pluvial flooding, and rainwater is always drained to the river by gravity. On the other hand, poor drainage is the condition with the maximum risk of pluvial flooding, and as with Yanagihara et al. (2022a), rainwater does not drain the river at all. The risk of pluvial flooding around rivers varies depending on the river water level, so we set the inundation conditions to minimize or maximize the risk of pluvial flooding as described above. In the flood analysis, we represented drainage to the river by always setting the depth of the river to 0. Additionally, we reproduced poor drainage to the river by setting the inflow to the river to 0. To calculate the inundation depth considering inland water drainage facilities, the rainfall amount according to the maintenance level of inland water drainage facilities was subtracted from the rainfall amount used in the flood analysis. We conducted a flood analysis assuming that all rainfall with a return period of 5 years in the baseline climate can be drained with reference to the target of the currently implemented measures (Yanagihara et al. 2022a, b).

13.2.2 Damage Amount Calculation

The damage amount was calculated on the basis of the Manual for Economic Evaluation of Flood Control Investment (Ministry of Land, Infrastructure, Transport and Tourism 2005). The damage amount was calculated by multiplying the asset value of the flooded land by the damage rate according to the inundation depth. The asset value was determined on the basis of the asset evaluation amount for each land use and prefecture. The amount of pluvial flood damage was evaluated by the annual expected damage amount while considering the occurrence probability. The damage calculation is based on Yanagihara et al. (2022a). Using extreme rainfall data of the future climate for multiple global circulation models (GCMs) (refer to Sect. 13.2.3), the amount of pluvial flood damage in the future climate was estimated. Therefore, the average amount of pluvial flood damage for each GCM was taken as the amount of pluvial flood damage in the future climate.

13.2.3 Estimation of Extreme Rainfall in the Future Climate

The distribution of extreme rainfall increase rates from the baseline climate to the future climate was estimated, and the future climate's extreme rainfall was estimated by multiplying the distribution of the increase rates by the baseline climate's extreme rainfall data. The method of estimating the distribution of extreme rainfall increase rates was the same as that used by Yanagihara et al. (2022a). Climate prediction data (Nishimori et al. 2019) for five GCMs and two representative concentration pathways (RCPs) were used to estimate the future climate's extreme rainfall data for each model and scenario. The GCMs were GFDL-CM3, HadGEM2-ES,

MIROC5, MRI-GCM3, and CSIRO-Mk3-6-0, and the RCP scenarios were RCP2.6 and RCP8.5.

13.2.4 Reflection of Adaptation Measures

In addition to the two adaptation measures examined by Yanagihara et al. (2022a), namely, the improvement in the maintenance level of inland water drainage facilities and the conversion of buildings to piloti buildings, the installation of flood prevention plates and the implementation of paddy field dams were added, and four adaptation measures were evaluated. The improvement of inland water drainage facilities is positioned as a basic measure against pluvial flooding. Specific measures include the improvement and development of rainwater pipes, pump stations, and storage infiltration facilities (Study Group on Countermeasures for Urban Inundation Considering Climate Change 2022). A piloti building is a building style where the first floor of the building is an external space with only columns. Therefore, when buildings are converted to piloti buildings, damage due to flooding on the first floor does not occur. Flood prevention plates are plates that prevent water from entering buildings. By installing flood prevention plates at the openings of buildings, flooding up to the height of the flood prevention plate can be prevented. A paddy field dam can reduce the outflow downstream by storing rainwater in paddy fields. The reflection methods of each adaptation measure are described below.

1. Improvement in the Maintenance Level of Inland Water Drainage Facilities (Plan 1)

By improving the maintenance level of inland water drainage facilities, it was assumed that the drainage capacity would be improved from rainfall with a return period of 5 years to rainfall with a return period of 10 years in the baseline climate. All rainfall with a return period of 10 years in the baseline climate was assumed to drain, and the damage amount was calculated.

2. Conversion of Buildings to Piloti Buildings (Plan 2)

Buildings were converted to piloti buildings to prevent floor flooding due to frequent pluvial flooding (frequency less than the service life of the house). The statutory service life of a wooden house used in Japan's asset valuation is 22 years (Real Estate Japan 2020). Therefore, we designated the cells of land for buildings where flooding of 45 cm or more (Ministry of Land, Infrastructure, Transport and Tourism 2005) occurred due to extreme rainfall with a return period of 10 years as piloti building zones. The inundation depth in the baseline climate was used as the standard for the piloti building zone. Additionally, to implement adaptation measures that assume the worst situation, the inundation depth under the poor drainage condition was used as the standard for the piloti building zone. To calculate the damage amount in the piloti building zone, we assumed the height of the piloti to be equivalent to the height of the first floor, 3 m, and calculated the damage amount by subtracting the inundation depth by 3 m.

3. Installation of Flood Prevention Plates (Plan 3)

Flood prevention plates were installed to prevent underfloor flooding due to frequent pluvial flooding (frequency less than the service life of the house). If underfloor flooding (flooding less than 45 cm) occurs due to extreme rainfall with a return period of 10 years under the poor drainage condition, that cell of land for buildings was designated as a flood prevention plate zone. The height of the flood prevention plate was set to be equivalent to the underfloor height, 45 cm. Because the installation of the flood prevention plate prevents flooding up to the underfloor height, it was assumed that no damage due to underfloor flooding occurred in the flood prevention plate zone. On the other hand, if the inundation depth was 45 cm or greater, the damage amount was calculated as if there were no flood prevention plates.

4. Implementation of Paddy Field Dams (Plan 4)

We applied the simple paddy field dam model of Chai et al. (2020) and investigated the effects when all the paddy fields in Japan exhibited a rainwater storage function in flood analysis. The modeled paddy field dam uses a free drain-type water fall control device. Chai et al. (2020) placed one water fall control device per 100 m^2. This study assumed six water fall control devices in one cell (approximately 250 m^2). Referring to the study of Kawagoe and Maruta (2021), we set the ridge height using landform classification data, considering that the storage effect is different in flat and sloping lands. Specifically, we set the ridge height to 0.30 m in flat lands, 0.10 m in sloping lands, and 0.20 m in alluvial fans in sloping lands. The average water depth during the heading and flowering periods, which coincides with the time when heavy rain is likely to occur, is 3 cm (Shimura 1982). On this basis, we set the initial paddy field water level and the height of the free drain pipe from the ground to 3 cm. For other quantities of the paddy field dam model, we used the values of Miyazu et al. (2017). The rainfall, which is the input data for the flood analysis, was set as the discharge amount of the paddy field dam model in the paddy field and the original rainfall outside the paddy field. This reflects the effect of the paddy field dam in the flood analysis.

13.3 Results and Discussion

13.3.1 Reduction Effect of Each Adaptation Measure on Pluvial Flood Damage in the Baseline Climate

In the baseline climate, the amount of pluvial flood damage across Japan under the drainage condition decreased by 52.8% due to the improvement in the maintenance level of inland water drainage facilities, by 55.5% due to the conversion of buildings to piloti buildings, by 16.2% due to the installation of flood prevention plates, and by 3.0% due to the implementation of paddy field dams. On the other hand, under the poor drainage condition, the amount of pluvial flood damage across Japan in the baseline climate decreased by 53.2% because of the improvement in the maintenance level of inland water drainage facilities, by 47.9% because of the conversion of buildings to piloti buildings, by 8.9% because of the installation of flood prevention plates, and by 5.7% because of the implementation of paddy field dams. The reduction rate of the amount of pluvial flood damage in each prefecture due to each adaptation measure in the baseline climate is shown in Fig. 13.1. The improvement in the maintenance level of inland water drainage facilities resulted in a generally equal reduction in the amount of pluvial flood damage in all prefectures. For other adaptation measures, regional characteristics were observed in the effect of reducing pluvial flood damage, so the factors involved are discussed below.

First, we consider the factors that affect the reduction effect of pluvial flood damage due to the conversion of buildings to piloti buildings. As shown in Fig. 13.2, the higher the proportion of the piloti building zone in the land for buildings is, the easier it is to reduce the amount of pluvial flood damage. However, even if the proportion of the piloti building zone in the land for buildings is the same, the reduction rate of the amount of pluvial flood damage varies. It is thought that in prefectures where flooding tends to spread due to topography, flooding tends to spread outside the piloti building zone, making it difficult to reduce the amount of pluvial flood damage. To confirm this, we examined the relationship between the ratio of the number of flooded cells for a return period of 100 years to the number of flooded cells for a return period of 10 years and the reduction rate of the amount of pluvial flood damage by prefecture and found a negative correlation (see Fig. 13.3).

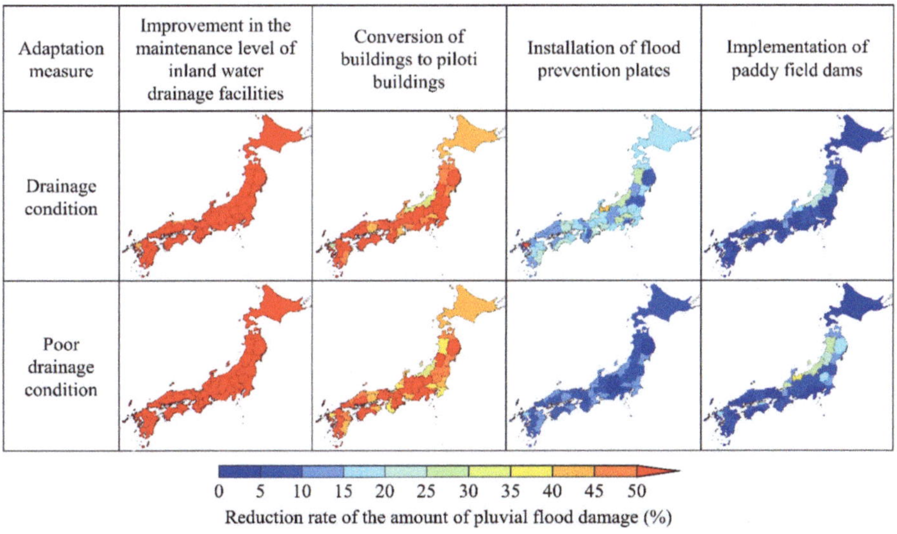

Fig. 13.1 Reduction rate of the amount of pluvial flood damage in each prefecture due to each adaptation measure in the baseline climate

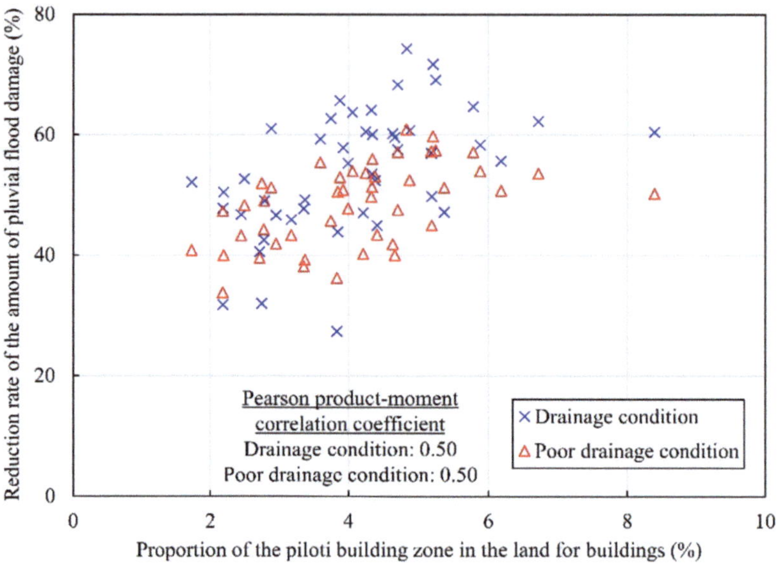

Fig. 13.2 Relationship between the proportion of the piloti building zone in the land for buildings and the reduction rate of the amount of pluvial flood damage by prefecture

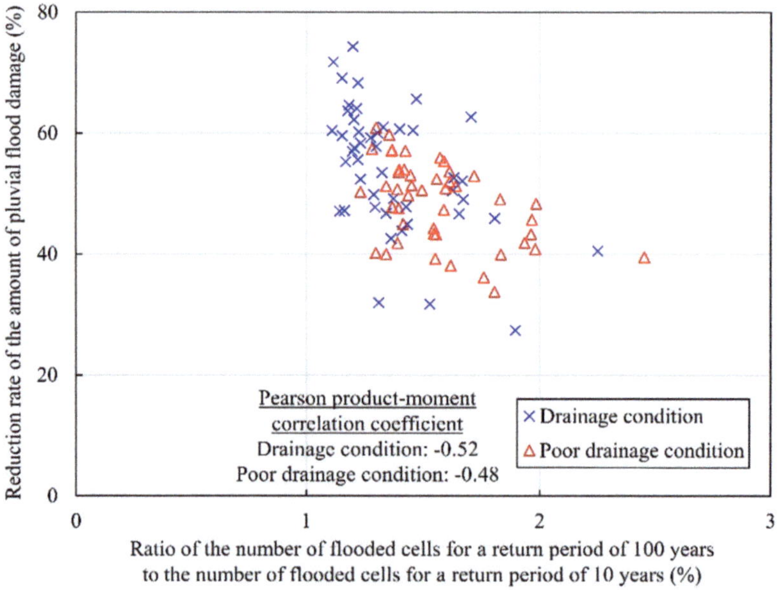

Fig. 13.3 Relationship between the ratio of the number of flooded cells for a return period of 100 years to the number of flooded cells for a return period of 10 years and the reduction rate of the amount of pluvial flood damage by prefecture

Therefore, reducing pluvial flood damage in prefectures where the range of flooding changes greatly depending on the scale of rainfall is difficult.

Next, we consider the factors that affect the reduction effect of pluvial flood damage due to the installation of flood prevention plates. There is a positive correlation between the proportion of the flood prevention plate zone in the land for buildings and the reduction rate of the amount of pluvial flood damage (see Fig. 13.4), but the variation in this relationship is large. In prefectures where the inundation depth tends to increase due to topography, the inundation depth may exceed the height of the flood prevention plate, making it difficult to reduce the amount of pluvial flood damage. To confirm this, we examined the relationship between the ratio of the average inundation depth in the flood area for a return period of 100 years to the average inundation depth in the flood area for a return period of 10 years and the reduction rate of the amount of pluvial flood damage by prefecture and found a negative correlation (see Fig. 13.5). Therefore, reducing pluvial flood damage in prefectures where the inundation depth changes greatly depending on the scale of rainfall is difficult.

Finally, we consider the factors that influence the effects of reducing pluvial flood damage by implementing paddy field dams. As shown in Fig. 13.6, the higher the proportion of paddy fields in all land use areas is, the easier it is to reduce pluvial flood damage. However, the size of paddy fields is not necessarily directly linked to a high degree of pluvial flood damage reduction. Focusing on the plot points in Fig. 13.6, there are prefectures where despite a high proportion of paddy fields compared with other prefectures (15% or more), the amount of pluvial flood damage is

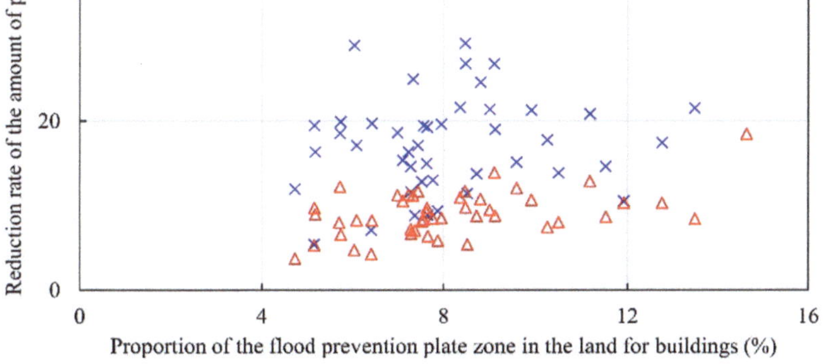

Fig. 13.4 Relationship between the proportion of the flood prevention plate zone in the land for buildings and the reduction rate of the amount of pluvial flood damage by prefecture

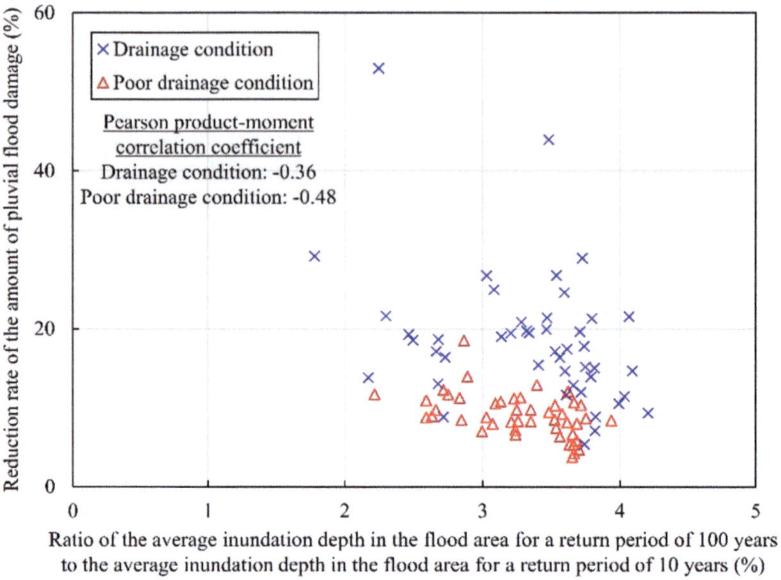

Fig. 13.5 Relationship between the ratio of the average inundation depth in the flood area for a return period of 100 years to the average inundation depth in the flood area for a return period of 10 years and the reduction rate of the amount of pluvial flood damage by prefecture

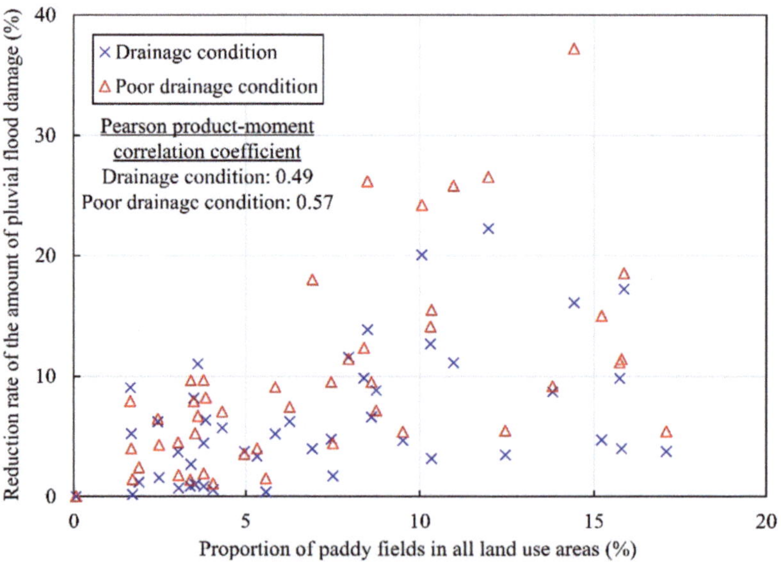

Fig. 13.6 Relationship between the proportion of paddy fields in all land use areas and the reduction rate of the amount of pluvial flood damage by prefecture

difficult to reduce (5% or less). These prefectures are thought to have regional characteristics that make it difficult to reduce pluvial flood damage with paddy field dams. To examine these regional characteristics, we focused on land use distribution. For comparison, we also focused on the land use distribution of the prefecture where the amount of pluvial flood damage was most reduced. As a result, compared with prefectures where pluvial flood damage is difficult to reduce, the prefecture where pluvial flood damage is most reduced tends to have densely located paddy fields and land for buildings. The denser the paddy fields and land for buildings are, the more likely the land for buildings is to benefit from the rainwater storage function of the paddy fields; thus, the pluvial flood damage reduction effect is likely greater.

13.3.2 Reduction Effect of Pluvial Flood Damage Across Japan Due to Adaptation Measures in the Future Climate

The change rates of the amount of pluvial flood damage across Japan from the baseline climate to each future climate when adaptation measures are implemented are shown in Table 13.1. In addition to the change rate of the amount of pluvial flood damage when adaptation measures are implemented alone, the change rate of the amount of pluvial flood damage when adaptation measures are combined is shown in Table 13.1. The red and blue cells in Table 13.1 represent increases and decreases from the pluvial flood damage in the baseline climate without implementing adaptation measures, respectively. First, we focus on the change rate of the amount of pluvial flood damage when adaptation measures are implemented alone. The amount of pluvial flood damage across Japan in the near-future climate and in the late twenty-first century under the poor drainage condition will increase even if adaptation measures are implemented alone. Additionally, under both inundation conditions, even if flood prevention plates are installed (Plan 3) or paddy field dams are implemented (Plan 4), the amount of pluvial flood damage across Japan in all future climates will increase from the baseline climate. The decrease in pluvial flood damage from the near-future climate to the late twenty-first century under the RCP2.6 scenario is likely due to the effects of mitigation measures.

Next, we focus on the change rate of the amount of pluvial flood damage when adaptation measures are combined in Table 13.1. There are combinations of adaptation measures that can keep the amount of pluvial flood damage across Japan below the baseline climate in all future climates. The four adaptation measures will keep the amount of pluvial flood damage across Japan below the baseline climate in all future climates. Except for the combination of the installation of flood prevention plates (Plan 3) and the implementation of paddy field dams (Plan 4), both of which are less effective at reducing the amount of pluvial flood damage alone, two or more adaptation measures reduce the amount of pluvial flood damage across Japan under

Table 13.1 Change rates of the amount of pluvial flood damage across Japan from the baseline climate to each future climate when adaptation measures are implemented (unit: %)

Inundation condition	Period	RCP scenario	No adaptation	Plan 1*	Plan 2*	Plan 3*	Plan 4*	Plan 1 & Plan 2	Plan 1 & Plan 3	Plan 1 & Plan 4
Drainage	Early 21st century	RCP2.6	35.6	-18.0	-32.6	17.8	31.8	-59.8	-29.4	-20.3
		RCP8.5	38.0	-13.5	-29.1	20.3	34.2	-56.1	-25.3	-16.1
	Near future	RCP2.6	69.1	15.7	-9.4	49.8	64.6	-41.0	0.6	12.6
		RCP8.5	73.1	18.5	-6.9	53.5	68.5	-38.7	2.9	15.4
	Late 21st century	RCP2.6	50.6	-7.4	-23.6	31.2	46.2	-54.2	-20.7	-10.2
		RCP8.5	97.6	44.1	9.8	77.4	92.7	-24.1	26.4	40.3
Poor drainage	Early 21st century	RCP2.6	45.6	-13.4	-10.4	37.6	37.3	-47.9	-18.7	-18.4
		RCP8.5	54.4	-3.7	-1.8	46.8	45.5	-40.2	-9.0	-9.4
	Near future	RCP2.6	94.6	30.4	30.0	86.9	83.6	-17.0	23.6	23.1
		RCP8.5	97.9	32.5	32.3	90.2	87.1	-15.3	25.6	25.3
	Late 21st century	RCP2.6	69.0	4.8	6.5	60.6	59.0	-36.7	-1.6	-1.7
		RCP8.5	133.4	65.9	62.1	126.3	120.6	8.7	58.4	56.7

Inundation condition	Period	RCP scenario	Plan 2 & Plan 3	Plan 2 & Plan 4	Plan 3 & Plan 4	Plan 1, Plan 2 & Plan 3	Plan 1, Plan 2 & Plan 4	Plan 1, Plan 3 & Plan 4	Plan 2, Plan 3 & Plan 4	Plan 1, Plan 2, Plan 3 & Plan 4
Drainage	Early 21st century	RCP2.6	-50.3	-35.5	14.4	-71.2	-61.5	-31.5	-52.9	-72.7
		RCP8.5	-46.8	-32.1	16.8	-67.8	-58.0	-27.6	-49.5	-69.5
	Near future	RCP2.6	-28.7	-13.0	45.5	-56.1	-43.4	-2.2	-32.2	-58.3
		RCP8.5	-26.5	-10.6	49.2	-54.2	-41.1	0.1	-30.0	-56.4
	Late 21st century	RCP2.6	-42.9	-27.0	27.1	-67.5	-56.3	-23.3	-46.1	-69.4
		RCP8.5	-10.5	5.5	72.6	-41.8	-27.2	22.8	-14.6	-44.7
Poor drainage	Early 21st century	RCP2.6	-18.4	-17.5	29.2	-53.2	-52.0	-23.7	-25.6	-57.4
		RCP8.5	-9.5	-9.6	37.7	-45.5	-44.9	-14.7	-17.4	-50.3
	Near future	RCP2.6	22.3	20.1	75.7	-23.8	-23.3	16.2	12.1	-30.2
		RCP8.5	24.5	22.5	79.1	-22.3	-21.5	18.3	14.5	-28.5
	Late 21st century	RCP2.6	-1.9	-2.3	50.4	-43.1	-42.1	-8.0	-10.9	-48.4
		RCP8.5	55.0	50.0	113.1	1.2	0.6	49.0	42.5	-7.1

*	
Plan 1	Improvement in the maintenance level of inland water drainage facilities
Plan 2	Conversion of buildings to piloti buildings
Plan 3	Installation of flood prevention plates
Plan 4	Implementation of paddy field dams

both inundation conditions of the late twenty-first-century climate/RCP2.6 scenario compared with the baseline climate. The four adaptation measures can reduce the amount of pluvial flood damage across Japan under both inundation conditions of the late twenty-first-century climate/RCP8.5 scenario to below the baseline climate. Under the poor drainage condition of the late twenty-first-century climate/RCP8.5 scenario, the amount of pluvial flood damage across Japan cannot be maintained below the baseline climate unless all four adaptation measures are implemented. These results indicate that the increase in the amount of pluvial flood damage across Japan in future climates can be suppressed by multiple adaptation measures. Additionally, if mitigation measures are taken, the amount of pluvial flood damage across Japan in the late twenty-first-century climate can be suppressed with fewer adaptation measures. Considering social conditions and financial constraints, there are limits to adaptation measures (Mimura 2015), and it is not always possible to implement all adaptation measures. For example, the implementation of paddy field dams using privately owned paddy fields requires the cooperation of agricultural workers. Therefore, it is conceivable that if an agreement of agricultural workers cannot be obtained, a paddy field dam cannot be implemented. Additionally, with the decrease in population, a decrease in tax revenue is expected (Mizuho Research Institute 2006), and public investment is expected to decrease. Therefore, it is conceivable that multiple adaptation measures cannot be implemented due to budget constraints. Considering these circumstances, it is necessary to implement both adaptation and mitigation measures simultaneously.

13.3.3 Reduction Effect of Pluvial Flood Damage by Prefecture Due to Adaptation Measures in the Future Climate

Figures 13.7 and 13.8 show the prefectures where the amount of pluvial flood damage in all future climates can be suppressed below the baseline climate for each combination of adaptation measures in the RCP2.6 scenario and the RCP8.5 scenario, respectively. In the combination of adaptation measures including the conversion of buildings to piloti buildings (Plan 2), many prefectures where the future climate's pluvial flood damage amount is less than that in the baseline climate are found. Additionally, even if the amount of pluvial flood damage across Japan increases relative to the baseline climate, there are prefectures where the amount of pluvial flood damage is less than that in the baseline climate. The increase rate of rainfall varies by prefecture, so the number and types of adaptation measures needed to suppress the increase in future pluvial flood damage amounts vary by prefecture. Even if all four adaptation measures are implemented, 5 out of the 47 prefectures cannot reduce the amount of pluvial flood damage to less than that in the baseline climate under both inundation conditions of the future climate and the RCP2.6 scenario. Even if all four adaptation measures are implemented, 17 out of the 47 prefectures cannot reduce the amount of pluvial flood damage to less than that in the baseline climate under both inundation conditions of the future climate and the RCP8.5 scenario. In these prefectures, the rate of increase in rainfall due to climate change is high. This means that the four adaptation measures alone cannot adapt to the impacts of climate change. Therefore, additional adaptation measures are needed in these prefectures.

13.4 Conclusion

This chapter quantitatively evaluated the reduction effect of pluvial flood damage amount by the four adaptation measures of the improvement in the maintenance level of inland water drainage facilities, the conversion of buildings to piloti

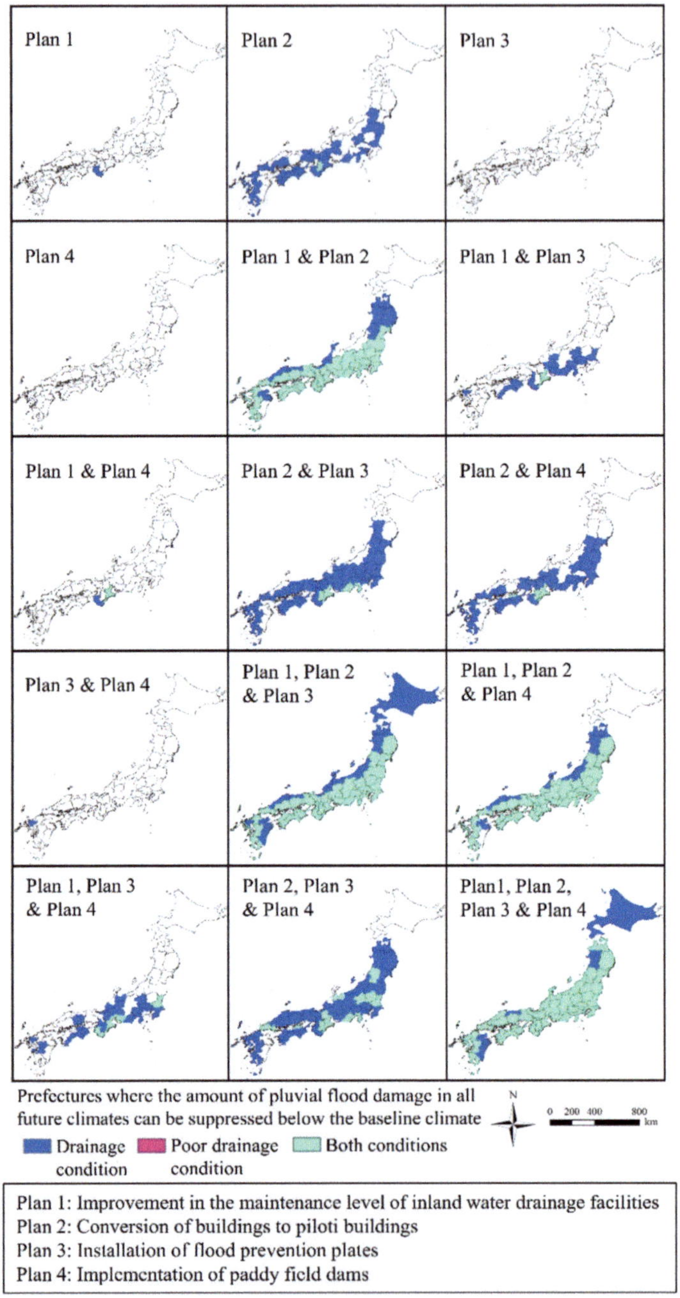

Fig. 13.7 Prefectures where the amount of pluvial flood damage in all future climates can be suppressed below the baseline climate for each combination of adaptation measures in the RCP2.6 scenario

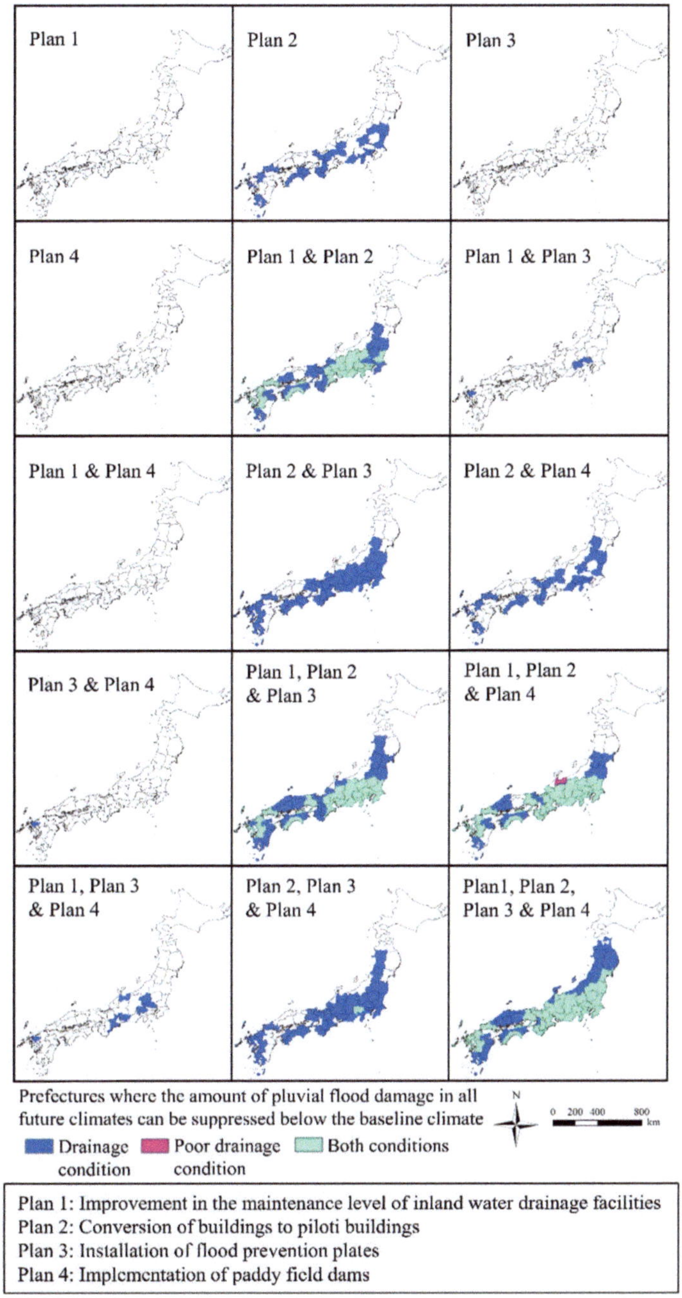

Fig. 13.8 Prefectures where the amount of pluvial flood damage in all future climates can be suppressed below the baseline climate for each combination of adaptation measures in the RCP8.5 scenario

buildings, the installation of flood prevention plates, the implementation of paddy field dams, and their combinations. Since it is possible to understand the measures necessary for adapting to the effects of climate change by prefecture, the results of this chapter are expected to be used as scientific information when formulating adaptation measures. However, because the GCM outputs were used in the impact assessment of climate change, it is necessary to be aware of the uncertainty in the amount of pluvial flood damage in the future climate. Additionally, although the ability of adaptation measures to reduce pluvial flood damage has been evaluated, the cost of adaptation measures has not yet been calculated. In the future, discussions on the efficiency of adaptation measures will be necessary.

Acknowledgments This research was supported by JHPCN (jh210001, jh220018, and jh230024), Joint Usage/Research Center for Interdisciplinary Large-scale Information Infrastructures. This study was also supported by JSPS KAKENHI Grant Number JP23KJ0119. Some of the calculation results were obtained using supercomputing resources at the Cyber Science Center, Tohoku University.

References

Chai Y, Touge Y, Shi K, Kazama S (2020) Evaluating potential flood mitigation effect of paddy field dam for Typhoon No.19 in 2019 in the Naruse River Basin. J Jpn Soc Civ Eng, Ser. B1 (Hydraul Eng) 76(1):295–303. https://doi.org/10.2208/jscejhe.76.1_295

Fujita M, Mizuta R, Ishii M, Endo H, Sato T, Okada Y, Kawazoe S, Sugimoto S, Ishihara K, Watanabe S (2019) Precipitation changes in a climate with 2-K surface warming from large ensemble simulations using 60-km global and 20-km regional atmospheric models. Geophys Res Lett 46(1):435–442. https://doi.org/10.1029/2018gl079885

Kawagoe S, Maruta S (2021) Development of water storage potential map based on modification of land structure and facilities in Japan. J Jpn Soc Civ Eng, Ser. G (Environ Res) 77(5):I_77–I_84. https://doi.org/10.2208/jscejer.77.5_i_77

Kawagoe S, Kazama S, Sarukkalige PR (2010) Probabilistic modelling of rainfall induced landslide hazard assessment. Hydrol Earth Syst Sci 14(6):1047–1061. https://doi.org/10.5194/hess-14-1047-2010

Löwe R, Urich C, Sto Domingo N, Mark O, Deletic A, Arnbjerg-Nielsen K (2017) Assessment of urban pluvial flood risk and efficiency of adaptation options through simulations—a new generation of urban planning tools. J Hydrol 550:355–367. https://doi.org/10.1016/j.jhydrol.2017.05.009

Mimura N (2015) Significance of climate change adaptation policy and its future directions. Jpn J Real Estate Sci 29(1):3–7. https://doi.org/10.5736/jares.29.1_3

Ministry of Land, Infrastructure, Transport and Tourism (1968) 1:500,000 scale landform classification maps. Digital National Land Information. https://nlftp.mlit.go.jp/kokjo/inspect/landclassification/land/l_national_map_50-1.html

Ministry of Land, Infrastructure, Transport and Tourism (2005) Manual for Economic Evaluation of Flood Control Investment (Draft). https://www.mlit.go.jp/river/kokusai/pdf/pdf06.pdf. Accessed 10 Jun 2024

Miyazu S, Yoshikawa N, Abe S (2017) Development of the function-independent runoff control device for the free-drain drainage system. Trans Jpn Soc Irrig 85(2):I_159–I_167. https://doi.org/10.11408/jsidre.85.I_159

Mizuho Research Institute (2006) The impact of population decline on local government finances: perspectives on the review of local taxation systems (Mizuho Report). https://dl.ndl.go.jp/view/prepareDownload?itemId=info%3Andljp%2Fpid%2F8900187&contentNo=1. Accessed 7 June 2024

Nishimori M, Ishigooka Y, Kuwagata T, Takimoto T, Endo N (2019) SI-CAT 1km-grid square regional climate projection scenario dataset for agricultural use (NARO2017). J Jpn Soc Simul Technol 38:150–154

Qiu Y, Schertzer D, Tchiguirinskaia I (2021) Assessing cost-effectiveness of nature-based solutions scenarios: integrating hydrological impacts and life cycle costs. J Clean Prod 329:129740. https://doi.org/10.1016/j.jclepro.2021.129740

Real Estate Japan (2020) Real estate Japan picture of the day—How long is the "life" of a house in Japan?. https://resources.realestate.co.jp/living/real-estate-japan-picture-of-the-day-how-long-is-the-life-of-a-house-in-japan/. Accessed 13 June 2024

Shimura H (1982) Evaluation of flood control function of paddy fields and rice paddies. J Agric Eng Soc, Jpn 50(1):25–29. https://doi.org/10.11408/jjsidre1965.50.25

Study Group on Countermeasures for Urban Inundation Considering Climate Change (2022) Promotion of countermeasures for urban inundation by sewage systems considering climate change, Proposal, Reference material (Partial revision). https://www.mlit.go.jp/common/001478480.pdf. Accessed 8 Jun 2024

Tezuka S, Takiguchi H, Kazama S, Sato A, Kawagoe S, Sarukkalige R (2014) Estimation of the effects of climate change on flood-triggered economic losses in Japan. Int J Disaster Risk Reduct 9:58–67. https://doi.org/10.1016/j.ijdrr.2014.03.004

Yanagihara H, Kazama S, Tada T, Touge Y (2022a) Estimation of the effect of future changes in precipitation in Japan on pluvial flood damage and the damage reduction effect of mitigation/adaptation measures. PLOS Clim 1(7):e0000039. https://doi.org/10.1371/journal.pclm.0000039

Yanagihara H, Yamamoto T, Kazama S (2022b) Estimation of inland flood damage based on extreme precipitation in Japan. Proc 39th IAHR World Congr:6303–6308. https://doi.org/10.3850/iahr-39wc252171192022214

Zhou Q, Leng G, Huang M (2018) Impacts of future climate change on urban flood volumes in Hohhot in northern China: benefits of climate change mitigation and adaptations. Hydrol Earth Syst Sci 22(1):305–316. https://doi.org/10.5194/hess-22-305-2018

Chapter 14
Evaluation of Flood Damage Reduction by Use of Irrigation Reservoirs Under Climate Change with the Worst Scenario

Atsuya Ikemoto, So Kazama, Hayata Yanagihara, and Takeo Yoshida

Abstract This study aims to evaluate the potential reduction in flood damage costs provided by irrigation reservoirs, assuming a worst-case scenario for damage reduction in Japan. We assessed the effects of irrigation reservoirs on flood damage reduction. When the storage rate was 0%, the damage cost reduction rate ranged from 1.1% to 1.7% by the end of the century. In this worst-case future climate scenario, the prefectures with the highest damage reduction rates included 16 prefectures for farmland out of 47 prefectures in Japan, 3 for residential areas, 2 for businesses, and 26 for golf courses. Notably, most prefectures with high damage reduction rates had golf courses. The findings from this chapter are expected to provide scientific insights into the use of irrigation reservoirs for flood control, particularly in light of climate change impacts across Japan, by prefecture and land use type.

Keywords Watershed management · Irrigation reservoirs · Flooding · Damage cost · Consensus building · Agricultural water use

14.1 Introduction

14.1.1 Overview of This Chapter

This chapter introduces the evaluation results of the flood damage reduction effect of agricultural reservoirs (hereafter, irrigation reservoirs) targeting flood inundation, which is a type of natural disaster. Irrigation reservoirs are storage facilities

A. Ikemoto (✉) · S. Kazama · H. Yanagihara
Tohoku University, Sendai, Japan
e-mail: ikemoto.atsuya.s1@dc.tohoku.ac.jp

T. Yoshida
National Agriculture and Food Research Organization, Tsukuba, Japan

N. Mimura, S. Takewaka (eds.), *Climate Change Impacts and Adaptation Strategies in Japan*, https://doi.org/10.1007/978-981-96-2436-2_14

dedicated to the aim of securing agricultural water; however, in recent years, they have been expected to be effective storage facilities against frequently occurring floods. Under the worst-case climate scenario, we evaluated the flood damage reduction effect of irrigation reservoirs throughout Japan, considering damage reduction on the safe side. The results of this chapter were obtained in Theme 3 "Prediction of Climate Change Impact and Evaluation of Adaptation Measures in the Field of Natural Disasters and Water Resources" of the S-18 Project, and some of the results of this chapter have previously been published in a study by Ikemoto et al. (2022, 2023).

14.1.2 Research Background

Watershed management efforts aimed at reducing flood damage are promoted as adaptation measures in Japan. Watershed management is an initiative to reduce flood damage by its stakeholders working together (Council for Social Infrastructure Development 2020), and includes the concepts of Ecosystem-based Disaster Risk Reduction (Eco-DRR) (Ministry of the Environment 2016) and Nature-Based Solutions (World Wide Fund for Nature 2022) (The sub-committee for research and proposition, The committee for ecosystem management, The Ecological Society of Japan 2023). Agricultural water-use facilities are mentioned in addition to flood control dams as storage facilities for rainwater and floodwater within a basin. Flood damage can be reduced by discharging the water in the storage facility before the occurrence of heavy rain and increasing the storage amount in the basin. Since ancient times, irrigation reservoirs have been built to secure water resources for agricultural use. Currently, there are approximately 150,000 to 210,000 reservoirs in Japan. Unlike dams, they can only be used for irrigation. Irrigation reservoirs play an important role in supporting agricultural activities in large rivers and areas with low rainfall. The demand for agricultural water increases from May to September.

The use of irrigation reservoirs for flood control can lead to water use risks and potential crop damage. Therefore, a framework for sharing the burden and compensation for damage across various sectors is effective. Consequently, a quantitative evaluation of the reduction effect on the flood damage costs of irrigation reservoirs by region and land use is required.

14.1.3 Issues of Previous Studies and Purpose of This Study

In previous domestic and foreign studies, flood damage reduction has been evaluated based on the amount of damage reduction (for example, Try et al. 2023; Choi et al. 2017). The effect of the peak discharge reduction using irrigation reservoirs in individual basins is shown as the impact of utilizing these reservoirs for flood control on river flow (for example, Yoshisako et al. 2013; Tanakamaru et al. 2022).

However, because the impact on flood inundation has not been quantitatively demonstrated, the sectors that benefit from the damage-cost-reduction effect of irrigation reservoirs are not clear. In addition, because the target basin area is small and the number of reservoirs to be examined is insignificant, its potential as a flood control measure is unclear. By evaluating the entire country as a target, we can discuss the reduction in flood damage costs by region. Furthermore, by evaluating land use, we can estimate the sectors that benefit from using irrigation reservoirs for flood control. Therefore, this study aimed to evaluate the potential reduction effect on flood damage costs by irrigation reservoirs, assuming a worst-case scenario for damage reduction and targeting Japan.

14.2 Dataset and Methodology

The study used the same dataset as Ikemoto et al. (2022, 2023). It consists of the rainfall distribution that causes extreme flows in each mesh (Tezuka et al. 2014), land use data, elevation data, ground slope data, and irrigation reservoir data. Following the methodology of Ikemoto et al. (2022, 2023), the reduction in flood damage using irrigation reservoirs for flood control was evaluated. The evaluation flow is shown in Fig. 14.1: (1) calculation of the inundation depth by flood inundation analysis, (2) calculation of the damage cost based on the inundation depth, and (3) calculation of the damage reduction rate. Details of this methodology are described below.

14.2.1 Irrigation Reservoir Data

The irrigation reservoir data were a partially edited version from the Irrigation Reservoir Disaster Prevention Support System created by the National Agriculture and Food Research Organization (NARO). To consider all the capacities of Japanese

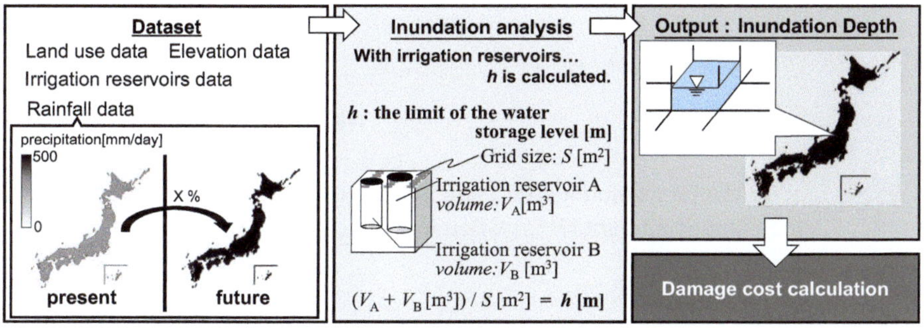

Fig. 14.1 Analysis flow

agricultural irrigation facilities, the dam data from the 2014 National Land Numerical Information was used, similar to Ikemoto et al. We added irrigation-specific dams not listed in the Irrigation Reservoir Support System data. In this study, the term "irrigation reservoir" includes agricultural dams in addition to the irrigation reservoirs listed in Japan's Irrigation Reservoir Disaster Prevention Support System.

14.2.2 Flood Inundation Analysis

In the inundation analysis, we applied a two-dimensional unsteady flow model (Tezuka et al. 2014) without distinguishing between rivers and floodplains and analyzed all of Japan simultaneously. This model was validated by Yanagihara et al. (2021). The grid size of the inundation analysis was approximately 250 m. For the analysis of the near-future climate (2031–2050) and end-of-century climate (2081–2100), we applied the extreme rainfall data of the future climate obtained using the method described in Sect. 17.2.4, for 24 h at a constant intensity. The parameters used for inundation analysis were the same as those used by Tezuka et al. (2014).

14.2.3 Calculation of Damage Amount and Reduction Rate of Damage Amount

The amount of damage was calculated based on the Flood Control Economic Survey Manual (Ministry of Land, Infrastructure, Transport and Tourism 2005, 2014). The land uses that caused damage were paddy fields, fields, residential areas, offices, and golf courses. We calculated the amount of damage by multiplying the asset value of the flooded plain by the damage rate according to the flood depth. We determined the asset value based on the asset evaluation value for each land use type and prefecture. The flood damage amount was evaluated based on the annual expected damage amount considering the occurrence probability. The amount of damage was calculated as described by Ikemoto et al. (2023). The damage reduction rate R [%] was calculated using Eq. (14.1).

$$R = \left(C_{\mathrm{no}} - C_{\mathrm{res}}\right) / C_{\mathrm{no}} \times 100 \tag{14.1}$$

where C_{no} is the damage amount [JPY] when the irrigation reservoir is not applied, and C_{res} is the damage amount [JPY] when using the irrigation reservoir.

14.2.4 Estimation of Extreme Rainfall in Future Climate

As in previous studies (Ikemoto et al. 2023; Yanagihara et al. 2024), we created and used a rainfall distribution that generated extreme flow rates in each mesh. We estimated the distribution of the increase rate of extreme rainfall from the reference climate to the future climate. The data of extreme rainfall in the future climate was estimated by multiplying this increase rate distribution by the data of extreme rainfall in the reference climate. The method for estimating the distribution of the rate of increase in extreme rainfall was the same as that used by Ikemoto et al. (2023) and Yanagihara et al. (2024). We used climate prediction data (Ishizaki et al. 2022) from five global climate models (GCMs) and one representative concentration pathway (RCP) scenario. The GCMs used were ACCESS-CM2, MIROC6, IPSL-CM6A-LR, MPI-ESM1-2-HR, and MRI-ESM2-0. We adopted SSP5-8.5 to consider the worst-case scenario.

14.2.5 Reflection of Irrigation Reservoirs

Using the same method as Ikemoto et al. (2023), we reflected the storage volume of irrigation reservoirs in the flood inundation analysis. This was estimated by applying the storage volume of the irrigation reservoirs as the storage height limit for the inundation analysis. The storage height limit was obtained by dividing the total storage volume of the irrigation reservoirs in each mesh by the mesh area. This indicated the flood depth that could be stored in each cell. In the inundation analysis, it was assumed that the water did not flow downstream until the water depth in the cells where the irrigation reservoirs were distributed exceeded the storage height limit, reflecting the storage capacity of the irrigation reservoirs. To evaluate the situation in which the storage capacity of the irrigation reservoirs was fully utilized, we set the storage rate to 0% as the initial condition for the inundation analysis. In addition, it is important not to cause water stress during the agricultural use of irrigation reservoirs, and it is necessary to maintain the storage volume without releasing all the water. Therefore, we set the storage rate of the irrigation reservoirs to 25%, 50%, and 75% as the initial conditions of the inundation analysis and applied them to the inundation analysis.

14.3 Results and Discussion

14.3.1 Reduction Rate of Damage Cost Nationwide in Japan by Flood Control Use of Irrigation Reservoirs

The nationwide reduction rate of flood damage costs in Japan for each GCM and storage rate is shown in Table 14.1. When the storage rate was 0%, the damage cost reduction rate was 1.4%–1.9% in the future climate and 1.1%–1.7% in the

Table 14.1 Rate of decrease in the EADC when flood control measures using reservoirs are taken

(a) For water storage rates of 0% and 25% (unit: %)

Rate of water storage	0%	25%	0%	25%	0%	25%	0%	25%	0%	25%	0%	25%
Period	ACCEESS-CM2		IPSL-CM-6A-LR		MIROC6		MPI-ESM1-2-HR		MRI-ESM2-0		Average	
Near future	1.4	1.0	1.9	1.5	1.7	1.4	1.6	1.3	1.9	1.5	1.7	1.4
Last twenty-first century	1.1	0.8	1.7	1.4	1.3	1.0	1.7	1.4	1.6	1.2	1.5	1.2

(b) For water storage rates of 50% and 75% (unit: %)

Rate of water storage	50%	75%	50%	75%	50%	75%	50%	75%	50%	75%	50%	75%
Period	ACCEESS-CM2		IPSL-CM-6A-LR		MIROC6		MPI-ESM1-2-HR		MRI-ESM2-0		Average	
Near future	0.8	0.4	1.1	0.6	1.0	0.5	1.0	0.5	1.1	0.6	1.0	0.5
Last twenty-first century	0.6	0.3	1.0	0.5	0.7	0.4	1.0	0.6	0.9	0.5	0.8	0.5

end-of-century climate. In addition, as the storage rate decreases, the damage cost reduction rate decreases. The nationwide reduction rate of flood damage costs in Japan in the end-of-century climate was smaller than the reduction rate of flood damage costs in the future climate. Therefore, it is suggested that the damage-cost reduction effect of irrigation reservoirs will decrease in the future. We evaluated the potential flood damage cost reduction effects of irrigation reservoirs by comparing them with other adaptation measures. Yamamoto et al. (2021) and Yanagihara et al. (2022) demonstrated the flood damage reduction rate due to piloti architecture and paddy field dams using a flow similar to that used in this study. The nationwide damage reduction rate in Japan was 14.0% for piloti architecture and 6.5% for paddy field dams. Compared with the nationwide damage reduction rate due to irrigation reservoirs, the results were lower for piloti architecture and paddy field dams. Hereafter, the maximum effect is discussed by introducing the analysis results with an irrigation reservoir storage rate set to 0%.

14.3.2 Evaluation of Annual Expected Damage Reduction Rate by Prefecture

To efficiently allocate budgets for adaptation measures, it is necessary to evaluate and compare the flood damage reduction rates of irrigation reservoirs by region. Fig. 14.2 shows a nationwide map of the flood damage reduction rate (average, maximum, and minimum values of five GCMs) for each prefecture. The maximum

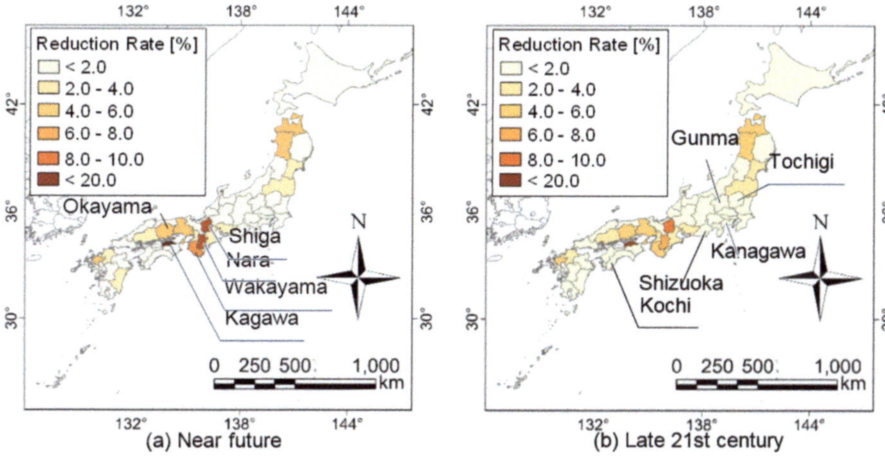

Fig. 14.2 Expected annual damage cost reduction rate in near future and late 21st century

and minimum national values have been mentioned above; therefore, in this section, we show the average value of five GCMs as a representative value. The top five prefectures with the highest damage reduction rates in the future were Kagawa (15.3%), Nara (9.4%), Shiga (8.8%), Wakayama (7.3%), and Okayama (5.8%). The last five prefectures are Kanagawa (0.1%), Tochigi (0.2%), Gunma (0.2%), Kochi (0.3%), and Shizuoka (0.3%) in terms of average values. At the end of the 21st century, they were Kagawa (15.5%), Shiga (8.9%), Nara (7.2%), Wakayama (5.6%), and Okayama (5.5%), with no change from the top five prefectures in the future climate, and high damage reduction rates were obtained mainly in prefectures facing the Seto Inland Sea and in the Tohoku region. In these prefectures, there is a strong possibility of a high damage reduction rate, owing to the large number of irrigation reservoirs and their location upstream of high-value land use. Therefore, budget allocation for adaptation measures for the flood control use of reservoirs is expected in these prefectures. The five lowest prefectures were Kanagawa (0.0%), Kochi (0.2%), Shizuoka (0.2%), Tochigi (0.2%), and Gunma (0.2%). There was no significant change in the ranking for either the maximum or minimum values in the five prefectures. The Kanto region picked up at the bottom, and the damage reduction rate was extremely low. In these prefectures and in the Kanto region, no effect is expected from budget allocation for the flood control use of irrigation reservoirs.

14.3.3 Evaluation of Flood Damage Reduction Rate by Land Use in Each Prefecture

Figures 14.3 and 14.4 indicates the damage reduction rates for each land-use type in the future climate and the climate at the end of the 21st century. The bar graph shows the average of five GCMs. The error bars indicate the maximum and minimum values of the five GCMs. For all land uses, the damage reduction rate is high

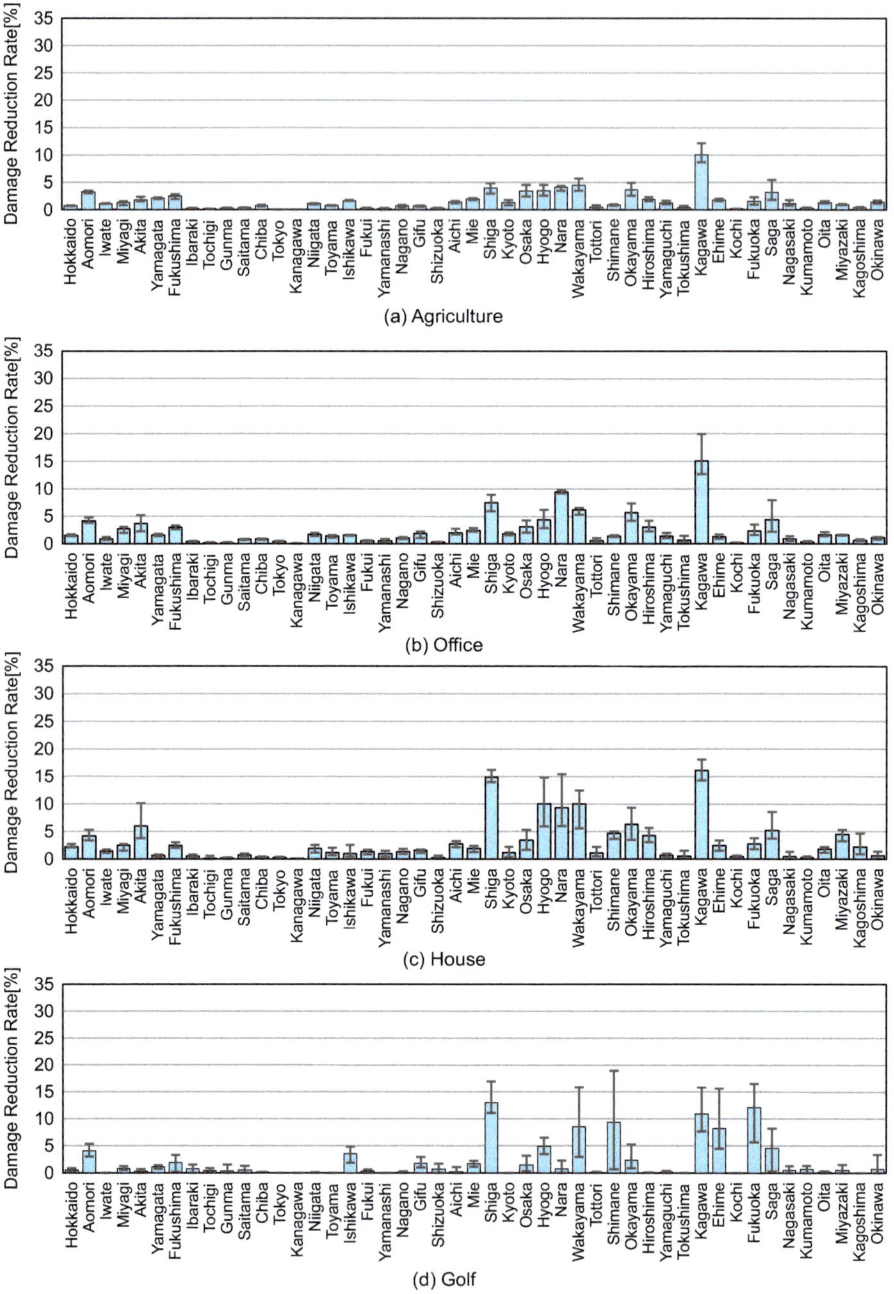

Fig. 14.3 Expected annual damage cost reduction rate of each land use in near future

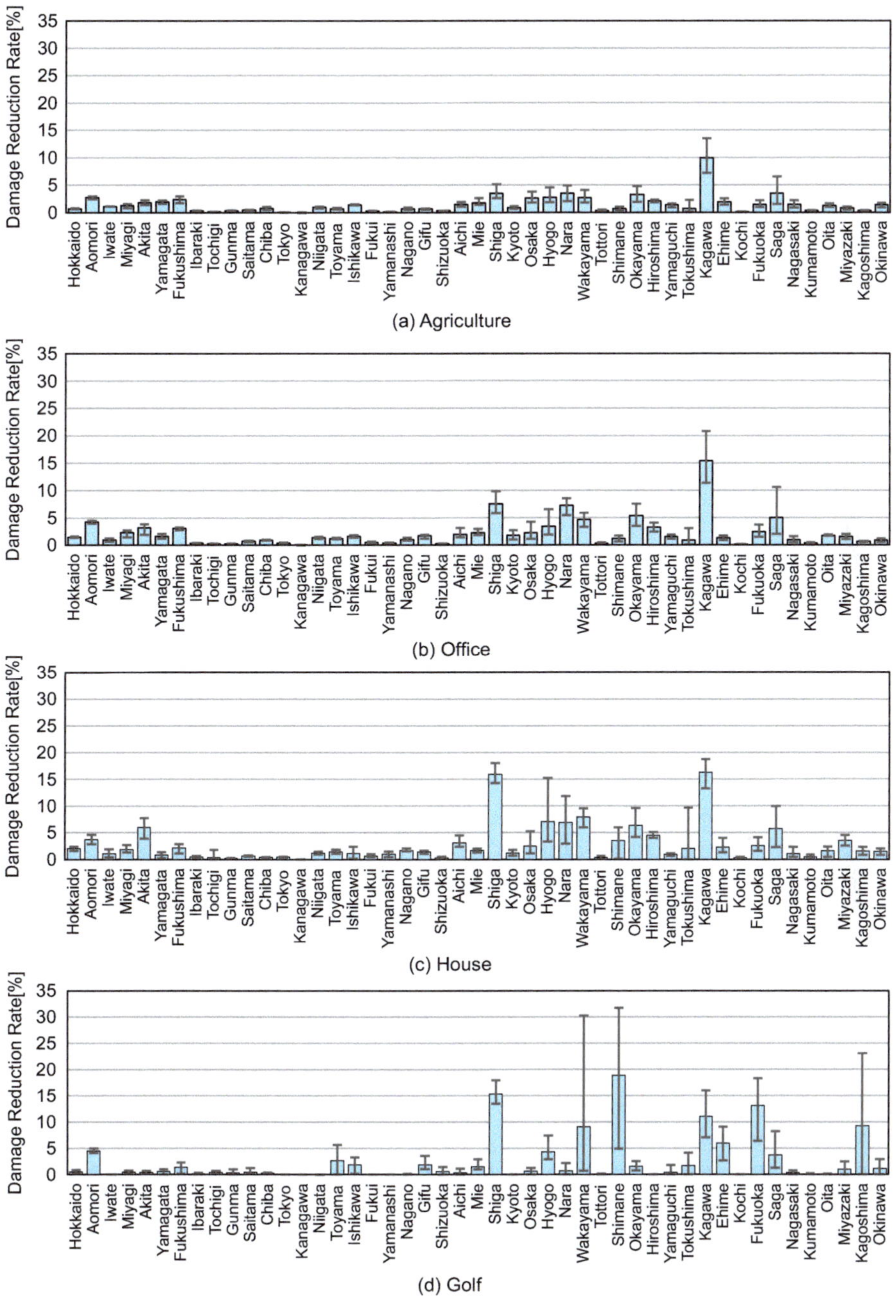

Fig. 14.4 Expected annual damage cost reduction rate of each land use in late 21st century

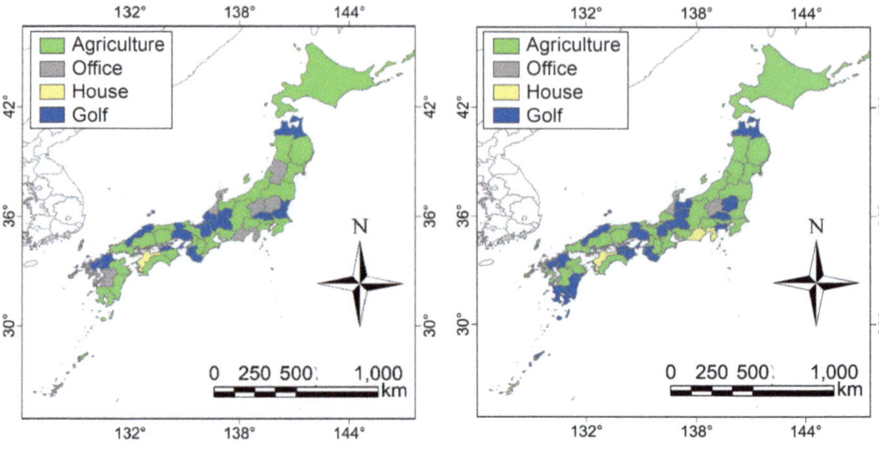

(a) Near future (b) Late 21st century

Fig. 14.5 Expected annual damage cost reduction rate of each land use

in western Japan, centered on the Seto Inland Sea region, in both the near future climate and the climate at the end of the 21st century. Observing the summarized damage reduction rates of all land uses, the damage reduction rate for farmland was the lowest, and the difference between the maximum and minimum values was also small. Therefore, in farmland, the impact of uncertainty is likely to be low, and farmland receives the damage-reduction effect of irrigation reservoirs compared to other land uses. However, the difference between the maximum and minimum damage-reduction rates in the golf courses was the largest. In Saga and Shimane prefectures in the near future climate, and in Wakayama and Kagoshima prefectures in the climate at the end of the 21st century, the minimum value is close to 0% because there is no damage reduction in golf courses according to uncertainty.

These results indicate that irrigation reservoirs reduce damage not only to farmland but also to land uses other than farmland. To understand the land uses that received the damage reduction effect, Fig. 14.5 displays the land use with the highest average damage reduction rate in each prefecture. In the future climate, the number of prefectures with the highest damage reduction rates will be 12 for farmland, 1 for residential areas, 8 for businesses, and 26 for golf courses. Additionally, in the climate at the end of the 21st century, the number of prefectures with the highest damage reduction rates was 16 for farmland, 3 for residential areas, 2 for businesses, and 26 for golf courses. Most prefectures with high damage reduction rates have golf courses. This is because the irrigation reservoirs upstream of golf courses cause significant damage reduction. However, residential areas and businesses indicate decreasing trends. In prefectures with the highest damage reduction rates in farmland, the use of irrigation reservoirs is prioritized. Alternatively, in prefectures with the highest damage reduction rate in land use other than farmland, sectors other than farmers were prioritized to receive the damage reduction effect of the irrigation reservoirs. Therefore, it is necessary to consider the water use risk and crop damage burden, including sectors other than farming.

14.4 Conclusion

This chapter evaluates the flood damage reduction effects of the irrigation reservoirs. The results of this chapter are expected to provide scientific information for the flood control use of irrigation reservoirs by understanding the damage reduction effect against the impact of climate change for Japan, by prefecture and land use. However, the flood damage reduction rate was evaluated by uniformly setting the water storage rate of the irrigation reservoirs, and the damage amount estimation for crops due to water use risk has not been reached. In the future, discussions that consider the trade-off between the benefits of flood control use of irrigation reservoirs and the damage caused by water use risk are required.

Acknowledgments To obtain the reservoir data, we received support from Professor Toshikazu Hori of the National Agriculture and Food Research Organization. Part of the calculation results from this study was obtained using the large-scale scientific calculation system of Tohoku University Cyberscience Center. We would like to extend our most sincere appreciation. In addition, this research was supported by JHPCN (jh220018, jh230024), Joint Usage / Research Center for Interdisciplinary Large-scale Information Infrastructures. This work was supported in part by MEXT WISE Program for Sustainability in the Dynamic Earth.

References

Choi H, Lee EH, Joo JG, Kim JH (2017) Determining optimal locations for rainwater storage sites with the goal of reducing urban inundation damage costs. KSCE J Civ Eng 21:2488–2500. https://doi.org/10.1007/s12205-016-0922-6

Council for Social Infrastructure Development (2020) How to prepare for water-related disasters in the light of climate change, https://www.mlit.go.jp/river/shinngikai_blog/shaseishin/kasenbunkakai/shouiinkai/kikouhendou_suigai/pdf/03_honbun.pdf. Accessed 18 July 2024

Ikemoto A, Kazama S, Yoshida T, Yanagihara H, Touge Y (2022) Potential of irrigation reservoirs for flood prevention and evaluation of the reduction rate of flood damage cost for each prefecture. J Jpn Soc Civil Eng Ser. B1 (Hydraul Eng) 78(2):I_265–I_270. https://doi.org/10.2208/jscejhe.78.2_I_265

Ikemoto A, Kazama S, Yoshida T et al (2023) Evaluation of an adaptation strategy for flood damage mitigation under climate change through the use of irrigation reservoirs in Japan. Water Resour Manage 37:4159–4175. https://doi.org/10.1007/s11269-023-03544-7

Ishizaki NN, Shiogama H, Hanasaki N, Takahashi K (2022) Development of CMIP6-based climate scenarios for Japan using statistical method and their applicability to heat-related impact studies. Earth Space Sci 9:e2022EA002451. https://doi.org/10.1029/2022EA002451

Ministry of Land, Infrastructure, Transport and Tourism (2005) Manual for Economic Evaluation of Flood Control Investment (Draft). https://www.mlit.go.jp/river/kokusai/pdf/pdf06.pdf. Accessed 10 Jun 2024

Ministry of Land, Infrastructure, Transport and Tourism (2014) National Land Numerical Information. Accessed 18 July 2024

Ministry of the Environment (2016) Ecosystem-based Disaster Risk Reduction in Japan—a Handbook for Practitioners. https://www.env.go.jp/content/900489554.pdf. Accessed 18 July 2024

Nsihida T, Iwasaki Y, Ohsawa T, Ogasawara S, Kamada M, Sasaki A, Takagawa S, Takamura N, Nakamura F, Nakashizuka T, Nishihiro J, Furuta N, Matsuda H, Yoshida T (2023) The subcommittee for research and proposition, the committee for ecosystem management, the ecological society of Japan. Jpn J Conserv Ecol 28:213–227. https://doi.org/10.18960/hozen.2211

Tanakamaru H, Kida N, Tada A (2022) Selection of irrigation ponds with large effect of flood mitigation by water release based on simple Estimation method. Water Land Environ Eng 90(6):401–404

Tezuka S, Takiguchi H, Kazama S, Sato A, Kawagoe S, Sarukkalige R (2014) Estimation of the effects of climate change on flood-triggered economic losses in Japan. Int J Disaster Risk Reduct 9:58–67. https://doi.org/10.1016/j.ijdrr.2014.03.004

Try S, Sayama T, Phy SR, Sok T, Ly S, Oeurng C (2023) Assessing the impacts of climate change and dam development on potential flood hazard and damages in the Cambodian floodplain of the lower mekong basin. J Hydrol 49. https://doi.org/10.1016/j.ejrh.2023.101508

World Wide Fund for Nature (2022) Nature Based Solutions—a review of current financing barriers and how to overcome these. https://www.wwf.org.uk/sites/default/files/2022-06/WWF-NBS-Public-Report-Final-270622.pdf. Accessed 18 July 2024

Yamamoto T, Kazama S, Touge Y, Yanagihara H, Tada T, Yamashita T, Takizawa H (2021) Evaluation of flood damage reduction throughout Japan from adaptation measures taken under a range of emissions mitigation scenarios. Climatic Change 165(60). https://doi.org/10.1007/s10584-021-03081-5

Yanagihara H, Yamamoto T, Kazama S, Touge Y, Chai Y, Tada T (2021) Regional evaluation of potential flood damage reduction by paddy field dam in Japan. J JSCE Ser G (Environ Res) 77(5):I_33–I_42. https://doi.org/10.2208/jscejer.77.5_I_33

Yanagihara H, Kazama S, Yamamoto T, Ikemoto A, Tada T, Touge Y (2024) Nationwide evaluation of changes in fluvial and pluvial flood damage and the effectiveness of adaptation measures in Japan under population decline. Int J Disaster Risk Reduct 110:104605. https://doi.org/10.1016/j.ijdrr.2024.104605

Yoshisako H, Koyama J, Ogawa S, Fukumoto M (2013) Flood peak mitigation effects by irrigation ponds group of valley type in the Mukunashi River Valley, Hiroshima Prefecture, Japan. Trans Jpn Soc Irrig Drain Rural Eng 81(3):205–214. https://doi.org/10.11408/jsidre.81.205

Chapter 15
Diffusion of Flood Adaptation Measures in Japan: An Agent-Based Model for Assessing Individual Behaviors and Policy Communication Effectiveness

Yoshiaki Nakagawa and Masayuki Yokozawa

Abstract Previous studies have statistically examined the social and psychological factors that influence the adoption of individual-level flood adaptation measures, focusing on the role of social networks in implementing these measures. Effective communication policies were explored in the Netherlands and France by simulating the diffusion of adaptation measures through social networks using empirical decision models and agent-based simulations. However, similar research still needs to be made available in Asia. This study investigated effective communication policies in Japan by simulating the diffusion of individual-level adaptation measures through social networks. Unlike the Netherlands and France, the results revealed that people-centered communication policies were less effective in Japan, while top-down policies proved more successful. At the same time, social networks were critical in adopting adaptation measures across all regions studied. Empirical decision models and agent-based approaches can provide valuable insights for developing effective communication policies to promote flood risk adaptation on a global scale.

Keywords Agent-based model · Protection motivation theory · Social network · Small-world · Flood risk communication · Adaptation diffusion simulation

15.1 Introduction

Climate change is increasingly affecting the world, with flooding as a major issue. While many adaptation measures have been proposed, O'Brien et al. (2006) noted that these often focus on technical solutions, neglecting the adaptation process itself. In Japan, flood risk assessment and adaptation research have mainly emphasized hydrological and engineering approaches. Sayama et al. (2008) evaluated

Y. Nakagawa (✉) · M. Yokozawa
Waseda University, Tokorozawa, Saitama, Japan
e-mail: nrj59355@nifty.com

N. Mimura, S. Takewaka (eds.), *Climate Change Impacts and Adaptation Strategies in Japan*, https://doi.org/10.1007/978-981-96-2436-2_15

flood risks under climate change using runoff models, and Tachikawa et al. (2011) predicted future flood frequency changes with integrated hydrological and climate models, providing insights into physical flood processes. However, these studies have not adequately considered social adaptation or the impact of human behavior on flooding. Numerical models assume static social systems, overlooking dynamic elements like the spread of social adaptation and the behavior change. Understanding how, by whom, and why adaptation is implemented is also crucial for effective policies, yet research on policy interventions promoting adaptation strategies is lacking.

Social factors are crucial in understanding stakeholders' adaptive behaviors. Agner et al. (2009) show that non-climatic influences can strongly determine adaptive actions. Risk perception, knowledge, and experience at individual and societal levels also shape adaptation measures (O'Brien et al. 2006; Agner et al. 2009).

Flood adaptation occurs at both societal and individual levels. Societal measures include levees, dams, flood retention basins, and building restrictions. Individual measures involve personal decisions, such as using water-resistant materials or storing valuables upstairs. This study examines the diffusion of individual adaptation measures.

Previous studies have explored the social and psychological factors influencing individual-level adaptation measures. Threat and coping appraisals are key psychological factors (Rogers 1975; Bubeck et al. 2013). Threat appraisal refers to an individual's assessment of the threat's probability and potential damage of the threat, while coping appraisal refers to an individual's assessment of his or her ability to cope with and avert damage caused by the threat and the associated costs. Recently, social networks have also been recognized as crucial in influencing adaptation decisions (Bubeck et al. 2013; Kunreuther et al. 2013; Lo 2013). Social networks refer to networks of individuals (such as friends and acquaintances) connected by interactions like information sharing and social contagion.

Effective communication policies on disaster risks and coping are vital for promoting individual adaptation measures diffusion (IPCC 2012). There are two types: top-down and people-centered (Haer et al. 2016; Erdlenbruch and Bonte 2018). Top-down policies involve government-led dissemination of flood-related information via guidelines, pamphlets, media, and websites. People-centered policies provide tailored information on risk perception and coping strategies based on individuals' needs.

Haer et al. (2016) developed an empirical decision model for adopting individual flood adaptation measures in the Netherlands, using it alongside an agent-based model with social networks to assess communication policy effectiveness. They found that people-centered policies, especially those targeting both threat and coping appraisals, were more effective in spreading adaptation measures than top-down policies. Additionally, social networks significantly impacted individuals' adoption of these measures. Erdlenbruch and Bonte (2018) conducted a similar study to Haer et al. (2016) in flood-prone areas in France, finding little difference between people-centered and top-down approaches. Communication policies targeting both threats and coping appraisals were effective, and social networks significantly influenced the adoption of adaptation measures, as also shown by Haer et al. (2016).

Effective communication policies can vary by region and culture, requiring local adjustments (Haer et al. 2016). However, few studies have been done in Asia. This study addresses this gap by investigating effective communication policies in Japan. We surveyed households in flood-prone areas, statistically estimated a decision model for adopting individual flood adaptation measures, and developed an agent-based model incorporating this decision model and social networks. We then simulated the diffusion of adaptation measures and assessed the effectiveness of various communication policies.

15.2 Methods

15.2.1 Overview

Figure 15.1 shows an overview of the method. First, we conducted a questionnaire survey on flood adaptation measures, and then estimated parameters of a decision model for implementing adaptation measure by using the data obtained from this survey. Furthermore, based on this data, we constructed small-world (SW) networks as social networks of individuals in the study area (see Sect. 15.2.4 for details on

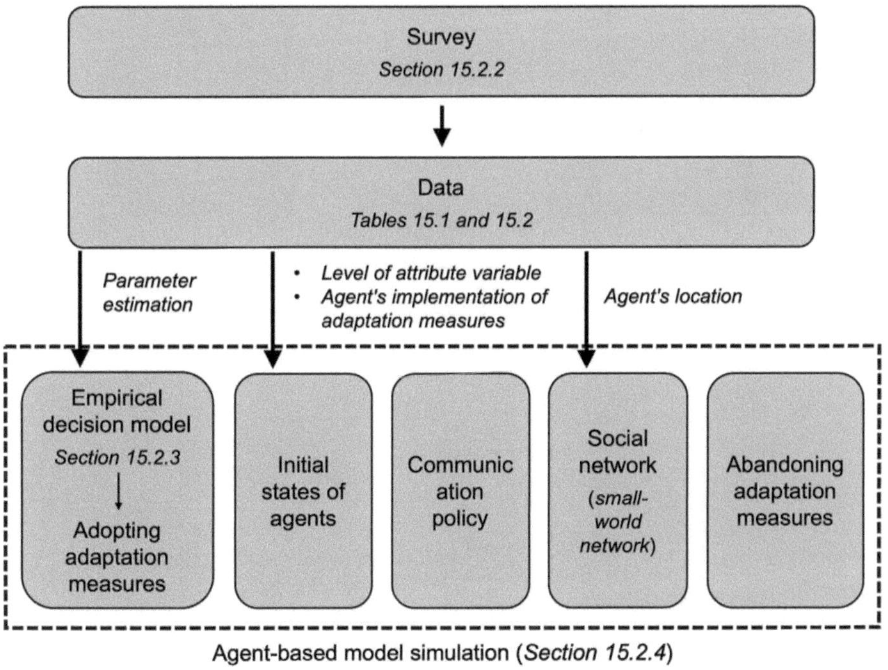

Fig. 15.1 Schematic overview of the method

SW networks). Finally, we developed an agent-based model to simulate the diffusion of adaptation measures, incorporating the decision model and the social network. The agent-based model also incorporated the influence of communication policies and the process of abandoning adaptation measures.

15.2.2 Survey

Study Areas In December 2022, we conducted an internet survey targeting 697 households in the municipalities of Japan, as shown in Fig. 15.2 (411 and 286 households in areas 1 and 2, respectively), through a research company (Macromill, Inc.). All the municipalities surveyed experienced severe flooding in the years preceding the survey. Area 1 includes municipalities in the Abukuma River Basin in Fukushima and Miyagi prefectures that suffered from flooding due to river flooding and pluvial flooding due to the 2019 East Japan Typhoon (TY1919, Hagibis). In these areas, approximately 10,500 houses were flooded above the floor level and approximately 3100 houses were flooded below the floor level (Ministry of Land, Infrastructure, Transport, and Tourism, Tohoku Regional Development Bureau,

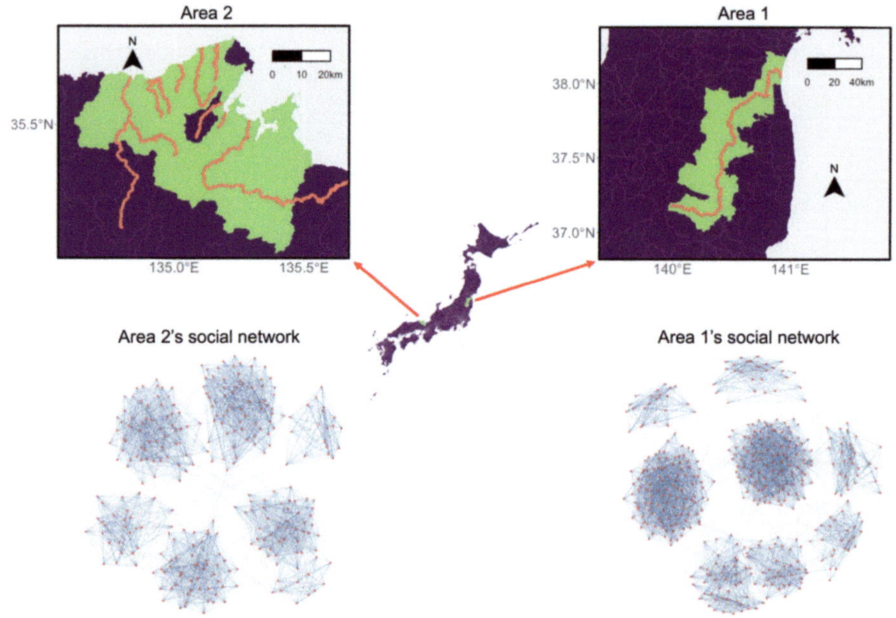

Fig. 15.2 Map of surveyed municipalities (upper panels) and social networks (lower panels). Orange lines represent rivers. Green areas represent the surveyed municipalities. Social networks between individuals are constructed as small-world (SW) networks of in-degree 15. The positions of the points in the graph are optimized to better visualize the network and are independent of an individual's geographic location

2020; Konami et al. 2020). The Abukuma River Basin has been hit by large-scale flood damage several times in its recorded history. In addition to 2019, large-scale floods occurred in 1991, 1998, 2002, and 2011. Area 2 includes the municipalities in Hyogo and Kyoto Prefectures that suffered from the flooding of rivers, such as the Maruyama, Izushi, and Yura Rivers, owing to Typhoon No. 23 in 2004 (Tokage). Approximately 3200 houses were flooded above the floor level, and 7200 houses were flooded below the floor level (Hyogo Prefecture Typhoon No. 23 Disaster Verification Committee, 2005; Kyoto Prefecture 2005). In addition to 2004, large-scale floods occurred in the Maruyama River in 1990, 2004, and 2009, as well as in the Izushi River in 2006, 2011, 2013, 2014, and 2017.

Sociodemographic Household Attributes The surveyed households were balanced in terms of age and sex. It is also well-diversified in terms of educational level. This study targeted only households that owned their homes and did not rent. This is because several measures listed in Table 15.1 can typically be implemented for home-owned households but not for rented households (Bubeck et al. 2013).

Adaption Measures, Survey Questions, and Attribute Variables This survey examined 11 major adaptation measures for flooding, as listed in Table 15.1. These adaptation measures were based on previous studies and individual-level measures for floods implemented in Japan (Bubeck et al. 2013; Richert et al. 2016; Erdlenbruch and Bonte 2018; Japan Federation of Housing Organizations 2021; Kiuchi and Nakano 2023). The survey questions and attribute variables are listed in Table 15.2. Among the attitude variables, perceived probability and perceived consequences ranged from 1 to 4, and the other variables ranged from 1 to 5.

Table 15.1 Adaptation measures, implementation rates, and coefficients of determination

		Implementation rate (%)		Nagelkerke R^2	
	Measure	Area 1	Area 2	Area 1	Area 2
M1	The ground of a house raised by embankment	16.30%	13.29%	0.21	0.14
M2	First floor with raised floor height or piloti	3.65%	6.99%	0.28	0.19
M3	Opening on the roof to facilitate evacuation	3.89%	2.80%	0.35	0.33
M4	Watertight door or water gate barrier	3.65%	4.90%	0.52	0.26
M5	Ventilation openings located at high locations with low risk of flooding	4.87&	3.85%	0.27	0.35
M6	Equipment to drain water in case of flooding (underfloor drain, etc.)	2.92%	3.85%	0.54	0.29
M7	Use of water-resistant materials (for the floor and/or the walls)	3.65%	2.10%	0.21	0.46
M8	Toilet, bath, and kitchen located upstairs	4.87%	4.55%	0.30	0.39
M9	Electrical wiring and systems and/or boiler installed higher up on the walls	3.41%	1.40%	0.28	0.51
M10	Sewer non-return valves	4.38%	2.45%	0.15	0.46
M11	Valuables stored upstairs	11.19%	7.34%	0.32	0.15

Table 15.2 Summary of PMT data: Attribute variable name corresponding to household attribute, mean value and standard deviation in the sample, question used in the questionnaire, and possible values for each attribute

Attribute variable	Area 1		Area 2		Question	Scale
	Mean	(Std Dev)	Mean	(Std Dev)		
Perceived probability	2.47	(1.09)	2.53	(1.15)	"How do you assess the following scenario: 'Your municipality will be flooded at least once in the next 10 years'?"	From 1 ("impossible") to 4 ("certain")
Perceived consequences	1.73	(0.94)	1.86	(0.97)	"In the case of flooding, how do you assess the following scenario: 'The water will reach your street'?"	From 1 ("impossible") to 4 ("certain")
Perceived self-efficacy	3.39	(1.09)	3.16	(1.16)	"To what extent do you agree with the following statement: 'I do not believe that I am able to avoid the consequences of floods in my household. I have no control over such events.'?"	From 1 ("strongly agree") to 5 ("strongly disagree")
Perceived efficacy of measure	3.47	(0.97)	3.44	(1.02)	"For each measure listed below, how effective do you think it will be in preventing the negative consequences of floods?"	From 1 ("not at all effective") to 5 ("very effective")
Past flood experience appraisal	1.37	(1.06)	1.56	(1.21)	"How do you assess the seriousness of the consequences of the reference flood for your household?"	From 1 ("not or for people who have not experienced a flood") to 5 ("extremely serious")
Perceived benefits	2.92	(0.84)	2.84	(0.78)	"How well do you feel in your municipality?"	From 1 ("not well at all") to 5 ("very well")
Perceived costs	2.25	(1.15)	2.3	(1.17)	"For each measure listed its use and its maintenance are they constraining?"	From 1 ("extremely serious") to 5 ("not at all")
Social network	1.31	(0.85)	1.27	(0.79)	"What percentage of neighbors in a social network implement each measure?"	1: 0%–20%, 2: 21%–40%, 3: 41%–60%, 4: 61%–80%, 5: 81%–100%

$N = 411$ (area 1), 286 (area2)

15.2.3 The Empirical Decision Model

The protection motivation theory (PMT) of Rogers (1975) has become an important social psychological model for individuals' decisions to implement flood adaptation measures (Bubeck et al. 2013; Richert et al. 2016). Based on PMT, in this study, we adopted "threat appraisal" and "coping appraisal" as variables in the model explaining adaptive behavior. The threat appraisal consists of three subcomponents: an individual's expectation of future flood occurrence, severity of its consequence, and the potential benefits of living in a hazardous area (i.e., "perceived probability," "perceived consequences," and "perceived benefits" in Table 15.2). Similarly, coping appraisal consists of three subcomponents: the effectiveness of the adaptation measures, the self-efficacy in adapting, and the cost of implementation of adaptation measures (i.e., "perceived efficacy of measure," "perceived self-efficacy," and "perceived costs" in Table 15.2). Furthermore, we introduced an individual's appraisal of past flood experience as a variable in the model to explain adaptive behavior (i.e., "past flood experience appraisal" in Table 15.2). Finally, we introduced the influence of the adaptive behavior of neighbors on social networks on an individual into this model (i.e., "social network" in Table 15.2). These variables were the same as those adopted by Erdlenbruch and Bonte (2018).

The probability to adopt adaptation measure i for each household, P_i, is described by the following equation:

$$P_i = \frac{1}{1+\exp\left(\alpha_{i,0}+\sum_{j=1}^{J}\alpha_{i,j}x_{i,j}\right)} = \frac{C_i\prod_{j=1}^{J}or_{i,j}^{\,x_{i,j}}}{1+C_i\prod_{j=1}^{J}or_{i,j}^{\,x_{i,j}}} \tag{15.1}$$

where j denotes the attributes, J denotes the number of the attributes, $\alpha_{i,j}$ denotes the estimated parameters by the logistic regression, $x_{i,j}$ denotes the level of attribute j for adaptation measure i, $or_{i,j}$ denotes the odds ratio of attribute j for adaptation measure i as estimated in a logistic regression, and C_i denotes the constant of the logistic regression (i.e., $\exp(\alpha_{i,0})$).

Furthermore, to examine whether the values of parameter $\alpha_{i,j}$ for each attribute and each adaptation measure differ between areas 1 and 2, we conducted the analysis of covariance (ANCOVA) with interaction terms (or with different slopes). A logistic regression analysis was performed using the following equation:

$$P_i = \frac{1}{1+\exp\left(\beta_{i,0}+\sum_{j=1}^{J}\beta_{i,j}x_{i,j}+\gamma_{i,0}t+\sum_{j=1}^{J}\gamma_{i,j}x_{i,j}t\right)} \tag{15.2}$$

where t denotes a dummy variable (in the case of area 1, $t = 0$ and in the case of area 2, $t = 1$); $\beta_{i,j}$ and $\gamma_{i,j}$ denote the estimated parameters by the logistic regression. If $\gamma_{i,j}$

is significantly different from zero, the parameter $\alpha_{i,j}$ for the target attribute and measure is significantly different between areas.

15.2.4 Simulation of the Diffusion of Adaptation Measures

The simulation model was implemented as an agent-based model, with households as agents. The agents were situated geographically and connected through social networks. Agents correspond to households, and the numbers of agents in areas 1 and 2 are 411 and 286, respectively. Agents linked to an agent were defined as neighbors of that agent. Agents obtained information regarding whether neighbors had implemented each adaptation measure. Each agent had two states: the level of the attribute variable and the implementation or non-implementation of each adaptation measure. The initial proportions of households that implemented each adaptation measure are listed in Table 15.1. The means and standard deviations of the initial values of the household attribute levels are listed in Table 15.2.

The general schedule for the simulation is as follows: Each year, in random order, each agent decides whether to adopt an adaptation measure. Subsequently, an annual communication policy is implemented in some agents to increase several attribute variables. Finally, every household observes its social network and updates its attribute level based on the number of neighbors adapted to the network. The time horizon is 30 years (or time steps). We performed 100 iterations for each simulation. We calculated the percentage of adapted households relative to the total number of agents.

Construction of the Social Network Following Erdlenbruch and Bonte (2018), we set the social network among agents as SW networks. In the SW networks, individuals being linked by a short chain of acquaintances. SW networks have been shown to be observed in many social interactions (Watts and Strogatz 1998). The steps for creating the SW network are as follows: First, we created a directed regular network connecting geographically close agents. Subsequently, based on the method described by Watts and Strogatz (1998), the links of this directed regular network were randomly rewired with a probability of 0.015 to create an SW network. The in-degree (i.e., the number of links entering an agent) was set to 5, 10, or 15. Figure 15.2 illustrates an example of the constructed SW networks.

Communication Policy Some agents are subject to communication policies. All agents have the probability P_{tar}, which is the product of the probability that communication reaches the agent and the probability that the communication policy changes the agent's attitude (Haer et al. 2016). P_{tar} was set to 0.3. According to Haer et al. (2016), communication policies for flood risk can take four forms: top-down policies on risk (td-r), top-down policies on risk and coping (td-rc), people-centered policies on risk (pc-r), and people-centered policies on risk and coping (pc-rc). The specific implementation of the communication policy in the simulation was explained by Haer et al. (2016) and is shown in Fig. 15.3. We examined the effects of five types of communication: td-r, td-rc, pc-r, pc-rc, and no communication (no-com).

Adopting and Abandoning Adaptation Measures We defined the probability that an agent adopts adaptation measure *i* each year as P_i, which is calculated using the empirical decision model (Eq. 15.1). The parameter value $\alpha_{i,j}$ of the empirical decision model to calculate P_i is the estimated value in Sect. 15.2.3 and is shown in

TD-R communication strategy

```
If communication reaches household [
        If communication is successful in changing attitude [
                increase attitude for perceived probability with 2 to a maximum of 4
                increase attitude for perceived consequence with 2 to a maximum of 4 ] ]
```

TD-RC communication strategy

```
If communication reaches household [
        If communication is successful in changing attitude [
                increase attitude for perceived probability with 1 to a maximum of 4
                increase attitude for perceived consequence with 1 to a maximum of 4
                increase attitude for self-efficacy with 4/3 to a maximum of 5
                increase attitude for response efficacy with 4/3 to a maximum of 5 ] ]
```

PC-R communication strategy

```
If communication reaches household [
        If communication is successful in changing attitude [
                If perceived probability < threshold
                        [ increase attitude for perceived probability with 2 ]
                Else [ increase attitude for perceived consequences with 2 to a maximum of 4]
                If perceived consequence < threshold
                        [ increase attitude for perceived consequence with 2 ]
                Else [ increase attitude for perceived probability with 2 to a maximum of 4 ]
```

PC-RC communication strategy

```
If communication reaches household [
        If communication is successful in changing attitude [
                If perceived probability < threshold
                        [ increase attitude for perceived probability with 1 ]
                Else [ If perceived consequence < threshold
                                [ increase attitude for perceived consequences with 1 to a maximum of threshold ]
                        Else [ increase self-efficacy and response efficacy with 4/3 to a maximum of 5 ] ]
                If perceived consequence < threshold
                        [ increase attitude for perceived consequence with 1 ]
                Else [ If perceived probability < threshold
                                [ increase attitude for perceived probability with 1 to a maximum of threshold]
                        Else [ increase self-efficacy and response efficacy with 4/3 to a maximum of 5 ] ]
```

Fig. 15.3 Algorithm of each communication policy. The algorithm proposed by Haer et al. (2016) was adjusted to the scale of the attribute variables in this study

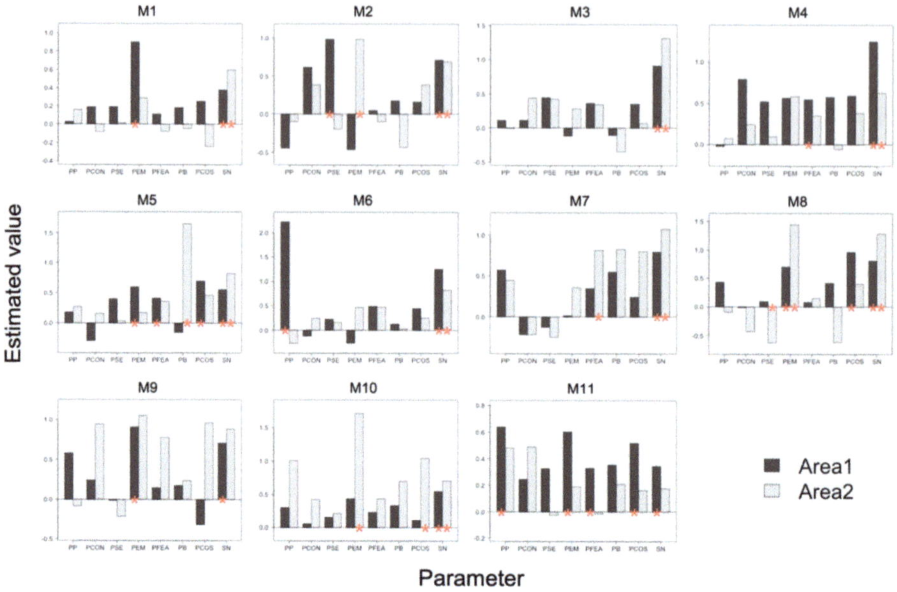

Fig. 15.4 Estimated parameters of decision model (Eq. 15.1). The bar represents the estimate of parameter. The red asterisk indicates that the parameter value is significantly different from zero (Wald's test). M1–M11 correspond to adaptation measures 1–11, as listed in Table 15.1. PP, PCON, PSE, PEM, PB, PCOS, and SN correspond to the attribute variables listed in Table 15.2

Fig. 15.4. Following Erdlenbruch and Bonte (2018), we assume that the average period of adoption of adaptation measures is *b* years. Therefore, the probability that an adapted agent abandons adaptation each year is 1/*b*. Abandoning adaptation resets the level of the perceived probability to zero. This corresponds to a feedback effect, in which risk perception decreases once an adaptation measure is adopted (Richert et al. 2016). Based on Erdlenbruch and Bonte (2018), *b* was set to 10. We also run simulations with different values of b (i.e., 1–30) and find that the order of effectiveness of communication policies remains the same regardless of the value of b.

15.3 Results and Discussion

15.3.1 Estimate of Empirical Decision Model

The estimated values of the parameters $\alpha_{i,j}$ of the empirical decision model are shown in Fig. 15.4. The Nagelkerke R^2 values for each adaptation measure are presented in Table 15.1. In a previous study, Nagelkerke R^2 was 0.34 (Erdlenbruch and Bonte 2018), and the value in this study was also around that. Therefore, the degree of fit of the model to the data is comparable to previous studies. Furthermore, we investigated the variance inflation factor (VIF) to evaluate multicollinearity. The VIF was less than 2.0 under all conditions (i.e., the conditions of area, adaptation measure, and attribute), indicating that there is hardly any multicollinearity between attributes.

In both areas, the parameters of the "social network" attribute were significantly different from zero and positive for most adaptation measures (Fig. 15.4). Therefore, for most adaptation strategies, an individual's motivation to implement the adaptation measure increases with the implementation rate of their neighbors in the social network. This result is consistent with prior studies (Haer et al. 2016; Erdlenbruch and Bonte 2018).

15.3.2 Comparison Between Coefficients of Empirical Decision Model in Different Areas

There were a few attributes for which the decision-making model parameters in the two areas differed significantly (Fig. 15.5). Therefore, a common decision model for the implementation of flood adaptation measures may be applicable to different regions in Japan.

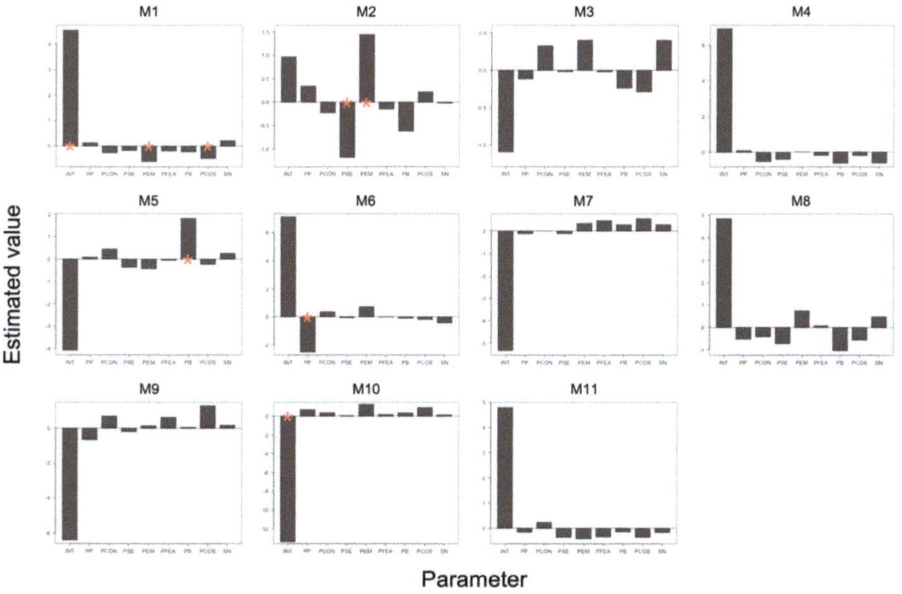

Fig. 15.5 Estimated values of parameter $\gamma_{i,j}$ of Eq. 15.2. INT corresponds to $\gamma_{i,0}$ in Eq. 15.2. See description of Fig. 15.4

15.3.3 Impact of Communication Policy

The implementation rates at time step 30 in areas 1 and 2 are shown in Figs. 15.6 and 15.7 (red boxplots), respectively. For each adaptation measure, the implementation rates were significantly different for most pairs of communication policies (Wilcoxon rank-sum test). In both areas, for many adaptation measures, "td-rc" had the highest implementation rate among all communication policies. This finding differs from the results of a previous study conducted in the Netherlands (Haer et al. 2016). In this previous study, "pc-rc" had the highest implementation rate among all communication policies.

15.3.4 Importance of Social Networks

The implementation rates at time step 30 for the case with no network in areas 1 and 2 are shown in Figs. 15.6 and 15.7 (blue boxplots), respectively. For all adaptation measures, the implementation rates were significantly different between the case of

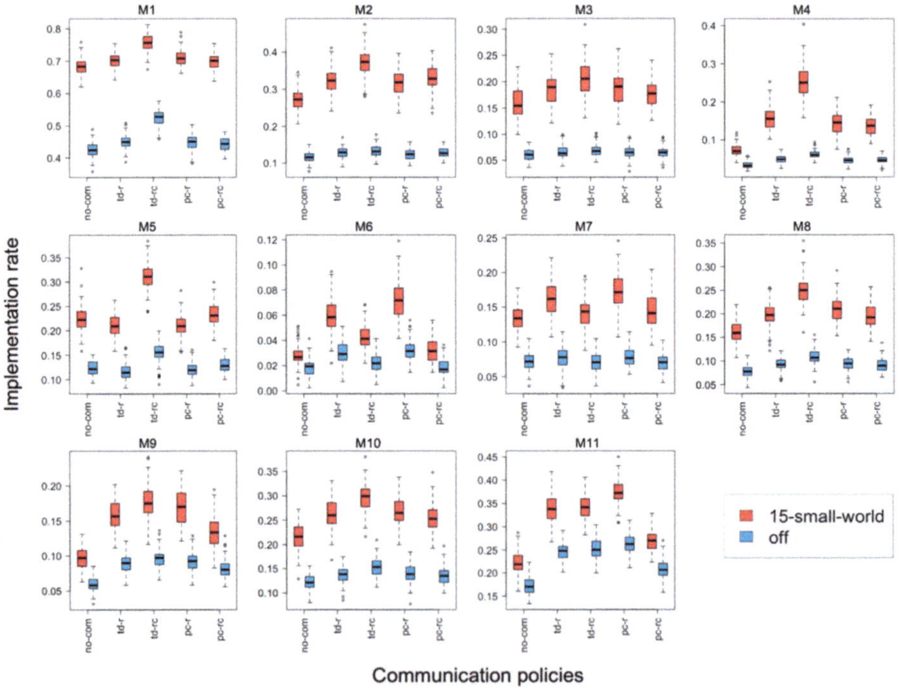

Fig. 15.6 Boxplots of implementation rate at time step 30 in the case of SW networks of in-degree 15 (red) and the case with no network (blue) in area 1. The x-axis is the type of communication policy

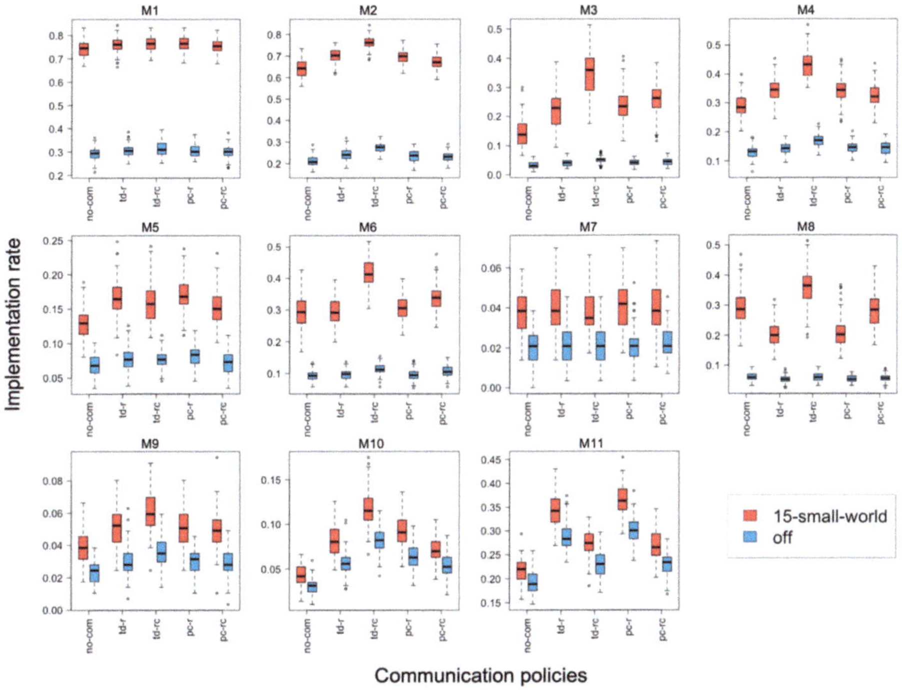

Fig. 15.7 Boxplots of implementation rate at time step 30 in the case of SW networks of in-degree 15 (red) and the case with no network (blue) in area 2. See description of Fig. 15.6

SW networks of in-degree 15 and the case with no networks (Wilcoxon rank-sum test). The implementation rates tended to be underestimated when social networks were not considered. Their results showed that social networks promoted the diffusion of flood adaptation measures. This also indicates that the design of flood risk communication strategies must consider the diffusion of information and behaviors through social networks.

Haer et al. (2016) highlighted that social media was effective in promoting the diffusion of flood adaptation measures in the Netherlands based on the finding that an individual's social network had a significant effect on whether or not an individual adopted flood adaptation measures. Our study yielded similar results. Therefore, social media can be effective in spreading flood adaptation measures in Japan. In addition, in our research, top-down communication policies were effective in Japan; therefore, a strategy based on the two-step flow model, which has been studied for many years in the field of media research (Lazarsfeld et al. 1944; Katz and Lazarsfeld 1955), can be effective in promoting the diffusion of flood adaptation measures in Japan. Based on the two-step flow model, information disseminated by the mass media was first transmitted to a small number of opinion leaders (also known as influencers) and then spread to many other people through social networks. Therefore, leveraging the influence of opinion leaders can effectively disseminate information and behavior (Rogers 1995). This method of promoting dissemination through the influence of opinion leaders will be effective in disseminating flood adaptation measures in Japan.

15.3.5 Effects of the Degree of Social Networks

The implementation rates of adaptation measures M1–M4 at time step 30 in areas 1 and 2 for different in-degrees of social networks are shown in Fig. 15.8. For adaptation measure M1 in area 1, the implementation rate barely changed with increasing in-degree (no significant difference in the Wilcoxon rank-sum test), which is consistent with Erdlenbruch and Bonte (2018). However, for adaptation measure M1 in area 2, the implementation rate increased as the in-degree increased. In addition, some adaptation measures (M2–M4 in area 1, and M3 and M4 in area 2) decreased as the in-degree increased. Under these conditions, the implementation rates were significantly different between different in-degrees (Wilcoxon rank-sum test). Therefore, social networks promote the spread of adaptation measures, as mentioned in the previous section, but denser social networks (i.e., social networks with higher degrees) do not necessarily facilitate a greater spread of adaptation measures.

The promotion of adaptation diffusion with an increase in in-degree occurs because the shortest path length between nodes in the network is reduced owing to an increase in in-degree. However, the impediment to the spread of adaptation measures owing to an increase in in-degree can be explained as follows: The flood adaptation diffusion model used in this study can be regarded as a type of linear threshold model (LT model; Centola and Macy 2005, 2007), which has been used in theoretical research on the diffusion dynamics of networks. In the LT model, the probability

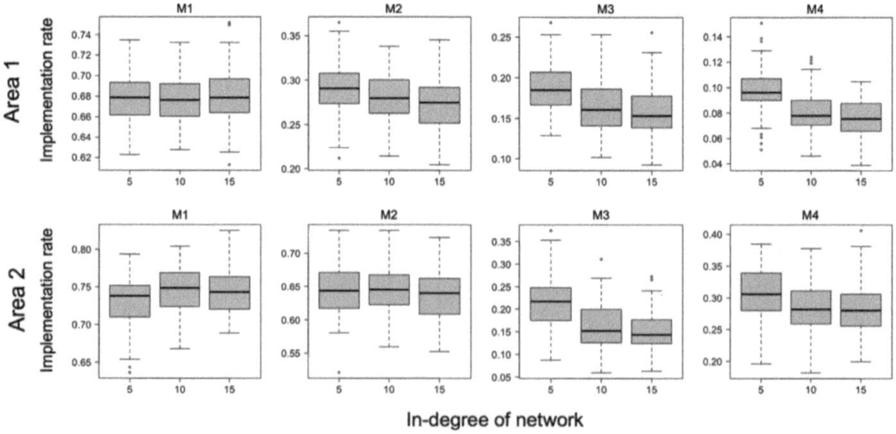

Fig. 15.8 Boxplot of implementation rate at time step 30 for each in-degree condition. The x-axis is the in-degree condition. These results are for no communication (no-com) conditions

that a node adopts information or behavior increases with the ratio of active neighboring nodes (individuals who adopt a specific behavior or have information) to the number of neighboring nodes. Previous studies have demonstrated that an increasing degree can inhibit diffusion in LT models (Watts 2002). This is because, as the degree increases (i.e., as the number of neighboring nodes increases), the ratio of the number of active neighbors to the number of neighboring nodes decreases, and the probability of adopting information or behavior also decreases. The spread of adaptation measures was hindered as the in-degree increased for the same reason.

15.4 Conclusion

This study investigated effective communication policies in Japan by simulating the diffusion of individual-level adaptation measures across social networks. Our findings revealed that, unlike in the Netherlands and France, top-down communication policies, especially those targeting both threat and coping appraisals, were more effective than people-centered approaches in Japan. However, the influence of social networks on the adoption of flood adaptation measures was significant across all studied regions. While our empirical decision model and agent-based approach provided valuable insights, several challenges for improvement remain. Future studies should focus on improving the representation of social networks in agent-based models. While our study used small-world networks, exploring other network structures and their impact on adaptation measure diffusion could yield valuable insights. Furthermore, the integration of real-world social media data could enhance the realism of simulated social interactions.

References

Agner WN et al (2009) Are there social limits to adaptation to climate change? Clim Chang 93:335–354

Bubeck P, Botzen WJW, Kreibich H, Aerts JCJH (2013) Detailed insights into the influence of flood-coping appraisals on mitigation behaviour. Glob Environ Chang 23:1327–1338

Centola D, Macy M (2005) Complex contagions and the weakness of long ties. Cornell University. https://courses.cit.cornell.edu/info435_2006sp/w13/WLT.pdf. Accessed 13 July 2024

Centola D, Macy M (2007) Complex contagions and the weakness of long ties. Am J Sociol 113(3):702–734

Erdlenbruch K, Bonte B (2018) Simulating the dynamics of individual adaptation to floods. Environ Sci Pol 84:134–148

Haer T, Botzen WW, Aerts JC (2016) The effectiveness of flood risk communication strategies and the influence of social networks-insights from an agent-based model. Environ Sci Pol 60:44–52

Hyogo Prefecture Typhoon No 23 Disaster Verification Committee (2005) Typhoon No 23 disaster verification report https://web.pref.hyogo.lg.jp/kk37/pa18_000000001.html. Accessed 10 Apr 2024 (in Japanese)

IPCC (2012) Managing the risks of extreme events and disasters to advance climate change adaptation. Cambridge University Press, Cambridge and New York
Japan Federation of Housing Organizations (2021) Guidance for designing flood countermeasures for houses Japan Federation of Housing Organizations, Tokyo (in Japanese)
Katz E, Lazarsfeld FP (1955) Personal influence; the part played by people in the flow of mass communications. Free Press, Glencoe, IL
Kiuchi N, Nakano T (2023) Studies on flood proofing plans of buildings and their cost-effectiveness, vol 153. Report of the Building Research Institute. (in Japanese)
Konami T, Koga H, Kawatsura A, Matsuyoshi K, Ida N (2020) Research on the proper flood information service based on the flood by typhoon Hagibis in 2019 in the upper Abukuma River Basin Japanese. J JSCE B1. (water engineering) 76(1):315–322 (in Japanese):315
Kunreuther HC, Meyer R, Michel-Kerjan EO (2013) Overcoming decision biases to reduce losses from natural catastrophes. In: Shafir E (ed) The behavioral foundations of public policy. Princeton University Press, Princeton, pp 398–414
Kyoto Prefecture (2005) Records of Typhoon No 23 disaster in 2004 https://wwwbousaiprefkyotolgjp/kikiweb/data/typhoonhtml. Accessed 10 Apr 2024 (in Japanese)
Lazarsfeld FP, Berelson B, Gaudet H (1944) The people's choice: how the voter makes up his mind in a presidential campaign. Columbia University Press, New York
Lo AY (2013) The role of social norms in climate adaptation: mediating risk perception and flood insurance purchase. Glob Environ Change 23:1249–1257
Ministry of Land, Infrastructure, Transport and Tourism Tohoku Regional Development Bureau (2020) Overview of the flooding caused by the 2019 East Japan typhoon and the Abukuma River emergency flood control project. https://www.thrmlitgojp/fukushima/kasen_seibi/2020dai14kai/06-14-siryou2-1pdf. Accessed 10 Apr 2024 (in Japanese)
O'Brien K, Eriksen S, Sygna L, Naess LO (2006) Questioning complacency: climate change impacts, vulnerability, and adaptation in Norway. Ambio 35(2):50–56
Richert C, Erdlenbruch K, Figuières C (2016) The determinants of households' flood mitigation decisions in France—on the possibility of feedback effects from past investments. Ecol Econ 131:342–352
Rogers RW (1975) A protection motivation theory of fear appeals and attitude change. J Psychol 91(1):93–114
Rogers E (1995) Diffusion of innovations. The Free Press, New York
Sayama T, Tachikawa Y, Takara K, Masuda A, Suzuki T (2008) Evaluating the impact of climate change on flood disasters and dam reservoir operation in the Yodo river basin. JSHWR 21(4):296–313. (in Japanese)
Tachikawa Y, Takino S, Fujioka Y, Yorozu K, Kim S, Shiba M (2011) Projection of river discharge of Japanese river basins under a climate change scenario. Proc JSCE B1 (water engineering) 67(1):1–15. (in Japanese)
Watts DJ (2002) A simple model of global cascades on random networks. PNAS 99(9):5766–5771
Watts DJ, Strogatz SH (1998) Collective dynamics of 'small-world' networks. Nature 393(6684):440–442

Chapter 16
Drought Risk and Agricultural Water Use: Changes in Water Resources and the Optimal Rice Growing Period

Takeo Yoshida and Asari Takada

Abstract Changes in hydrological processes due to climate change will have various potential impacts on water resources and related human activities. To prevent crop failure caused by high temperatures, farmers are likely to adopt adaptation measures that could alter regional water use patterns. However, the implementation of these measures may be hindered by other factors, even when their benefits are recognized: "soft adaptation limits." We introduce a novel framework for predicting the occurrence of soft adaptation limits in typical Japanese watersheds, where irrigation water use is predominant. The framework integrates two process-based models—a crop model and a watershed hydrology model—designed to evaluate water resources and rice production. This framework successfully predicted potential soft adaptation limits that may arise when implementing adaptation measures. We applied this framework to other regions, which enhanced our understanding of soft adaptation limits, yielding more robust insights based on varying climate and regional characteristics.

Keywords Soft adaptation limit · Irrigation · Rice cultivation · Crop failure · Water resources · Process-based models

16.1 Introduction

Changes in hydrological processes due to climate change will have various potential impacts on water resources and associated human activities. Agricultural water use accounts for approximately 70% of water use worldwide, and its proper management is key to sustainable development. Projections of the impacts of climate change on water resources indicate that heavy snowfall areas in the temperate zone of Japan are markedly vulnerable to temperature increases; they are expected to experience a large reduction in water resources in earlier snowmelt (Kudo et al.

T. Yoshida (✉) · A. Takada
National Agriculture and Food Research Organization, Tsukuba, Japan
e-mail: takeoys@naro.affrc.go.jp

N. Mimura, S. Takewaka (eds.), *Climate Change Impacts and Adaptation Strategies in Japan*, https://doi.org/10.1007/978-981-96-2436-2_16

2017). However, these approaches have overlooked some critical points. Because high temperatures have direct impacts on crops, to avert crop failure under climate change, farmers are likely to implement various adaptive actions that may alter regional water use patterns. Various adaptation measures have been proposed to moderate the negative impacts of climate change, including water management, fertilizer management, transplantation date shift, switching of crop type, and agricultural insurance. Among the options, changing the rice transplantation date is deemed relatively easy to implement; however, implementation of these measures may be hampered by other factors even if the benefits are acknowledged—a situation termed "soft adaptation limits" by the IPCC (IPCC 2022).

We therefore built a framework for predicting the occurrence of soft adaptation limits in typical Japanese watersheds dominated by irrigation water use. The framework, which consists of two process-based models, a crop model and a watershed hydrology model, was designed to evaluate the water resources and rice production. Throughout this work, we ask the following questions: How does an effective measure in one sector (agriculture) influence the other sector (water resources)? How does climate change affect the relationship between agricultural outcomes and water resources? and Does the proposed framework help us understand under what circumstances a soft adaptation limit will occur?

16.2 Water–Rice Coupled Systems in Japan

16.2.1 Relationship Between Water Resources and Rice Cultivation

Paddy rice cultivation expanded rapidly in the seventeenth century, when local governments attempted to develop as many rice paddies as possible wherever irrigation water and suitable land were available (Satoh and Ishii 2021). Irrigation methods using open channels were developed at this time and are still common today. When open channels are used to carry or divert water to branch channels, farmers must keep water levels in the open channels high, and as a result, extra water is required, above the actual water requirements of the crop. Current irrigation systems using open channels therefore have a high proportion of return flow from irrigated areas back to rivers (Yoshida et al. 2016). During drought, irrigation water is generally restricted by 10–30% before any restrictions are placed on other water users. In response to such restrictions, farmers adjust flows to the branch channels every day to change water diversion routes. As a result, each paddy field is supplied with water at intervals of several days, preventing water stress on the paddy rice even during the restriction period. Thus, although drought generally causes water stress and yield loss of paddy rice globally, these negative effects are less likely to occur because of these efforts by farmers.

Overall, the tightly coupled system of water and rice cultivation is based on mature irrigation systems (hereafter, water–rice coupled systems) and an established administrative framework for water resources management. Therefore, the impact of a water deficit is not direct (e.g., water stress leading to decline of rice yield) but indirect (i.e., difficulties in setting appropriate water rights for rice cultivation). Changes in meteorological conditions under climate change may increase the amount of water that needs to be withdrawn for paddy rice cultivation. Because of substantial amounts of water for withdrawal, changes in the water use period may lead to conflicts with other water users and disrupt the availability of streamflow. Therefore, because the water–rice systems are limited not by the availability of sufficient water for withdrawal but by the period of water withdrawal for irrigation, a soft adaptation limit may occur if water is not available for withdrawal during the optimum rice-growing period.

16.2.2 Impacts of Climate Change on Rice Cultivation and Water Resources

Rice cultivation and water resources are both affected by climate change, and efforts to assess the individual impacts of climate change on both have been carried out. High temperatures during the heading period can reduce the appearance quality of rice by increasing the proportion of white immature grains (Ishigooka et al. 2011). The proportion of white immature grains in a given number of grains is one criterion for determining the rice grade (i.e., first- or second-grade rice). Adaptation measures to moderate the negative impacts of climate change on rice quality are attracting considerable attention from farmers and government. Shifting the transplantation date is both relatively inexpensive and easier to implement than other adaptation measures. However, the shifting the transplantation date has assumed that sufficient irrigation water will be available and that it will be possible to set withdrawal periods freely if transplantation dates are shifted under climate change. However, as mentioned in Sect. 16.2.1, whether it is possible to change the period of water withdrawal to implement a shift in the transplantation date is dependent on river conditions. Therefore, the feasibility of adaptation measures for maintaining rice quality must consider the availability of water resources.

16.3 Materials and Methods

16.3.1 Evaluation of the Soft Adaptation Limit

We present a novel framework for predicting soft adaptation limits under climate change based on the assumption that shifts in the transplantation date water–rice coupled systems are related to the occurrence of the soft adaptation limits. We focused two stakeholders of water–rice coupled systems: farmers and river administrators. We assumed that the only measure taken by farmers to adapt to climate change was a shift in the transplantation date and that the area occupied by rice paddies and the cultivars used would mostly remain unchanged. We compiled two process-based models, one to evaluate the risk of a water deficit and the other to evaluate the impacts of the transplantation date shifts on rice production (e.g., rice yield and quality). Then, we conducted an experiment in which we shifted the transplantation date applied by each model by 1 week for up to 5 weeks before and after the current date under the same climate change scenarios. The period of water withdrawal was also shifted in accordance with the transplantation date shift. Lastly, we calculated the rice production and drought risks for each transplantation date.

We examined the adaptation options from the relationships between benefits and risks were synergistic or trade-off relationships (Fig. 16.1). If the adaptation option leads to a relationship whereby the risk decreases as the benefit increases (i.e., the slope of the relationship is negative), then the relationship can be described as "synergistic" (Fig. 16.1a). If, however, the adaptation option increases both the risk and the benefits, the result is a "trade-off" relationship (Fig. 16.1b). At a predefined threshold level of risk (e.g., water deficit level), the river administrators do not allow actions that lead to a further increase in the risk (horizontal dashed orange line). Thus, of these two relationship types, the trade-off relationship would very likely face a soft adaptation limit (Takada et al. 2024a). To predict soft adaptation limits in water–rice coupled systems, we hypothesized two possible patterns of transplantation date changes based on the past data: One pattern was a trend toward earlier transplantation dates, as in the data from the 1950s to about 2000 (i.e., a trend toward higher rice yields). The other was a trend toward later transplantation dates, as in the data from the 2000s and later (i.e., a trend toward a greater appearance quality of rice).

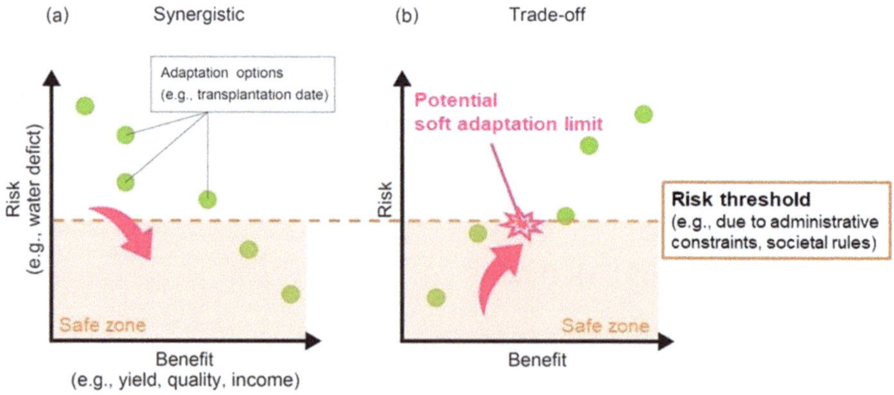

Fig. 16.1 Two types of agriculture–water resource relationships resulting from adaptation options. (Reproduced from Takada et al. (2024a))

16.3.2 Climate Change Scenarios

We used general circulation model (GCM)-based historical or scenarios that projected outcomes up to 2050 based on Representative Concentration Pathways (RCPs) 2.6 and 8.5. We chose to limit our future projections to up to the year 2050 because a shift in the transplantation date is an incremental adaptation measure (Iizumi 2019), and we assumed that other, more transformative adaptation measures would be implemented after 2050. We used the outputs of three GCMs, MIROC5, MRI-CGCM3, and HadGEM2-ES, for the case study in the Shinano River Watershed (Takada et al. 2024a). For application of the model to the entire Japanese archipelago, we used the data of the five GCM of CMIP6 (MIROC6, MRI-ESM2-0, ACCESS-CM2, IPSL-CM6A-LR, MPI-ESM1-2-HR) for the historical (1980–2014) and future (2015–2050) with three RCP scenarios (2.6, 4.5, and 8.5).

16.3.3 Process-Based Model Used to Evaluate the Risk of Water Deficit

We used the distributed water circulation model developed by Yoshida et al. (2016) to evaluate the risk of water deficit. The model represents a watershed on a 1-km grid-cells and simulates the streamflow in each grid cell. Furthermore, the model simulates water circulation processes in paddy irrigation districts by water-use facilities such as reservoir management, water withdrawal from rivers, and water allocation, enabling the simulation of streamflow that is greatly disturbed by irrigation systems (Yoshida et al. 2016). A hydrological drought can be evaluated as occurring when the flow rate falls below the minimum flow requirement. The cumulative amount of streamflow below this threshold during the irrigation period (hereafter water deficit) was calculated annually.

16.3.4 Process-Based Model to Evaluate the Rice Production

We used the process-based rice growth model developed by Ishigooka et al. (2011) to assess the rice production benefits resulting from a shift in the transplantation date. This model has three major components: phenological development, biomass production, and yield formation. We assumed that rice growth was not affected by the availability of water resources, because water stress on rice growth is relatively unlikely to occur owing to human-controlled mechanisms that are implemented to prevent water stress on paddy rice (see Sect. 16.2.1).

We used two indices to evaluate the rice production benefit: total yield and appearance quality. Total yield was calculated by the rice growth model. Appearance quality was estimated based on the heat stress index for rice quality, defined by Ishigooka et al. (2011). The heat stress index is related to the emergence of chalky grains due to high temperatures—that is, to deterioration of the appearance quality of the rice. Among the three classes based on the heat stress index, we used the yield of the lowest class (hereafter Class A yield) as the indicator of appearance quality in our evaluation.

16.4 Case Study I: Shinano River Watershed

The Shinano River is the longest and the third largest watershed in Japan (Fig. 16.2). The lower part of the watershed is one the heaviest snowfall areas. We included the operation of 29 major reservoirs (multipurpose, municipal, hydropower, and irrigation reservoirs) and 88 irrigation districts in the watershed. The amounts and periods of water withdrawal in the irrigation districts were modeled based on the recorded withdrawal rates over 5 years (2011–2015).

We used the Ojiya gauging station as the reference point for water use in the middle and lower areas of the Shinano River watershed. The minimum flow requirement defined by the river administrator during the irrigation period (from 28 April to 15 September) is 145 m^3/s. This minimum flow requirement takes account of the aggregated water rights of all downstream water users (domestic and industrial users, irrigation, fisheries, and the environment). The rice yield and cultivation

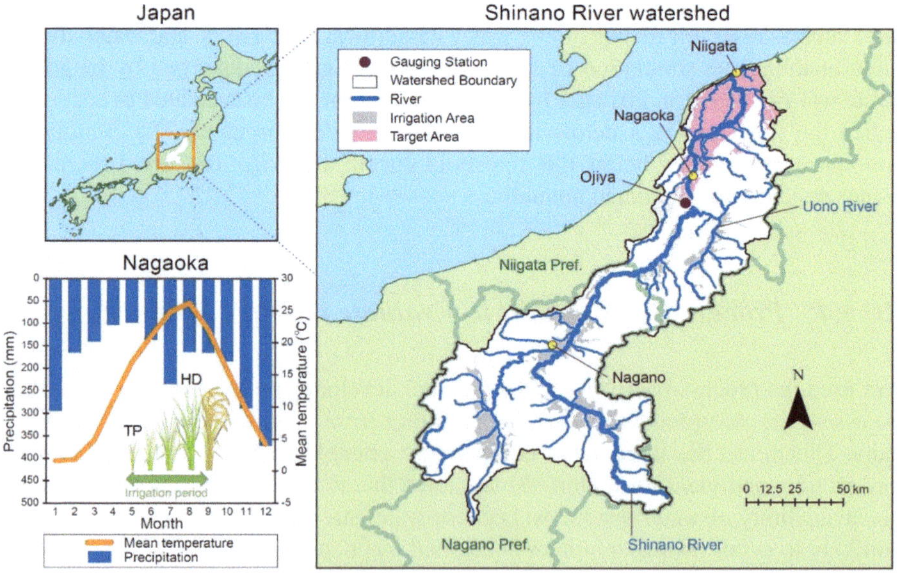

Fig. 16.2 The Shinano River watershed and the target irrigation area. The bottom-left panel shows the monthly mean temperature (orange line) and precipitation (blue bars) and the stage of the rice-growing cycle at Nagaoka. TP and HD denote transplanting and heading, respectively. (Reproduced from Takada et al. (2024a))

schedules (dates of sowing, transplantation, heading, and harvesting) are summarized in an annual statistics (hereafter crop statistics, published by Ministry of Agriculture, Forestry and Fisheries) at the prefectural level until 1968 and by sub-administrative regions after 1969. The transplantation date in our target area, 9 May, is used as the "current transplantation date" (Takada et al. 2024a).

16.4.1 Water–Rice Relationship Under the Historical Scenario

We first examined the relationships between the water deficit risk and the rice production benefit (i.e., total and Class A yields) in the historical scenario (Fig. 16.3). The relationship between the water deficit and the total yield is synergistic (Fig. 16.3a), while the relationship between the water deficit and the Class A yield shows a trade-off (Fig. 16.3b).

16.4.2 Soft Adaptation Limits

The relationships between the water deficit and total yield under the RCP 2.6 and 8.5 climate change scenarios for 2011–2030 and 2031–2050 are compared with the results under the historical (1981–2000) scenario in Fig. 16.4. Under all future

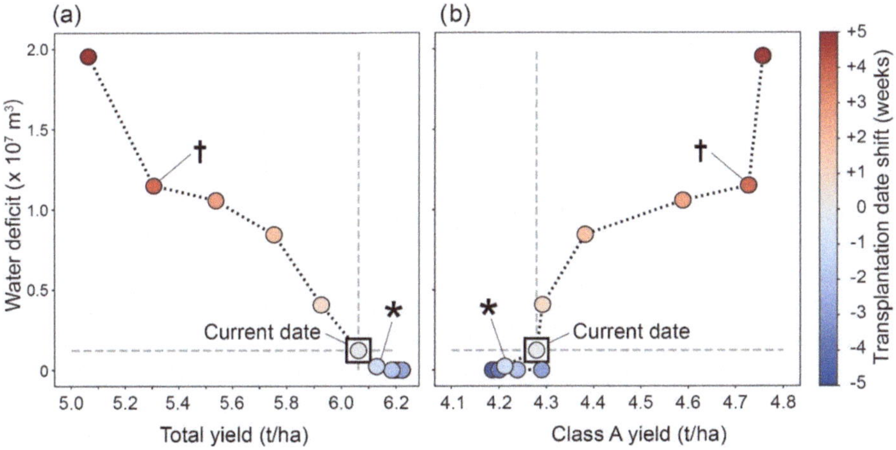

Fig. 16.3 Relationships under the historical scenario (1981–2000) between (**a**) the water deficit and total yield and (**b**) the water deficit and the Class A yield resulting from shifts in the current transplantation date (square). The vertical and horizontal dashed gray lines indicate the total or Class A yield and the water deficit, respectively, on the current transplantation date. † and * indicate the peak transplantation date in the 1950s and in 2000, respectively. The former corresponds to a transplantation date shift of +4 weeks, and the latter corresponds to a shift of −1 week. (Reproduced from Takada et al. (2024a))

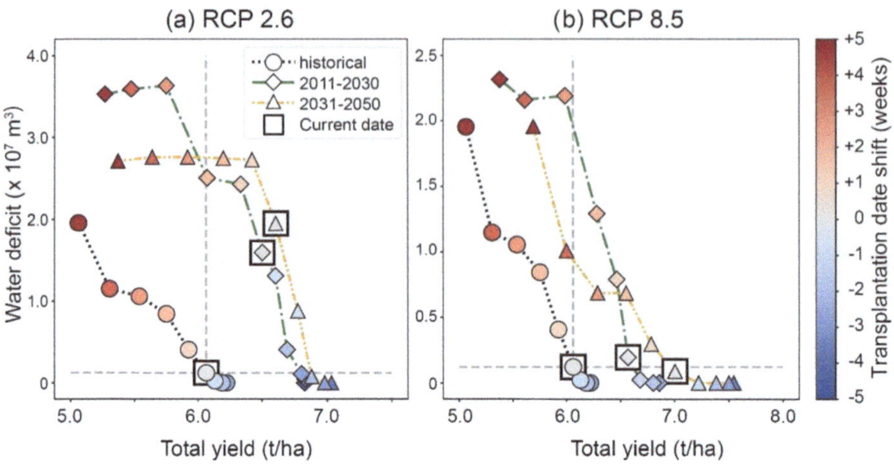

Fig. 16.4 Relationships between the water deficit and total yield for 2011–2030 and 2031–2050 with shifts of the current transplantation date (squares) under the (**a**) RCP 2.6 and (**b**) RCP 8.5 scenarios, and that under the historical scenario (1981–2000). (Reproduced from Takada et al. (2024a))

scenarios, the relationships between the water deficit and total yield were synergistic as in the historical scenario. With the current transplantation date, the total yields in all future scenarios were higher than that in the historical scenario. In the RCP 2.6 scenario for both 2011–2030 and 2031–2050, water deficits were higher than in the historical scenario (Fig. 16.4a), whereas the water deficits in the RCP 8.5 scenario did not differ substantially from that in the historical scenario (Fig. 16.4b). These results indicate that transplantation date shifts to increase total yield in the future may not be limited by the water deficit risk.

We also plotted the relationships between water deficit and Class A yield under the climate change scenarios (Fig. 16.5). In all future scenarios, a trade-off relationship was observed as in the historical scenario. Class A yields with the current transplantation date decreased in the future under both scenarios, but in 2011–2030, the same yield as the historical one could be ensured by shifting the transplantation date; a 5-week delay in the transplantation date under both the RCP 2.6 and RCP 8.5 scenarios for 2011–2030 († in Fig. 16.5) would ensure yields, which are comparable to the yield with the current transplantation date under the historical scenario. However, delaying the transplantation date by 5 weeks in 2011–2030 caused the water deficits to be approximately 2.2 and 11.8 times greater, respectively, than those with the current transplantation date. Furthermore, during 2031–2050, a 5-week delay from the current transplantation date would not ensure a Class A yield as large as that with the current transplantation date under the historical scenario, and the water deficit would also increase relative to that under the historical scenario (* in Fig. 16.5). These results indicate that in the future delaying the transplantation date to increase the Class A yield may be hampered by an increased water deficit.

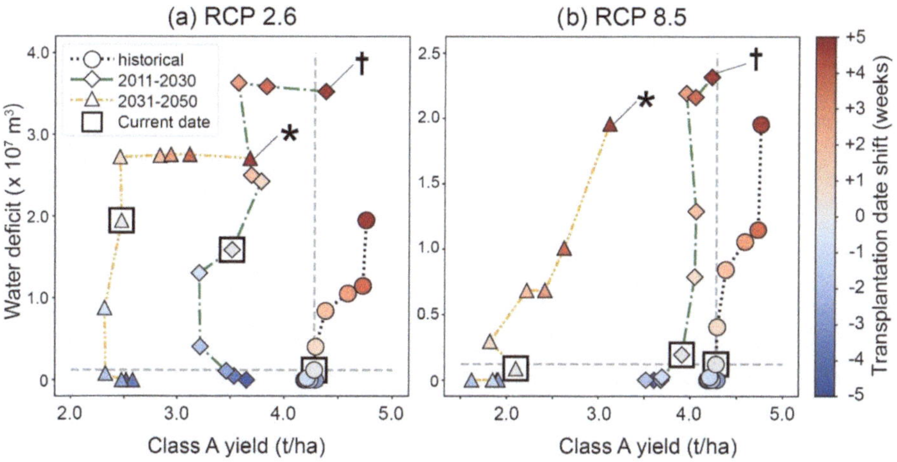

Fig. 16.5 Relationships between water deficit and Class A yield with shifts of the current transplantation date (squares) under the (**a**) RCP 2.6 and (**b**) RCP 8.5 scenarios in 2011–2030 and 2031–2050. The dashed lines indicate the water deficit and Class A yield with the current transplantation date under the historical scenario (1981–2000). † and * indicate a 5-week delay in the current transplantation date in 2011–2030 and 2031–2050, respectively. (Reproduced from Takada et al. (2024a))

We posit that there are soft adaptation limits in water–rice coupled systems, whereby the current yield or quality of rice can be ensured by shifting the transplantation date, but the shifted transplantation date is not acceptable from the perspective of the water deficit. If farmers decide the optimal transplantation date with the aim of maximizing yield, those shifts would not be restricted by water resources (Fig. 16.4). However, if the farmers decide to aim at maximizing quality, then soft adaptation limits would occur in 2011–2030 (Fig. 16.5). During 2031–2050, the shifted transplantation date cannot ensure the same level of rice quality as at present (Fig. 16.5), which can be categorized as a hard adaptation limit. Our predictions suggest that whether soft adaptation limits occur depends on the farmers' decisions regarding the transplantation date (Takada et al. 2024).

16.5 Case Study II: 77 Watersheds in Japan

We evaluated the agriculture–water resource relationship at 77 reference gauging stations for water use and sub-administrative regions. Water deficit was calculated based on stream flow during the irrigation period. Data on peak transplanting and heading periods in 2000 for the sub-administrative regions were obtained from crop statistics. We evaluated the changes in the water deficit in a 10-year return period for historical (1980–2014) and projected (2015–2050) scenarios. In Fig. 16.6, the change ratio in the water deficit is mapped onto the 77 watersheds: watersheds with

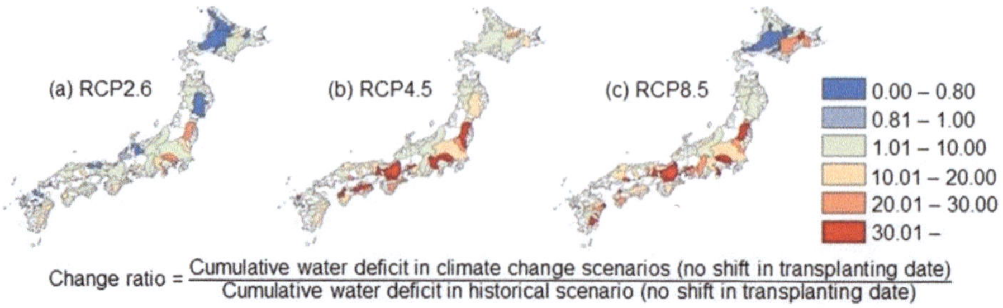

Fig. 16.6 Change ratio of the cumulative water deficit in the climate change scenarios (2015–2050) relative to that in the historical scenario (1980–2014) with no shift in the transplantation date for (**a**) RCP2.6, (**b**) RCP4.5, and (**c**) RCP8.5. (Reproduced from Takada et al. (2024b))

negative values (blue) show a decreasing drought risk, whereas those with positive values (red) show an increasing drought risk. Without a change in the transplantation date, the simulation with RCP2.6 scenario resulted in a decreased water deficit in 15 watersheds, whereas the RCP4.5 and RCP8.5 scenarios resulted in an increased drought risk in almost all watersheds.

To assess the impact of changes in the transplantation date on drought risk, we selected the transplantation date resulting in the highest yield under the climate change scenarios (2015–2050). Then, we compared the water deficit under the projected scenarios between the simulations with the shifted transplantation dates and those with the current dates. Figure 16.7 shows the changes in water deficit simulated with the selected transplantation dates relative to those simulated with the current dates for the period from 2015 to 2050. Hot colors indicate increased drought risk, while cool colors indicate lower risk. These results indicate that transplantation date shifts are relatively easier to implement in regions where the water–rice relationship is synergistic, whereas their implementation may be hindered in regions with a trade-off relationship.

We classified watersheds into synergistic and trade-off based on the slope of linear regression lines fitted to the water rice plots (Fig. 16.8).Watersheds for which the slope was negative (i.e., synergistic relationship; n = 38) are colored light blue and those for which the slope is positive (i.e., trade-off relationship; n = 37) are colored red. A clear geographic distinction can be seen between watersheds with synergistic and trade-off relationships, where the former correspond to regions with heavy snowfall. Advancing the transplantation date generally offers for the crop more chance to grow and is likely to result in higher yields. In regions with heavy snowfall, the snowmelt provides ample streamflow from March to May, thereby reducing the risk of a water deficit during the crop growing period, resulting in synergistic relationships. In contrast, in less snowy regions without ample snowmelt, a trade-off relationship occurs, likely because of the longer water-use periods and higher evaporation loss in summer.

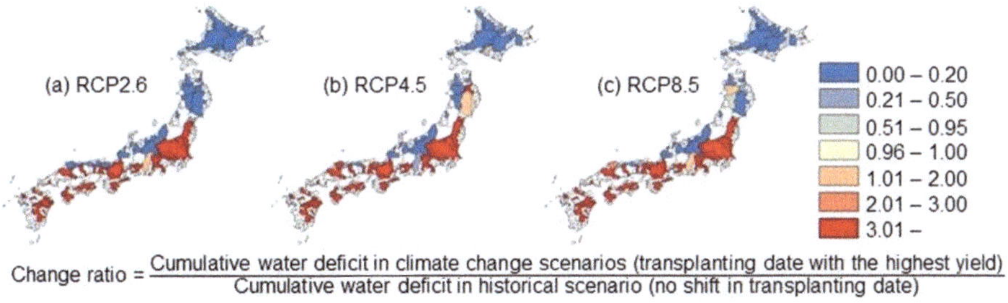

Fig. 16.7 Change ratio of the cumulative water deficit for the transplantation dates with the highest yield under the climate change scenarios (2015–2050) relative to that under the historical scenario (1980–2014) with no shift in the transplantation date for (**a**) RCP2.6, (**b**) RCP4.5, and (**c**) RCP8.5. (Reproduced from Takada et al. (2024b))

Fig. 16.8 Classification of 77 watersheds into synergistic and trade-off based on water–rice relationships. (Reproduced from Takada et al. (2024b))

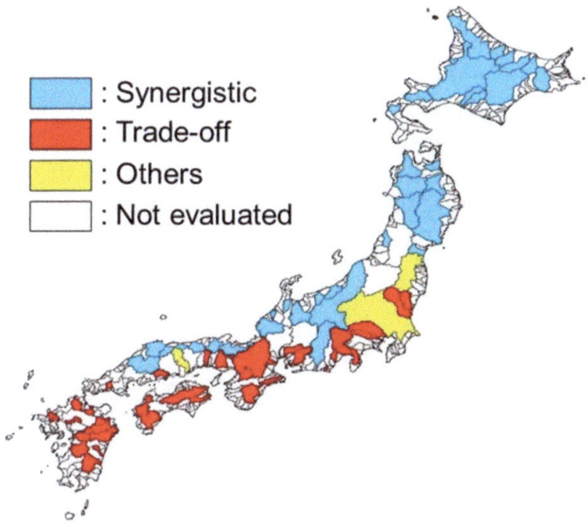

16.6 Conclusion

When we applied the new framework to the Shinano River watershed, a typical Japanese watershed where streamflow is highly disturbed by irrigation systems, we found that with the current transplantation date, the water–rice relationship was synergistic because the water deficit decreased when the rice yield increased. This synergistic relationship between the water deficit and rice yield was apparent under all climate scenarios; therefore, in this case, shifts in transplantation dates to increase the total yield in the future may not be limited by the water deficit risk. In contrast, we found that shifts in the transplantation date to benefit rice quality would result in an increased water deficit risk under all climate change scenarios. Under the RCP 2.6 and RCP 8.5 scenarios, delaying the transplantation date by 5 weeks during

2011–2030 would ensure rice quality comparable to that with the current transplantation date under the historical scenario, but it would result in a greater risk of a water deficit. We posit that this situation represents a soft adaptation limit. Our results thus revealed two contrasting development pathways in the study watershed. Soft adaptation limits imposed by water availability will occur by 2030 if farmers optimize for quality, whereas mutual benefits to farmers and river administrators will be achieved if farmers prioritize yield.

The application of the framework to 77 Japanese watersheds revealed that the water–rice relationship in response to changing transplantation dates differs by region. A synergistic relationship between a water deficit and rice yield was apparent in regions with heavy snowfall; therefore, in these regions, shifts in transplantation dates to increase the total yield in the future may not be limited by the risk of water deficit.

Our proposed framework successfully predicted the soft adaptation limits that may be faced for the implementation of adaptation measures. The novelty of the framework is that it is based on process-based models that use an actual adaptation measure, a shift of the transplantation date. Moreover, societal rules that explicitly govern institutional decisions are employed in the framework as constraints on adaptive actions. Applying this framework to other regions has the potential to enhance our understanding of the occurrence of soft adaptation limits based on more robust results using different climate and regional characteristics.

References

Iizumi T (2019) Emerging adaptation to climate change in agriculture. Adaptation to Climate Change in Agriculture, pp 3–16

IPCC (2022) Climate change 2022: impacts, adaptation and vulnerability. In: Pörtner H-O, Roberts DC, Tignor M, Poloczanska ES, Mintenbeck K, Alegría A et al (eds) Contribution of working group II to the sixth assessment report of the intergovernmental panel on climate change. Cambridge University Press, Cambridge, UK and New York, NY, p 3056

Ishigooka Y, Kuwagata T, Nishimori M, Hasegawa T, Ohno H (2011) Spatial characterization of recent hot summers in Japan with agro-climatic indices related to rice production. J Agric Meteorol 67(4):209–224

Kudo R, Yoshida T, Masumoto T (2017) Nationwide assessment of the impact of climate change on agricultural water resources in Japan using multiple emission scenarios in CMIP5. Hydrol Res Lett 11(1):31–36

Satoh M, Ishii A (2021) Japanese irrigation management at the crossroads. Water Alternatives 14:413–434

Takada A, Yoshida T, Ishigooka Y, Maruyama A, Kudo R (2024b) Effects of shifting rice transplanting date on agricultural water use under climate change in Japan. Jpn J JSCE, 80(16) (in Japanese with English abstract)

Takada A, Yoshida T, Ishigooka Y, Maruyama A, Kudo R (2024a) Potential barriers to adaptive actions in water–rice coupled systems in Japan: a framework for predicting soft adaptation limits. Water Resour Res 60(4):e2022WR034219

Yoshida T, Masumoto T, Horikawa N, Kudo R, Minakawa H, Nawa N (2016) River basin scale analysis on the return ratio of diverted water from irrigated paddy areas. Irrig Drain 65:31–39

Part V
Quality of Life, Human Health, and Urban Systems

Chapter 17
Climate Change and Quality of Life: What Affects the Happiness of Citizens?

Kiyo Kurisu, Kosuke Shirai, and Yoko Imai

Abstract Local municipalities' perceptions of climate change and people's subjective evaluation of quality of life (QoL) are discussed. Climate change has an impact on QoL, and there is a need to address the negative impacts of climate change on QoL. There are various definitions of QoL, including objective conditions and subjective evaluations. The objective conditions are perceived by local municipalities as the severity of climate change impacts and reflect their mitigation and adaptation measures. In this chapter, we use a questionnaire survey to clarify how local municipalities perceive the impacts of climate change and to what extent they are implementing adaptation measures. In addition, we clarify how residents perceive the local environment and how this is related to residents' recognition of happiness. Then, finally, we discuss the relationship between the local municipalities' perceptions of climate change and people's subjective evaluation of QoL.

Keywords Quality of Life (QoL) · Local municipality · Perceived climate change severity · Cluster analysis · Subjective happiness · Structural equation modeling · Questionnaire survey

17.1 Introduction

The IPCC's sixth assessment report (AR6) addresses the impact of climate change on quality of life (QoL). It states that climate change will have a negative impact on QoL due to meteorological disasters, loss of access to nature due to coastal erosion, heat, declining air quality, changes in behavioral patterns, etc. The World Health Organization (WHO) has also pointed out the need to respond to the negative impact

K. Kurisu (✉) · K. Shirai
The University of Tokyo, Tokyo, Japan
e-mail: kiyo@env.t.u-tokyo.ac.jp

Y. Imai
The University of Tokyo, Tokyo, Japan

National Institute of Environmental Studies, Tsukuba, Japan

N. Mimura, S. Takewaka (eds.), *Climate Change Impacts and Adaptation Strategies in Japan*, https://doi.org/10.1007/978-981-96-2436-2_17

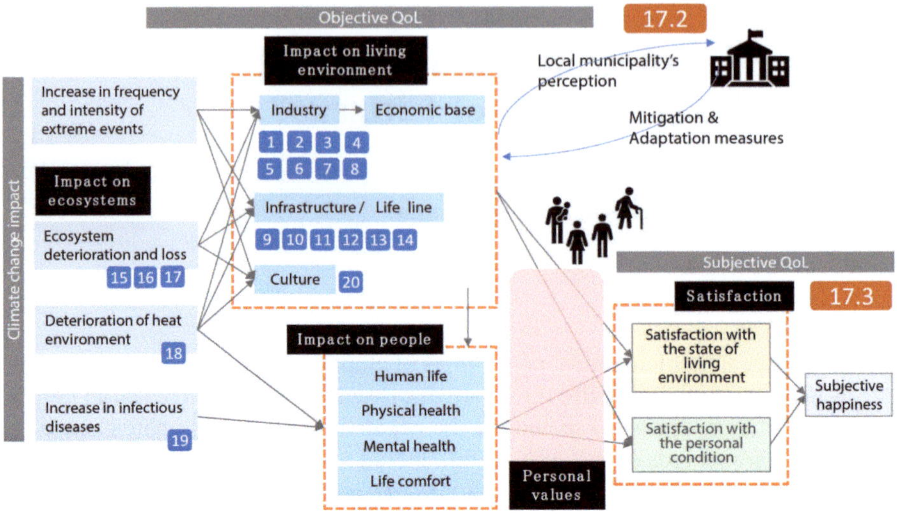

Fig. 17.1 Climate change influence on QoL
The numbers represent the 20 categories affected by climate change that are focused in Sect. 17.2: (1) Agriculture, (2) Forestry, (3) Fisheries and marine products, (4) Manufacturing, (5) Construction, (6) Other local industries, (7) Tourism, (8) Other commerce, (9) Water supply, (10) Sewage, (11) Other urban infrastructure, (12) River disasters, (13) Coastal disasters, (14) Mountain disasters, (15) Terrestrial ecosystems, (16) Freshwater ecosystems, (17) Coastal ecosystems, (18) Heat stroke, (19) Transmitting infections animal increase, and (20) Traditional events.

of climate change on QoL. Figure 17.1 is a schematic diagram showing how climate change threatens people's QoL. Various impacts, such as "increase in frequency and intensity of extreme events," "deterioration of heat environment," "increase in infectious diseases," and "ecosystem deterioration and loss" mentioned in the AR6 WGII SPM, directly affect human life, physical health, mental health, and life comfort, and can also have indirect effects through their impacts on industry, infrastructure, culture, etc.

There are various definitions of QoL, but broadly speaking, it includes objective conditions such as the living environment and personal condition, the subjective evaluation of the degree of satisfaction with those conditions, and the overall evaluation of these, the degree of happiness. The objective conditions are perceived by local municipalities as the severity of climate change impacts and reflect to their mitigation and adaptation measures. Therefore, in the next section, 17.2, we use a questionnaire survey to clarify how local municipalities perceive the impacts of climate change and to what extent they are implementing adaptation measures. In addition, in Sect. 17.3, we use a questionnaire survey to clarify how residents perceive the local environment and how this is related to residents' recognition of happiness. Finally in Sect. 17.4, we discuss the relationship between the two.

17.2 Local Municipality and Climate Change (Imai and Kurisu 2022)

17.2.1 Local Municipalities' Perceptions About Climate Change

In Japan, the Climate Change Adaptation Act was enacted in December 2018, setting out the goal of establishing regional climate change adaptation centers in local municipalities (including prefectures and cities and villages) and formulating regional climate change adaptation plans. Here, local municipalities are considered important actors in promoting climate change adaptation. In this context, it is important to know how each local municipality perceives the impacts of climate change. Through a questionnaire survey of local municipalities, we clarify how each local municipality perceives the impacts of climate change on various fields and their current actions and limitations.

17.2.2 Questionnaire Survey to Local Municipalities

To understand local municipalities' perceptions about climate change impacts, we designed a questionnaire consisting of four parts: (1) perceptions about the severity of climate change impacts, (2) awareness of adaptation and the status of formulating climate change adaptation plans, (3) the status of implementation of adaptation measures, and (4) resources for implementing adaptation measures.

In Japan, there are 47 prefectures under the nation, and 1,741 cities and villages under them. We sent out the questionnaire to these 1,788 local municipalities and requested their responses in December 2020. Of the 1,123 responses collected, 1,098 were valid responses excluding deficiencies (34 from prefectures and 1,064 from cities and villages), giving a response rate of 61.4%.

17.2.3 Survey Results

17.2.3.1 Perceptions About the Climate Change Impacts

Figure 17.2 shows the scores for the perception of the severity of climate change impacts using a six-point scale (six being perceived as the most severe). As shown in Fig. 17.2a, "intensive rainfall," "annual average temperature rise," and "heat waves" were recognized as being highly severe nationwide. In addition, as shown in Fig. 17.2b, the perception of severity was high in four fields: "agriculture," "river disasters," "mountain disasters," and "heat stroke." On the other hand, "drought, water shortage," "sea level rise," and "sea temperature rise" in Fig. 17.2a and

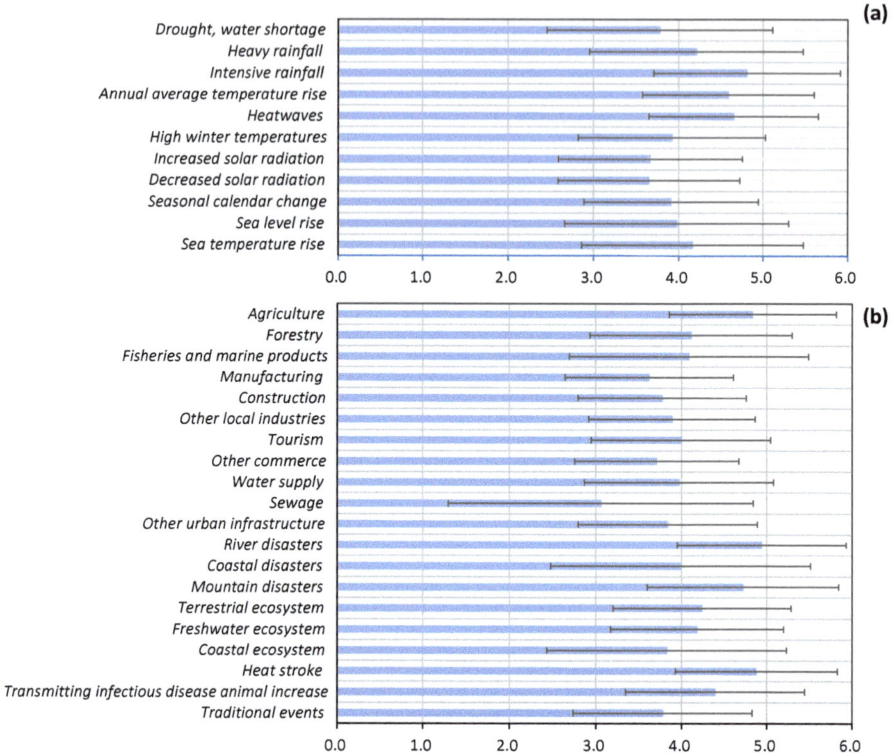

Fig. 17.2 Perception of the severity of climate change. (Adapted from Imai and Kurisu 2022) Responses were asked on a six-point scale ranging from very serious to not at all serious, with very serious scored as 6.0 and not at all serious scored as 1.0. Bars indicate standard deviation.

"fisheries and marine products," "sewage," "coastal disasters," and "coastal ecosystems" in Fig. 17.2b were heavily influenced by geographical factors, such as whether the area faces the sea or is a rainy area, so the standard deviation was large at over 1.3, and there was a wide variation in the scores.

To identify different response trends by municipality, the results of the severity perception by field were subjected to cluster analysis (Ward method). The local municipalities ($n = 1098$) were categorized into six clusters (C1 – C6). The number of local municipalities and the average perception score for each cluster are shown in Fig. 17.3. C3 showed the highest severity for all sectors. Especially for "agriculture," C3 showed the highest severity followed by C4. As seen in Fig. 17.4a, b, the local municipalities categorized in these clusters have higher total agricultural production and agricultural production per capita than other clusters. This high dependency on agriculture would cause their higher perceptions about climate change impact severity on agricultural sector. C3 and C4 showed almost the same severity perception scores for agriculture, river disasters, and mountain disasters, but C4 was slightly lower in other sectors.

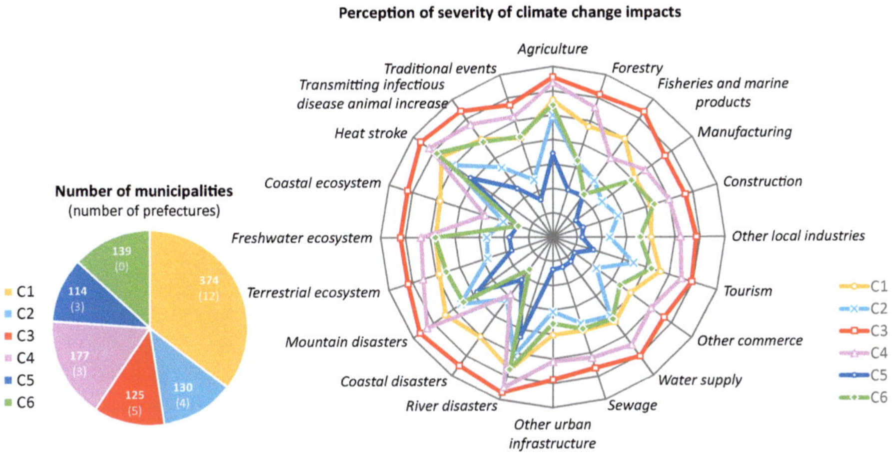

Fig. 17.3 Municipality categorization based on perception of the climate change severity. (Adapted from Imai and Kurisu (2022))
Responses were asked on a six-point scale ranging from very serious to not at all serious, with very serious scored as 6.0 and not at all serious scored as 1.0. The outermost scale line on the radar chart is 6.0, which is the highest score for perceived severity.

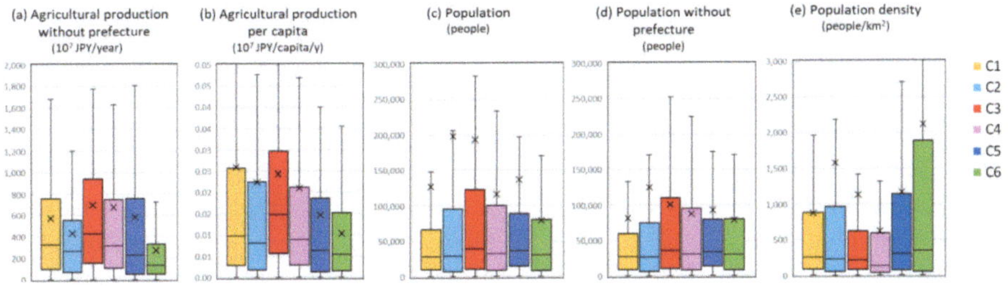

Fig. 17.4 Agricultural production and population data in each cluster
X shows the average value of each cluster. Data for prefectures are not included in (**a**) and (**d**) because the prefectures are large and the scale effects appear in the data.

As many inland local municipalities were classified into C4, the recognition of the severity was lower in coastal sectors. Similar effects of geographical conditions were also seen in C2, C5, and C6. These clusters are also located in inland areas. C2 and C5 are clusters showing the lower severity recognition than other clusters. As seen in Fig. 17.4, C2 has higher total population, higher population density, and relatively lower agricultural production. This indicated that these municipalities are relatively urbanized areas. C5 showed the lowest scores for all sectors especially for industries and infrastructure.

17.2.3.2 Knowledge About Climate Change Adaptation and the Status of Adaptation Planning

Figure 17.5 shows the surveyed results about the knowledge of climate change adaptation and the status of adaptation plan formulation for each cluster. Regarding adaptation to climate change, 9.7% of the local municipalities responded that they had never heard of it before or were not aware of it, and 37.3% responded that they were not very familiar with it, indicating that approximately half of the local municipalities do not have sufficient knowledge about adaptation as of 2020. In C3 and C4, the proportion of the local municipalities that responded that they were "well known" or "somewhat known" exceeded 60%, showing a high proportion. In addition, regarding adaptation plans, C3 and C4 had a higher proportion of local municipalities that had "already formulated" or "in preparation" than the other clusters, indicating that these clusters, which have a high perception of the severity of the climate change impacts, also have a high level of knowledge about adaptation and have formulated adaptation plans. On the other hand, C5 showed the lowest adaptation knowledge and the lowest implementation of adaptation center, which corresponded to their lowest recognition about the climate change severity.

The implementation status of adaptation measures was asked by a six-point scale using "already implemented," "planned to be implemented within this year," "planned to be implemented next year or later," "no plan but under consideration," "no plan," and "don't know," and each response was given a score from 6.0 to 1.0. Figure 17.6 shows the implementation status of the 15 adaptation measures, divided into the local municipalities that have formulated or are preparing adaptation plans and those that have not. While soft-side measures such as "development of warning systems," "development of hazard maps," and "raising public awareness about disaster prevention" are being implemented, many local municipalities have not implemented hard-side measures such as "facility improvement related to drought," "introduction of green infrastructure," and "facility improvement related to heat environment." When comparing by the status of adaptation plan formulation, the implementation rates of all measures were high in the local municipalities that had already formulated or were in the process of formulating adaptation plans.

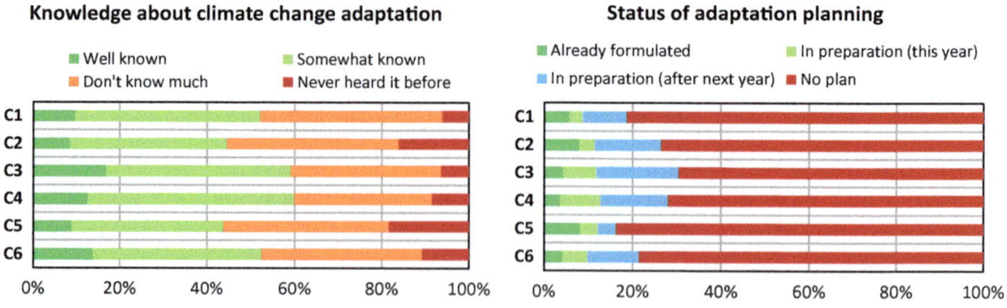

Fig. 17.5 Knowledge about climate change adaptation and the status of adaptation planning. (Adapted from Imai and Kurisu (2022))

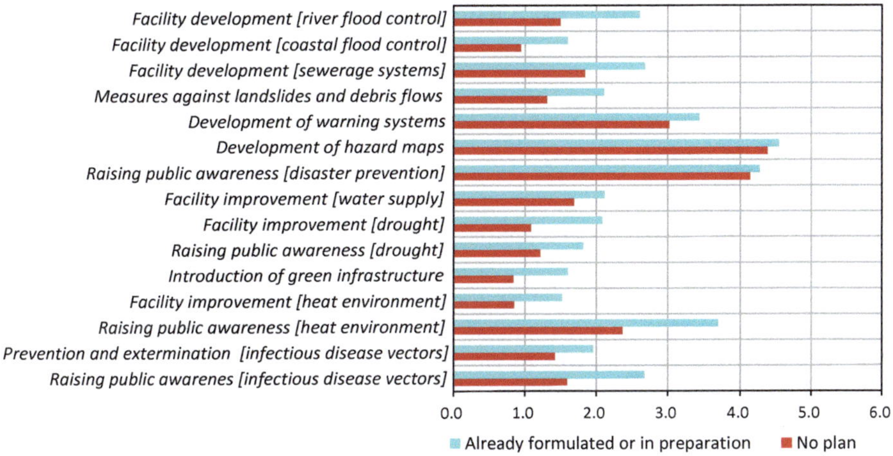

Fig. 17.6 Relationship between adaptation plan formulation status and adaptation measure implementation. (Adapted from Imai and Kurisu 2022)

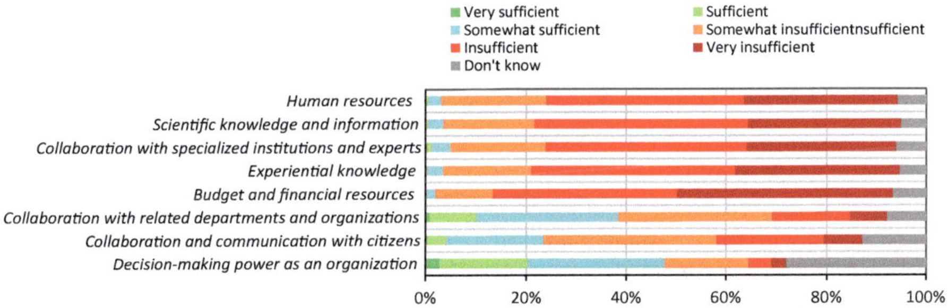

Fig. 17.7 Recognition about resource sufficiency. (Adapted from Imai and Kurisu (2022))

Figure 17.7 shows the results of a question about the availability of various resources required for adaptation. Of the eight resources, "collaboration with related departments and organizations," "cooperation and communication with citizens," and "decision-making power as an organization" tended to be answered as sufficient, while more than 80% of the local municipalities answered that "budgets and funds for implementation" were "insufficient" or "very insufficient." Looking at the resources by cluster, C3 and C4, who recognize the severity of the impacts of climate change more, tended to answer that each resource was sufficient compared to other clusters. In contrast, C2 and C5, who recognize the severity less, recognized the sufficiency of resources less than the other clusters, with C2 recording the lowest values in multiple items.

17.3 Quality of Life (QoL) and Climate Change (Shirai et al. 2023)

17.3.1 Questionnaire Survey About Subjective Evaluation of QoL

To understand the resident's perceptions about QoL related to living environment, we summarize the QoL categories and indicators for living environment as shown in Table 17.1. Five categories can be considered for QoL related to the state of the living environment: "local environment," "comfort," "convenience," "vitality," and "safety." The possible indicators to measure the categories are shown in the same table and aspects that are possibly affected by climate change are marked with a ✓.

Many previous studies have evaluated only the objective state of the living environment as QoL. However, as shown in Fig. 17.1, the state of the living environment ultimately leads to a sense of satisfaction and subjective happiness, and these subjective cognitions are also considered to be an important element of QoL (Felce and Perry 1995). Therefore, here we develop a model based on data obtained from a questionnaire survey to show how subjective happiness is connected to other subjective evaluations.

Figure 17.8 shows a hypothetical model to describe the relationships among the subjective QoL elements. "Subjective happiness" is determined by three variables, such as "satisfaction with life" (Shin 2015), "emotional experiences" (Diener 2006), and "eudemonic happiness" (Waterman et al. 2008). In this model, "satisfaction with life" is determined by "satisfaction with the state of living environment" and "satisfaction with the personal condition." Five categories summarized in Table 17.1 are involved in the "satisfaction with the state of living environment." For the personal condition, four aspects, namely economic condition, health, human relationship, and local community, are involved.

To evaluate the model developed in Fig. 17.8, data for measuring each variable were collected through a questionnaire survey. The satisfaction of each aspect was asked using a six-point scale ranging from "very satisfied" to "not satisfied at all."

An online questionnaire survey was conducted from March 24th to 29th, 2022, targeting people in their 20 s to 70 s across Japan. The survey targeted two to three local municipalities from each of the 47 prefectures. The prefectural capitals of each prefecture and the local municipalities selected from each cluster categorized in Sect. 17.2 (C1-C6) as well as areas highly dependent on tourism that are thought to have different industrial structures, were selected. The survey was designed to collect equal numbers of respondents by age group and gender, and only when samples were insufficient, they were filled in from adjacent categories. As a result, a total of 11,880 responses were obtained.

Table 17.1 QoL categories related to the state of the living environment and possible indicators

QoL categories based on previous studies			Possible indicators		
Category	**Sub-category**	**Ref.**	**Aspect**	**Indicator candidates**	**Climate change influence**
Local environment	Natural environment	1a, 8a, 9c	Natural abundance	Forestry area	
	Climate	9c	Temperature	Discomfort index (summer)	✓
	Water environment	1a, 6a, 9c	Water resources	Water resource reserves	✓
	Local culture	5a, 6b, 8b, 9b	Regional uniqueness	Number of tourism resources	✓
			Regional attractions	Number of visitors	
	Environmental consciousness	2a, 2b, 5b	Waste	Recycle rate	
			Environmental loads	CO_2 emission	
			Energy	Renewable energy introduction	
Comfort	Residence	1b, 4a, 4b, 5c, 6c, 7a, 8a	Living environment	Living floor area	
			Surrounding environment	Park and green area	
	Parks, green spaces	1a, 2c, 4a, 5b, 5c, 7a			
	Landscape	3a, 6c, 8a			
	Environmental quality	1a, 2a, 7a, 8c	Air quality	Number of photochemical oxidant warning days	✓
			Water quality	Water quality standard achievement rate	✓
	Health		Mortality	Life expectancy	
			Mortality	Low temperature deaths	✓
			Disease	Number of people transported for heat stroke	✓
			Disease	Dengue outbreak probability	✓
			Symptom	Hay fever prevalence rate	✓

(continued)

Table 17.1 (continued)

QoL categories based on previous studies			Possible indicators		
Convenience	Consumption facilities	3b, 5a, 6b, 6d, 7b, 8d, 9b	Consumption	Number of retail stores	
			Consumption	Number of large retail stores	
			Consumption	Number of restaurants	
	Transportation	2d, 4c, 5b, 6a, 6c, 6d, 8d, 9b	Public transportation	Number of stations	
			Public transportation	Number of bus stops	
			Public transportation	Existence of airport	
	Education	1c, 1d, 2e, 3b, 4d, 5a, 7b	Education	Number of nursery schools	
			Education	Number of high schools	
	Medical care	2f, 3b, 5a, 6e, 7b, 8d, 9e	Medical care	Number of hospitals	
	Culture, entertainment	2 g, 3b, 4d, 6d, 7b	Culture	Number of culture facilities	
	Infrastructure, public service	2 h, 3b, 4d, 6a			
	Telecommunication	2i, 4c			
	Price of commodities	1e, 9a			
Vitality	Local finance		Productivity	Regional GDP	✓
			Finance	Local government current account balance ratio difference	
			Infrastructure	Infrastructure maintenance and operation costs	
			Employment	Unemployment rate	
	Local industry	1e, 2j, 4b, 5d, 6b, 7b, 9a			
	Local activity	1f, 8b	Human relationships	Rate of people engaging in dating/ relationship activities	
			Community activity	Volunteer activity rate	
			Sports	Sports participation rate	

(continued)

Table 17.1 (continued)

QoL categories based on previous studies			Possible indicators		
Safety	Disaster safety	1 g, 2 k, 3c, 4d, 6e, 7c, 8c, 9a, 9d	Water disaster	Expected annual flood depth	✓
			Other disaster	Number of designated non-flood danger areas	
	Transportation & public safety	1 g, 2 l, 4d, 5e, 6e, 7c, 8c, 9d	Crime	Number of criminal offences	✓
			Accident	Number of traffic accidents	

1: Cabinet Office—Indicators of satisfaction and quality of life (2022), 1a: natural environment, 1b: housing, 1c: ease of raising children, 1d: educational environment and educational level, 1e: employment and wages, 1f: social connections, 1 g: safety in the surroundings
2: ISO 37120 (Sustainable Development of Communities: Indicators for City services and Quality of Life, 2018), 2a: environment, 2b: waste, 2c: urban planning, 2d: transportation, 2e: education, 2f: medical care and health, 2 g: entertainment, 2 h: sewage treatment, 2i: communications, 2j: economy, 2 k: firefighting and emergency response, 2 l: public safety
3: Myers (1998), 3a: quality of views and scenery, 3b: social capital and public and private facilities, 3c: probability of disasters
4: Ülengin et al. (2001), 4a: physical environment, 4b: economic environment, 4c: transportation and communication, 4d: social environment
5: Hayashi et al. (2004), 5a: opportunities for life services, 5b: environmental impact reduction, 5c: comfort, 5d: opportunities for economic activity, 5e: safety and security
6: Doi et al. (2006), 6a: environmental sustainability, 6b: opportunities for economic activity, 6c: spatial comfort, 6d: opportunities for lifestyle and culture, 6e: safety and security
7: Kachi et al. (2006), 7a: residential comfort, 7b: transportation convenience, 7c: disaster safety
8: Ishikawa and Asami (2012) (Excerpt of satisfaction items only), 8a: satisfaction with residential environment, 8b: satisfaction with image, 8c: satisfaction with safety and security, 8d: satisfaction with convenience
9: Togawa et al. (2020), 9a: economic efficiency, 9b: convenience, 9c: comfort, 9d: safety, 9e: health

17.3.2 Results

Satisfaction with the state of the local living environment was initially envisaged into five categories, but when the responses were subjected to factor analysis (maximum likelihood method with Promax rotation), six factors were extracted: "f1: convenience," "f2: local environment and culture," "f3: vitality," "f4: safety," "f5: local climate," and "f6: educational services." Part of the "local environment" was integrated with "comfort" to extract "local environment and culture," while "educational services" within "convenience" and "local climate" within the "local environment" were extracted as separate factors.

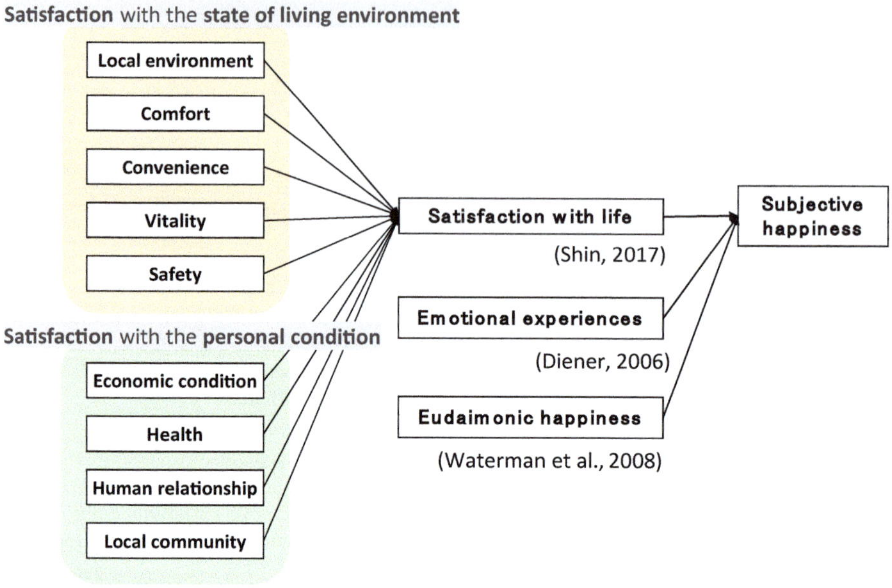

Fig. 17.8 Hypothetical model to determine the subjective happiness. (Adapted from Shirai et al. (2023))

Using the factor scores for these six factors for the variables in the developed model, the hypothetical model was evaluated and the final model was gained as shown in Fig. 17.9. The "subjective happiness"is mainly determined by "satisfaction with life"directly and indirectly through "positive emotion." The "satisfaction with life" is strongly influenced by satisfactions with "economic condition" and "human relationship," while the influence is a little bit smaller than personal conditions but the satisfactions with "convenience" and "local environment and culture" also showed positive and significant influence on the "satisfaction with life."

The local municipalities were categorized into nine clusters (G1 – G9) by the score of satisfaction with each local environmental aspect. Figure 17.10a, b, c show the average factor score of each cluster for "satisfaction with local environment," "satisfaction with personal condition," and "aspects related to subjective happiness," respectively. As shown in Fig. 17.11, the clusters of G2, G3, G5, G7, G8, and G9 were clusters which have relatively lower population densities. G9 has the lowest population density, showed the lowest scores for satisfaction with the local environment in "convenience," "vitality," and "educational services," and the lowest scores in all categories for satisfaction with personal conditions. The scores of G7 are not as low as those of G9, but show a similar trend. On the other hand, G8, which has the same low population density as G9 and low satisfaction with "convenience," "vitality," and "educational services," has relatively high scores in "local environment and culture," personal "economic condition," and "human relationships," which is thought to bring about higher life satisfaction of G8 than G7 and G9.

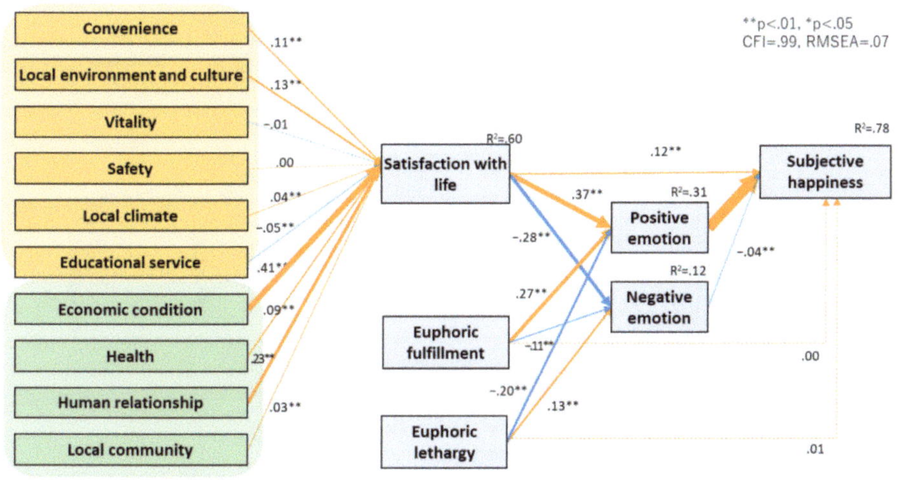

Fig. 17.9 Model to describe subjective happiness. (Adapted from Shirai et al. (2023)) ***$p < 0.01$, *$p < 0.05$, *The numerical values of path coefficients indicate standardized estimates. Covariance between exogenous variables is omitted in the figure.

G4 and G6 are densely populated areas. Both show high satisfaction with "convenience," "vitality," and "educational services." G4 also has high satisfaction with personal conditions, which translates into high life satisfaction. On the other hand, G6 includes large cities, but satisfactions with the "living environment and culture," and "local community" are particularly low, which is the reason why life satisfaction and happiness of G6 are lower than G1 and G4.

17.4 Summary

Figure 17.12 shows the relationship between clusters in Sect. 17.2 (C1-C6) and Sect. 17.3 (G1 – G9). We can see that C3, where the perception of the severity of climate change was high in all areas, contained many G7 and G9 local municipalities. These regions also had low overall QoL satisfaction and were subject to serious additional climate change impacts. On the other hand, C5, where the perception of severity was the lowest, contained many G4 and G1 local municipalilties, and their living conditions were relatively good, which led to the perception that the severity of the impacts of climate change was not high.

As we saw in Sect. 17.3, the current state and evaluation of QoL varies greatly from region to region. Small local municipalities in particular have a low current QoL, and as we saw in Sect. 17.2, local industries, especially agriculture, will also be affected by climate change. Even if they are affected by the same climate change, the risks vary greatly depending on the basic QoL situation of each local municipality. We need to evaluate the current state of QoL and the risks of climate change in each region based on it and proceed with adaptation.

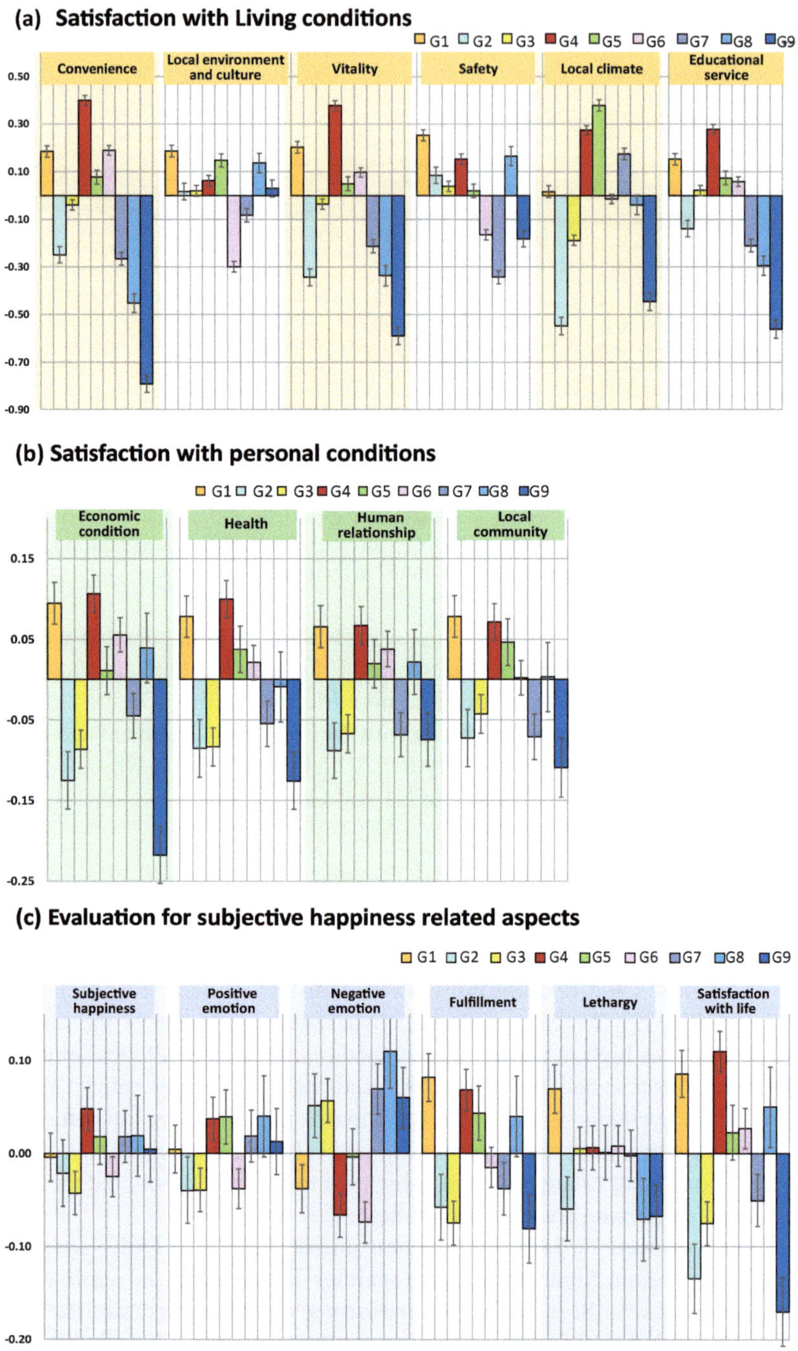

Fig. 17.10 Factor scores for (**a**) living conditions, (**b**) personal conditions, and (**c**) subjective happiness-related aspects in each cluster. (Adapted from Shirai et al. (2023))

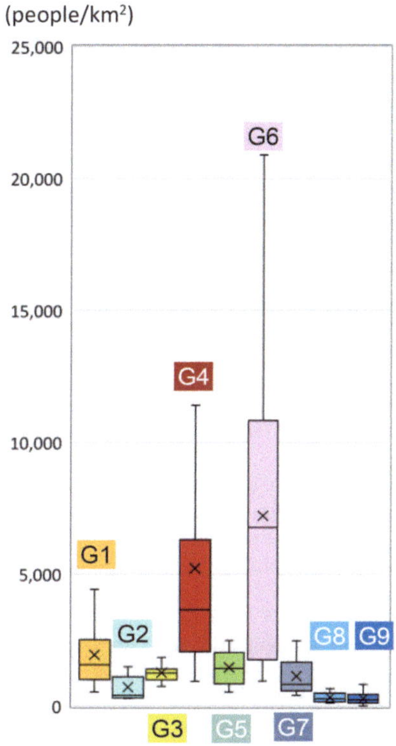

Fig. 17.11 Population density of each cluster. (Adapted from Shirai et al. (2023))

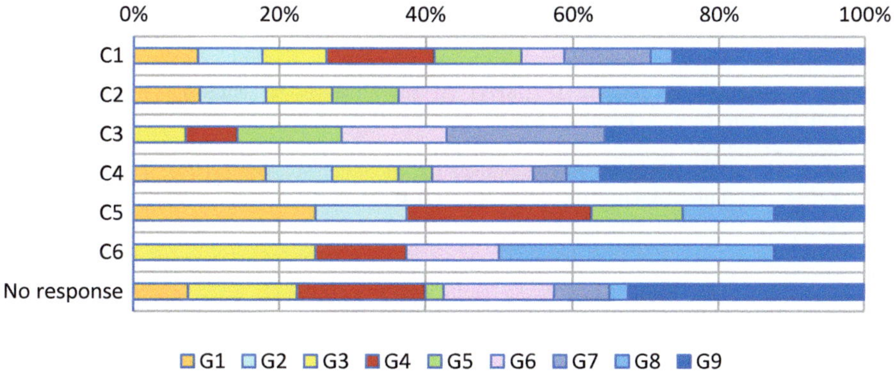

Fig. 17.12 Relationship between the clusters in Sects. 17.2 (C1-C6) and 17.3 (G1 – G9) "No response" is the municipality which did not respond to the questionnaire in Sect. 17.2.

References

Cabinet Office, Japan (2022) Indicators of satisfaction and quality of life. https://www5.cao.go.jp/keizai2/wellbeing/manzoku/index.html. Accessed Aug 2024
Diener E (2006) Guidelines for national indicators of subjective well-being and ill-being. Appl Res Qual Life 1:151–157. https://doi.org/10.1007/s11482-006-9007-x
Doi K, Nakanishi H, Sugiyama I, Shibata H (2006) Development of a QoL-based multi-dimensional evaluation system for urban infrastructure planning. J Jpn Soc Civil Eng D 62(3):288–303. (in Japanese) https://doi.org/10.2208/jscejd.62.288
Felce D, Perry J (1995) Quality of life: its definition and measurement. Res Dev Disabil 16(1):51–74. https://doi.org/10.1016/0891-4222(94)00028-8
Hayashi Y, Doi K, Sugiyama I (2004) Evaluating infrastructure projects by means of measures of the quality of life. J Jpn Soc Civil Eng 2004(751):55–70. (in Japanese). https://doi.org/10.2208/jscej.2004.55
Imai Y, Kurisu K (2022) Local government's perceptions about climate change impacts and their severities. J Jpn Soc Civil Eng G 78(6):II_99–107. (in Japanese). https://doi.org/10.2208/jscejer.78.6_II_99
Ishikawa T, Asami Y (2012) Perception of the quality of urban living and residential satisfaction—in relation to residential characteristics, human values, and physical environments. J City Plan Instit Japan 47(3):811–816. (in Japanese). https://doi.org/10.11361/journalcpij.47.811
ISO 37120 (2018) Sustainable development of communities: indicators for city services and quality of life
Kachi N, Kato H, Hayashi Y, Morisugi M (2006) A quality of life index measured by life year for evaluating residential areas and its application to examining policies to control urban sprawl. J Jpn Soc Civil Eng D 62(4):558–573. (in Japanese). https://doi.org/10.2208/jscejd.62.558
Myers D (1998) Building knowledge about quality of life for urban planning. J Am Plan Assoc 54:347–358. https://doi.org/10.1080/01944368808976495
Shin DC (2015) How people perceive and appraise the quality of their lives: recent advances in the study of happiness and wellbeing. UC Irvine CSD Working Papers. https://escholarship.org/uc/item/0hq2v2wx. Accessed in Aug 2024
Shirai K, Kurisu K, Fukushi K (2023) Evaluation of the relationship between satisfaction with regional conditions and subjective well-being for the prediction of future climate change impact. J Jpn Soc Civil Eng 79(26):23–26005. (in Japanese). https://doi.org/10.2208/jscejj.23-26005
Togawa T, Takano T, Morita H, Ooba M, Estoque RC, Kondo M (2020) Quantitative assessment framework of the impact of climate change on QoL. J Jpn Soc Civil Eng G 76(5):I_461–I_470. (in Japanese). https://doi.org/10.2208/jscejer.76.5_I_461
Ülengin B, Ülengin F, Guvenc U (2001) A multidimensional approach to urban quality of life: the case of Istanbul. Eur J Oper Res 130(2):361–374. https://doi.org/10.1016/S0377-2217(00)00047-3
Waterman AS (2008) Reconsidering happiness: a eudaimonist's perspective. J Posit Psychol 3(4):234–252. https://doi.org/10.1080/17439760802303002

Chapter 18
Implementing Urban Design Workshops for Climate Change Adaptation at the District Scale

Junya Yamasaki and Akito Murayama

Abstract This chapter introduces advanced examples of workshops (WS) that have been conducted multiple times to design districts for climate change adaptation as part of area-based management activities in Japan. These activities are aimed at understanding the potential impacts of climate change on the district and generating ideas for adaptation measures, primarily emphasizing physical and environmental solutions. The themes of the WS primarily focus on heat-related measures, and the program is characterized by its use of scientific knowledge, including thermal environment measurements and simulation technologies, to understand the current situation and future challenges within the district. These WS were held a total of four times, with the first having a theme of "Inspire," the second of "Envision," the third of "Evaluate," and the fourth of "Experience and Plan." This chapter presents content on the series of WS such that urban planners can replicate it, and these examples can be widely referenced and adapted to global urban development, accounting for local characteristics.

Keywords Urban design · Area-based management · District scale · Thermal environment simulation · Computational fluid dynamics · Surface temperature · Wet Bulb Globe Temperature

J. Yamasaki (✉)
Nagoya University, Nagoya, Japan
e-mail: yamasaki.junya.i5@f.mail.nagoya-u.ac.jp

A. Murayama
The University of Tokyo, Tokyo, Japan

N. Mimura, S. Takewaka (eds.), *Climate Change Impacts and Adaptation Strategies in Japan*, https://doi.org/10.1007/978-981-96-2436-2_18

18.1 Introduction

Climate change adaptation is not only a policy issue at the national or municipal level but also a matter that should be examined individually and concretely at smaller scales, such as the district level. It is crucial for relevant stakeholders to collaborate in preparing a district's physical environment. For example, Just Communities, an organization established in 2021, promotes racially equitable and climate-resilient local environments by building a certification system for district-scale development processes (Just Communities 2024). Additionally, district-scale sustainability assessment systems, such as LEED for Cities and Communities, BREEAM Communities, and the Living Community Challenge, are advancing initiatives related to climate actions (BREEAM 2024; International Living Future Institute; 2024; U.S. Green Building Council 2024). In Japan, "area-based management" activities aimed at enhancing the value of specific districts are becoming widespread, and the organizations involved are expected to play key roles in addressing local issues (Yasui and Izumiyama 2021). As of September 2024, there are 117 organizations nationwide that have been officially designated as "Urban Renaissance Corporations" by local governments (MLIT 2024), and these organizations collaborate through networks such as the "National Area Management Network" (Area Management Network 2024). While social and economic aspects, such as revitalizing local economies and creating vibrant districts, have often been emphasized in Japan's urban development, it is now essential to actively address environmental aspects, including climate change mitigation and adaptation.

Therefore, this chapter introduces advanced examples of workshops (WS) that have been conducted multiple times to design districts for climate change adaptation as part of such area-based management activities. These activities are aimed at understanding the potential impacts of climate change on the district and generating ideas for adaptation measures, primarily focusing on physical and environmental solutions. The development and operation of the WS were led by the Urban Land Use Planning Unit of the University of Tokyo (Urban Land Use Planning Unit 2024), in collaboration with local area-based management organizations. The themes of the WS primarily focus on heat-related measures, and the program is characterized by its use of scientific knowledge, including thermal environment measurements and simulation technologies, to understand the current situation and future challenges within the district. This chapter presents content on the series of WS such that urban planners can replicate it, and these examples can be widely referenced and adapted to global urban development, accounting for local characteristics.

This chapter was conducted as part of the research theme of Team 4, titled "Projection of Climate Change Impacts on Quality of Life (QoL) of People and Their Associated Infrastructure and Local Industries and Evaluation of Adaptation Options," within the S-18 Project. The Urban Land Use Planning Unit of the University of Tokyo has also conducted research on climate change impact assessments, with the results summarized by Yamasaki et al. (2024). This chapter presents an example of developing practical processes based on these research outcomes and attempting social implementation.

18.2 Case Study Site

The case study site for these WS was the Nishiki 2 District in Nagoya City, Aichi Prefecture, Japan. The city is located in the middle of mainland Japan and functions as the regional center of the Nagoya metropolitan area, which is one of Japan's three major metropolitan areas. The Nishiki 2 District is located in a commercial and business area approximately 1 km from Nagoya Station, which is the central station of Nagoya City. It has a grid-like layout and spans approximately 400 × 400 m (Fig. 18.1). Once a thriving wholesale textile district, the Nishiki 2 had experienced a gradual decline since the 1990s owing to changes in its social structure, leading to issues such as the hollowing out of local industries and population decline. In the 2000s, a community development council was established to address this situation with efforts to revitalize the district, including attracting an outdoor venue for an international art festival. In 2018, an area-based management corporation was established to actively promote a community development project (Nishiki 2 Area Management 2024), and Nagoya City designated it as an Urban Renaissance Corporation in 2021. As a result, the Nishiki 2 is now recognized as a district characterized by advancing community development activities.

In 2020, the area-based management corporation established the "Nishiki 2 Area Platform (N2/LAB)" as a platform for experimentation to advance the conception, research, and co-creation towards realizing a sustainable district (N2/LAB 2024). Furthermore, the "Nishiki 2 Future VISION," which outlines the activity policies of N2/LAB, was formulated in 2021, with the promotion of climate change measures being positioned as one of its key policies. These WS were designed and conducted in alignment with the district trends.

18.3 Contents of the WS

These WS were held a total of four times, with the first having a theme of "Inspire," the second of "Envision," the third of "Evaluate," and the fourth of "Experience and Plan" (Fig. 18.2). Participants were primarily recruited through websites that

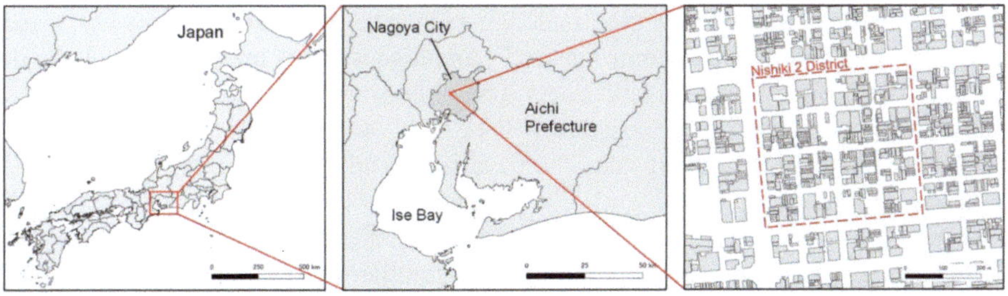

Fig. 18.1 Location of the Nishiki 2 District in Nagoya City, Aichi Prefecture, Japan. (Reproduced from Yamasaki et al. (2024))

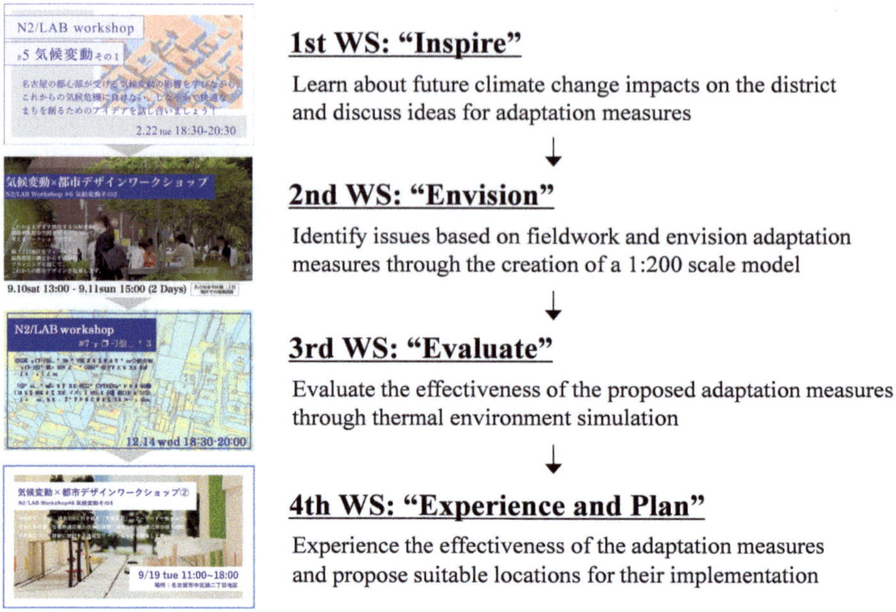

Fig. 18.2 Flow of the series of WS

included local residents, university members, company employees, and administrative officials (including those who were not experts on climate change).

18.3.1 First WS: "Inspire"

In the first WS, participants aimed to learn about scientific knowledge related to climate change impacts and discuss ideas for future impacts and adaptation measures in the Nishiki 2. This WS was held online due to COVID-19, and the program was structured into two parts. Participants discussed designated topics twice after lectures by experts (Table 18.1). In the first lecture, titled "How does climate change affect our lives?", experts introduced climate change impacts both domestically and internationally. Based on this, the first discussion asked, "What problems will arise if the heat and heavy rain become severe?", and discussed its impact on our lives in general, not limited to the Nishiki 2. As a result, opinions were expressed, such as "The beauty of Japan's four seasons will be lost" and "People will refrain from going out, reducing restaurant sales," with regard to heat, and "Land prices will fall in towns with frequent floods" and "Traffic congestion and accidents will increase," with regard to heavy rain.

In the second lecture, titled "What will the summer heat in the Nishiki 2 be like in the future?", the future environment of the district was introduced using technologies such as thermal environment simulations. The Urban Land Use Planning Unit at the University of Tokyo has focused on simulation analysis based on Computational

Table 18.1 Outline of the first WS: "Inspire"

Purpose	Learn about future climate change impacts on the district and discuss ideas for adaptation measures
Date	February 22, 2022, 18:30–20:30
Location	Online (due to COVID-19)
Participants	40 people
Program	Part 1: Lecture 1 "How does climate change affect our lives?" (10 min) Discussion 1 "What problems will arise if the heat and heavy rain become severe?" (35 min) Part 2: Lecture 2 "What will the summer heat in Nishiki 2 be like in the future?" (20 min) Discussion 2 "What can we do to overcome the heat? From now until the 2090s" (40 min)

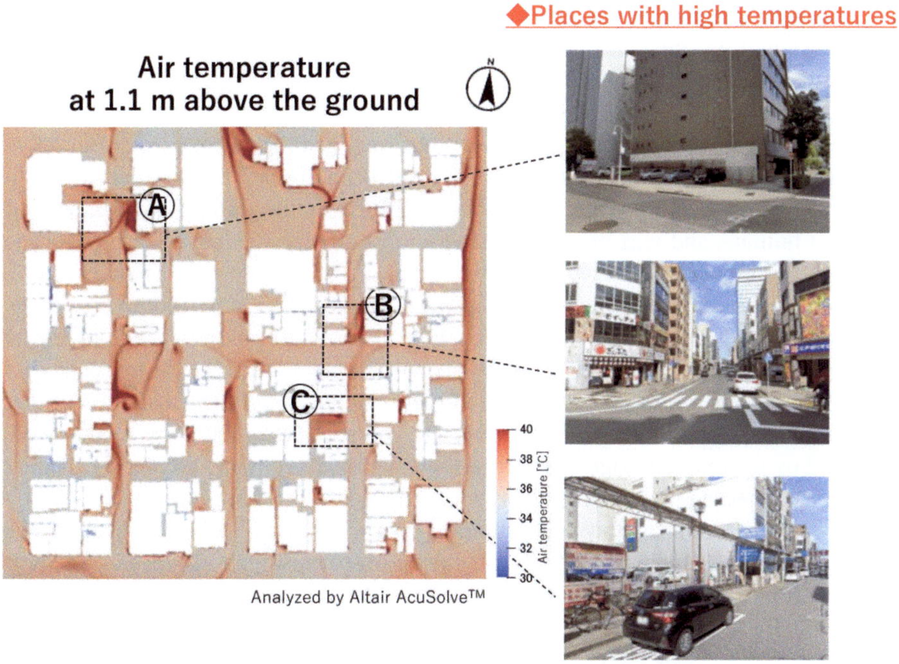

Fig. 18.3 Simulation results of the thermal environment under current weather conditions. (Using CFD software Altair AcuSolve™)

Fluid Dynamics (CFD) theory, allowing for the visualization of the thermal environment. The simulation results for the Nishiki 2 were shared, explaining which places were particularly hot under both current and future weather conditions (Fig. 18.3). Based on this, the second discussion asked, "What can we do to overcome heat? From now until the 2090s," and discussed adaptation measures in the district. Physical adaptation measures were categorized by spatial position in relation to the human scale (e.g., up, down, and sideways), with ideas such as "Install cooling mist along main routes (up)" and "Green many parking lots in the district (down)."

18.3.2 Second WS: "Envision"

In the second WS, the participants aimed to understand the issues through fieldwork in the Nishiki 2 and envision future adaptation measures based on ideas from the first WS by creating a 1:200 scale model. The program on the first day included expert lectures on climate change and urban planning, fieldwork on thermal environment surveys, and discussions on the survey results. The program on the second day included group work on urban design based on discussions from the first day, followed by presentations of outcomes and feedback (Table 18.2). The design of this WS focused on three areas within the district: the public spaces on the ground floor of a high-rise building on one block and north–south and east–west roads spanning two blocks (each approximately 200 m).

On the first day, three experts delivered lectures to provide the participants with scientific knowledge. The lectures covered the following topics: (1) the history of urban development in Nagoya City and the Nishiki 2, (2) urban design for climate change adaptation, and (3) concepts and methods for heat-related measures. Second, to identify issues in the district's outdoor public spaces, participants were divided into six groups, and fieldwork was conducted to survey the thermal environment in their assigned areas. The primary survey items included the surface temperatures of buildings and the ground, and the WBGT (Wet Bulb Globe Temperature) at 1.1 m above the ground. The surface temperature influences the radiant environment of adjacent humans and was measured using a thermographic camera. The WBGT is an index used to assess the risk of heatstroke based on factors such as temperature, humidity, and radiation, measured using specialized devices and tripods (Fig. 18.4). The positions and heights of street trees, plants, and sidewalk roofs that affect the thermal environment were surveyed at the same time. Third, the participants discussed the issues while presenting the survey results on a 1:200 scale model that depicted the current situation of the district. The surface temperature results were obtained by printing photographs from a thermographic camera and placing them at their corresponding positions on the model. The WBGT results were represented by placing colored pins at the corresponding positions, with colors indicating the measured values (Fig. 18.5). Finally, each group presented the issues in their assigned areas and shared their findings.

Table 18.2 Outline of the second WS: "Envision"

Purpose	Identify issues based on fieldwork and envision adaptation measures through the creation of a 1:200 scale model
Date	September 10, 2022, 13:00–17:00, and September 11, 10:00–15:00
Location	The Nishiki 2 District
Participants	22 people
Program	Day 1 Part 1: Expert lectures on climate change and urban planning (70 min) Part 2: Fieldwork on thermal environment surveys (80 min) Part 3: Discussions on the survey results (80 min) Day 2 Part 1: Group work on urban design (180 min) Part 2: Presentations of the outcomes and feedback (60 min)

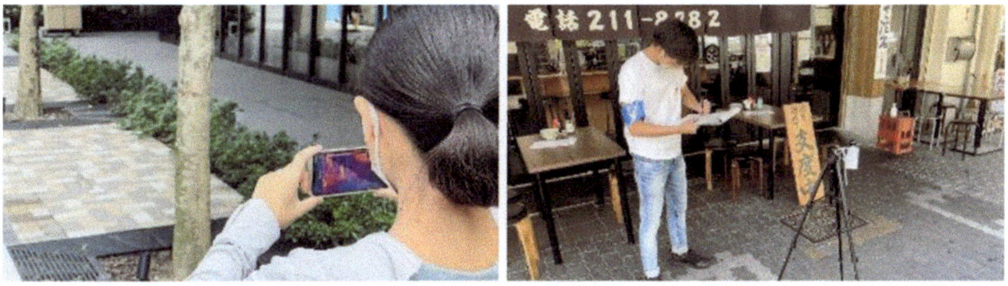

Fig. 18.4 Fieldwork on thermal environment surveys

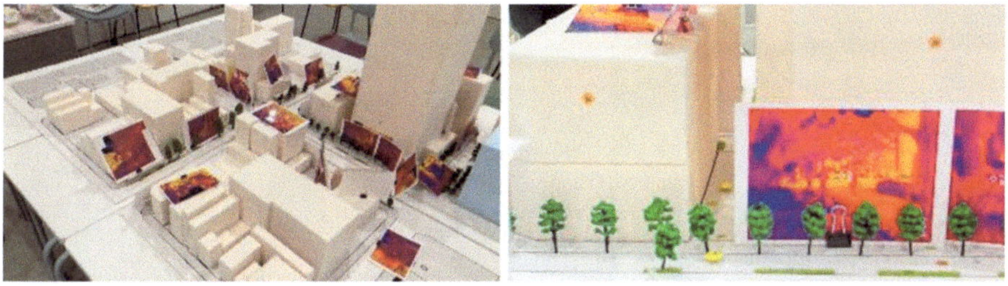

Fig. 18.5 Presentation of survey results

Fig. 18.6 Group work on urban design

On the second day, participants discussed adaptation measures for the district and designed them through group work by creating a 1:200 scale model. The six groups on the first day were reorganized into three new groups, and each new group took responsibility for the fieldwork areas from the first day: the public spaces on the ground floor and the north–south and east–west roads. Adaptation measures were represented by incorporating new structures into the existing model, altering the surface materials, and making other modifications (Fig. 18.6). Finally, each group presented their outcomes and received feedback from the experts. Each group included both short- and long-term adaptation measures. Short-term measures included wall greening, mist showers, and installing sidewalk roofs. Long-term measures included redistributing and woodifying road spaces, installing biotopes, and designing new public spaces through redevelopment (Fig. 18.7).

Fig. 18.7 Outcomes of urban design

Table 18.3 Outline of the third WS: "Evaluate"

Purpose	Evaluate the effectiveness of the proposed adaptation measures through thermal environment simulation
Date	December 14, 2022, 18:30–20:00
Location	The Nishiki 2 District
Participants	15 people
Program	Part 1: Sharing thermal environment simulation results (30 min) Part 2: Discussion (60 min)

18.3.3 Third WS: "Evaluate"

The purpose of the third WS was to evaluate the proposed urban design of the Nishiki 2 through a thermal environment simulation and discuss the effects of adaptation measures (Table 18.3). The simulation was conducted in advance by the Urban Land Use Planning Unit at the University of Tokyo, as done for the first WS. The procedure first involved modeling the proposed urban design, including physical elements, such as new buildings, street trees, sidewalk roofs, and biotopes, using CAD software. Second, a simulation was conducted for both the current and proposed conditions of the district using weather data from 10:00 to 18:00 on a representative summer day. By simultaneously analyzing the airflow, temperature, and radiation, it is possible to understand the wind flow, air temperature, and shadow distribution caused by the new structures (Fig. 18.8).

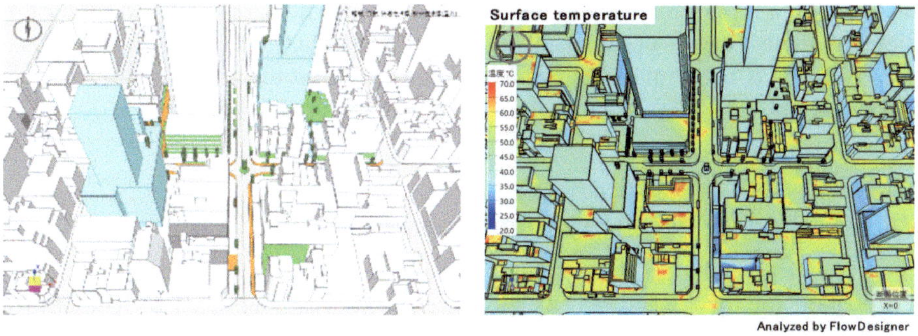

Fig. 18.8 Thermal environment simulation for urban design outcomes. (Using CFD software FlowDesigner)

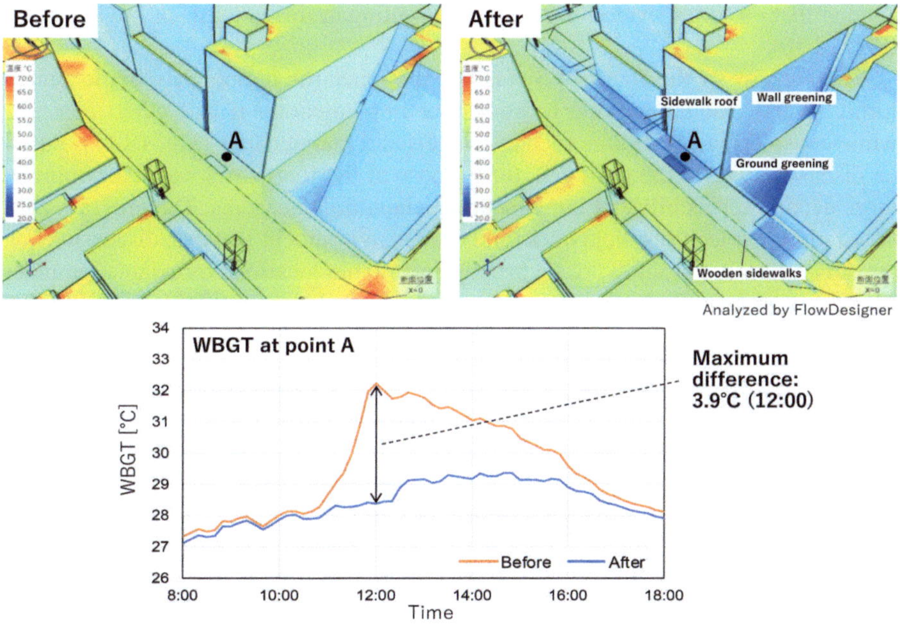

Fig. 18.9 Evaluation of the effectiveness of adaptation measures. (Using CFD software FlowDesigner)

In the WS, the experts first visually explained the differences in the thermal environment before and after the implementation of adaptation measures in the three areas. As an example, the effectiveness of the adaptation measures in the north–south road area is shown in Fig. 18.9. This area had a problem with few structures blocking direct sunlight at noon; however, the walking environment was significantly improved by proposals such as sidewalk roofs, wall greening, and woodifying road spaces. The daytime WBGT under the sidewalk roof decreased by 3.9 °C at 12:00 on a representative summer day. Second, participants engaged in discussions on three topics: "(1) Impressions of the WS so far and the analysis results," "(2) New ideas for climate change adaptation based on the analysis results," and "(3) How can WS and thermal analysis be utilized in urban planning?"

18.3.4 Fourth WS: "Experience and Plan"

To prepare for the fourth WS, a field experiment was conducted to temporarily implement the adaptation measures proposed up to the third WS. Two types of outdoor furniture were implemented in public spaces, in collaboration with two companies that developed products for outdoor heat-related measures. The first is a bench with a roof and misting function, and the second is a movable standalone awning that provides shade for pedestrians waiting at traffic lights or buying lunch (Fig. 18.10).

The purpose of the fourth WS was to experience the temporarily implemented furniture and discuss suitable locations for implementing it and other measures while considering constraints such as budgets and maintenance (Fig. 18.11). Based on discussions up to the third WS, the purpose of the fourth WS was to examine in more detail the feasibility of implementing adaptation measures. The program consisted of expert lectures, fieldwork on thermal environment surveys, discussions on the survey results, group work on proposing suitable locations for adaptation measures, and presenting these outcomes and receiving feedback (Table 18.4). In this

Fig. 18.10 Field experiment on the implementation of outdoor furniture

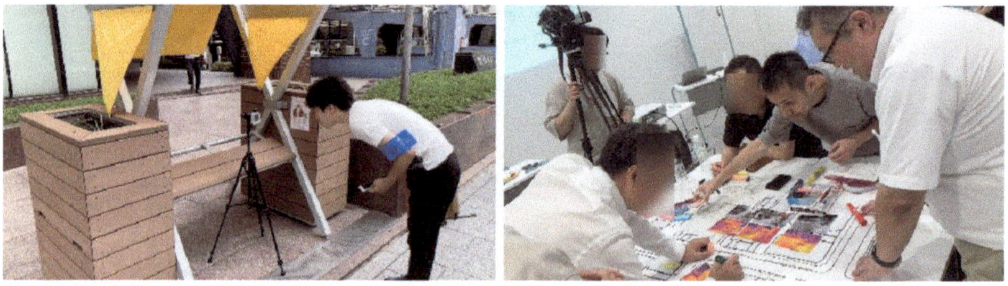

Fig. 18.11 Fieldwork and group work on the implementation of adaptation measures

Table 18.4 Outline of the fourth WS: "Experience and Plan"

Purpose	Experience the effectiveness of the adaptation measures and propose suitable locations for their implementation
Date	September 19, 2023, 11:00–17:30
Location	The Nishiki 2 District
Participants	20 people
Program	Part 1: Expert lectures on climate change and urban planning (45 min) Part 2: Fieldwork on thermal environment surveys (60 min) Part 3: Discussions on the survey results (45 min) Part 4: Group work on proposing suitable locations of adaptation measures (60 min) Part 5: Presentations of the outcomes and feedback (30 min)

WS, the Nishiki 2 road space was divided into five areas, with participants forming five groups, each responsible for surveying and discussing one area.

Group work was conducted in two steps: first, by considering locations for implementing adaptation measures without budget constraints, and second, by considering locations with budget constraints. Information on the implementation costs of each measure was distributed to each group. This led to a process in which many measures were first considered for implementation and then selected based on cost constraints, allowing for the determination of suitable locations based on the cost-effectiveness of each measure. Five groups conducted these discussions and a concrete direction for future urban development was obtained.

18.4 Conclusions

A total of 97 people participated in the four WS, and they contributed to the promotion and dissemination of climate change adaptation through urban development activities. Adaptation is an abstract concept, and its effects are difficult to quantify. As a result, there are fewer opportunities for the inclusion of concrete measures in the plans of local governments and companies than for mitigation efforts. In response to this situation, the main feature of these WS was their clear focus on the theme of adaptation, and the participants recognized this focus as a new direction in urban planning. As there is no panacea for urban development or adaptation and both processes take extended time, it is expected that fostering a common understanding with stakeholders will have long-term effects. The policies and ideas for adaptation measures developed in these WS will be incorporated into guidelines for future area-based management activities. The content of each WS is introduced on the N2/LAB website in Japanese (N2/LAB 2024).

In realizing this activity, the contributions of university members and local organizations that have been addressing issues for an extended period were significant. Therefore, it is possible to consider the potential of promoting similar activities in areas where such foundations have already been established. We hope these cases will be shared through international networks and partnerships related to urban development.

References

Area Management Network (2024) What is area management? (in Japanese). https://areamanagementnetwork.jp. Accessed 1 Sep 2024

BREEAM (2024) BREEAM communities. https://breeam.com/standards/communities. Accessed 1 Sep 2024

International Living Future Institute (2024) Living community challenge. https://living-future.org/lcc. Accessed 1 Sep 2024

Just Communities (2024) Just communities | The Evolution of Ecodistricts. https://justcommunities.info. Accessed 1 Sep 2024

MLIT (Ministry of Land, Infrastructure, Transport and Tourism, Japan) (2024) Urban renewal corporation (in Japanese). https://www.mlit.go.jp/toshi/toshisaisei. Accessed 1 Sep 2024

N2/LAB (2024) Nishiki 2 are platform (in Japanese). https://n2-lab.jp. Accessed 1 Sep 2024

Nishiki 2 Area Management (2024) NISHIKI2 (in Japanese). http://nishiki2areamanagement.co.jp. Accessed 1 Sep 2024

U.S. Green Building Council (2024) LEED for cities and communities. https://www.usgbc.org/leed/rating-systems/leed-for-cities-communities. Accessed 1 Sep 2024

Urban Land Use Planning Unit (2024) Urban land use planning unit, The University of Tokyo. https://up.t.u-tokyo.ac.jp. Accessed 1 Sep 2024

Yamasaki J, Wakazuki Y, Iizuka S, Yoshida T, Nitanai R, Manabe R, Murayama A (2024) Microclimate Simulation for Future Urban District under SSP/RCP: Reflecting changes in building stocks and temperature rises. Urban Clim 57:102068. https://doi.org/10.1016/j.uclim.2024.102068

Yasui M, Izumiyama R (2021) Area management case method. Textbook of regional management through public-private collaboration (in Japanese). Gakugei shuppansha

Chapter 19
Urban Metabolism and Adaptation Options for Climate Change

Hiroki Tanikawa and Naho Yamashita

Abstract This chapter examines the sustainability of cities by exploring urban material metabolism and climate change adaptation. It emphasizes the importance of using urban structures to mitigate environmental impacts like greenhouse gas emissions and using the Japanese concept of "mottainai" (avoiding waste). We explore the dynamics of material stock and flow, particularly the accumulation of material stocks in industrialized nations and emerging economies. The analysis employs indicators such as material productivity, service utilization, and stock retention time to assess urban sustainability. Adaptive measures are necessary to address climate risks, including reinforcing building structures and managing material circulation.

Keywords Socioeconomic metabolism · Material stock and flow · Climate change adaptation · Building and infrastructure · Urban mine

19.1 Introduction

This chapter discusses the sustainability of cities from the perspective of material metabolism systems and climate change adaptation. Photo 19.1 shows the center of Tokyo, the capital of Japan, where building reconstruction and enhancement of social infrastructures are ongoing. Commercial buildings and high-rise apartments made of reinforced concrete under construction, the steel-framed Tokyo Tower, and infrastructures are lined up. In this area, large-scale redevelopment projects centered around Tokyo Station are also underway. To continue urban activities over a long period of time, it is necessary to adapt to the potential impact of climate change. Otherwise, it will not be possible to continue living safely and securely.

H. Tanikawa (✉)
Nagoya University, Nagoya, Japan
e-mail: tanikawa.hiroki.z7@f.mail.nagoya-u.ac.jp

N. Yamashita
The University of Tokyo, Tokyo, Japan

N. Mimura, S. Takewaka (eds.), *Climate Change Impacts and Adaptation Strategies in Japan*, https://doi.org/10.1007/978-981-96-2436-2_19

Photo 19.1 Urban material stock (Central Tokyo, 2024, photographed by the author)

The concept of using things for a long time is widely understood from the Japanese sense of "mottainai," and this is an important idea from the perspective of sustainability and environmental load reduction. The long-term use of urban structures has a mitigating effect of suppressing the amount of material inflow and greenhouse gas (GHG) emissions. If the material stock of buildings and infrastructures is used only for a short period of time, the impact on climate change through the increase in resource and energy consumption and the associated environmental load (waste generation and GHG emissions) for frequent reconstruction and renewal becomes a concern. In addition, the long-term use of materials has the aspect of adaptation measures leading to urban resilience, such as improving durability and earthquake resistance through structural reinforcement, and changes in structure such as pilotis. However, unintended effects such as a temporal increase in material input and in environmental load due to structural reinforcement and construction material substitution may occur. In this chapter, we discuss the long-term use of materials and the impact of climate change, considering socioeconomic factors and the geographical distribution of major urban structures, from both adaptation and mitigation perspectives. The results shown in this chapter are based on the research results on urban climate change adaptation measures in the S-18 project.

19.2 Material Stock Supporting Human Activities

In the early stages of urban development, a large amount of resources and energy are consumed to support rapid economic growth, which often turns out to be a flow-type society. Continuous material flows are also required in mature societies if short-lived products and structures are prevalent. Resources and energy are required for frequent replacements and updates in a flow-type society. To clarify the dynamics of material flow and stock that support human activities, it is necessary to know the current state of the materials existing in society. However, it is not easy to grasp the trends of a vast and diverse range of materials over the years and comprehensively. There is relatively a lot of research on material flow in each country. On the other hand, information on material stock, such as when, where, and in what materials were accumulated, is limited domestically and internationally. Therefore, material stocks have been treated as a black box that adjusts between inflow and outflow, and their quantity and quality have not been sufficiently discussed.

19.2.1 Increase in Material Stock in the Twentieth Century

We are advancing material stock analysis in collaboration with researchers from around the world (Krausmann et al. 2017). The amount of material stock that supports our human activities has increased about 23 times worldwide during the twentieth century, especially in industrialized countries where manufacturing capital has greatly expanded. As shown in Fig. 19.1a, the material stock increased from 35 billion tonnes to 792 billion tonnes from 1900 to 2010, growing at an average rate of 2.9% per year. Considering the direct and indirect consumption of resources and energy, it indicates that we disturb the natural environment several times to several hundred times the mass of the material stock itself (Matthews 2000).

According to Fig. 19.1b, the per capita material stock in 2010 is shown to be an average of 115 tonnes worldwide. The per capita material stock in industrialized countries is an average of 335 tonnes/person, whereas 136 tonnes/person in China, and 38 tonnes/person in other countries (Rest of the World: RoW). There is a large difference between industrialized countries and China and other countries. People's lives in industrialized countries are supported by 335 tonnes per person of material stocks in the forms of urban structures and durable goods. In the near future, if emerging countries uniformly establish nearly 300 tonnes of material stock per person equivalent to industrialized countries, a huge amount of resources and energy will be needed to maintain them. Suppose that a material stock of 200 tonnes per person is established in China in the future, 1.4 billion people (2020) × 200 tonnes/person = 2800 billion tonnes of material will be needed. Namely, nearly half of the total material stocks in the world will be newly established in China alone.

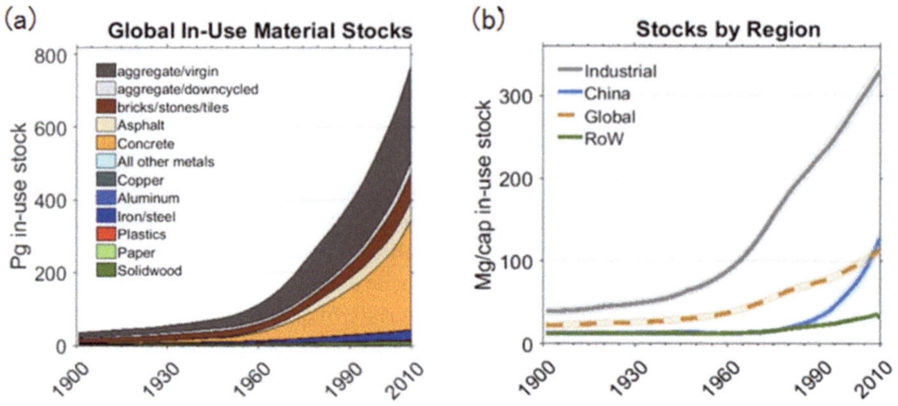

Fig. 19.1 Increasing global material stocks (**a**) In-use material stock by type of material, (**b**) per-capita material stocks by region. (Source: Reproduced from Krausmann et al. (2017) with permission)

19.2.2 Current Status of Material Flow and Stock in Japan

Figure 19.2 shows the material flow and stock for the entire Japan. The upper left side of the figure represents the inflow into society that year, while the right side represents the outflow that is consumed in production activities and energy conversion, as well as discharged as waste and recycling. The total material input (the total amount of material input into society in one year) in Japan in 1990 was about 2.4 billion tonnes, of which about 65% was accumulated into infrastructures, buildings, durable goods, etc. In 2015, the total material input significantly decreased compared to 1990, and human activities are supported by an annual inflow of about 1.6 billion tonnes. Looking in detail at each item that constitutes the inflow, it turns out that almost no change has occurred except for one item. The most significant change is in an item called the net addition to stock (NAS), which has decreased by more than half over about 30 years. NAS is the net increase in materials in society, subtracting the amount of outflow from the amount of material newly accumulated in material stocks that year. In Japan, the inflow still exceeds the outflow, and the overall material stock in society is increasing, but it is moving towards a mature society in terms of stock saturation. The lower part of the figure is the material stock, and the left and right sides show the breakdown by application, such as infrastructures and buildings, durable goods, and materials such as concrete, metal, and timber. Material stock data is yet limited in terms of the range of applications and the components of materials, but the figure is useful to understand how much materials are actually required to support our lives (Tanikawa et al. 2021).

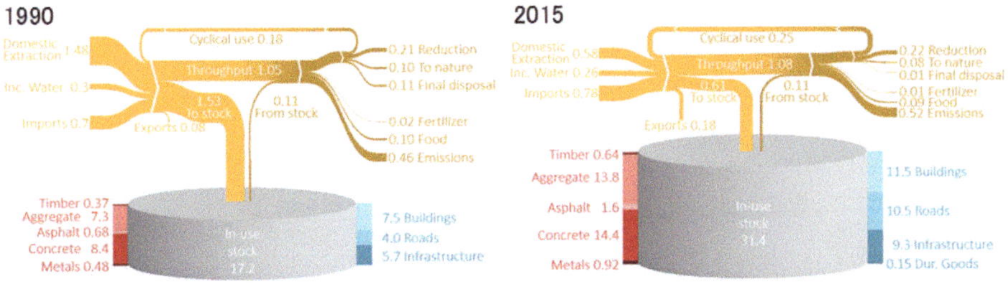

Fig. 19.2 Material flow and stock in Japan in 1990 and 2015 (billion tonnes) (Source: Reproduced from Tanikawa et al. (2021) with permission from Elsevier)

19.2.3 Spatiotemporal Distribution of Material Stock in Japan

It is important to know the situation of existing material stocks, whether they are effectively used or not, and where and how long they remain in society. By explicitly showing the temporal and spatial distribution of the materials accumulated in each urban structure using GIS (Geographic Information System), it contributes to environmental measures that are more in line with the actual situation of the region (Tanikawa et al. 2015). Figure 19.3 shows a distribution of Japan's building and infrastructure stocks estimated by identifying the congruency of structures between generations. That process mechanically distinguishes the change in shape of structures over the years, considering structures of the same shape as those that have remained and others as newly constructed or demolished. It enables us to understand the dynamics of urban structures in time and space. Japan's building stock was estimated to be 9.6 billion tonnes in 2003 and 11.8 billion tonnes in 2020, increasing by about 1.2 times in 17 years. Spatially, it is concentrated in urban areas such as the 23 wards of Tokyo, Osaka, and Nagoya. The building stock in the 23 wards of Tokyo and government-designated cities accounts for about 32% of the total material stocks. Infrastructure stock (roads, railways, airports, ports, fishing ports, dams) was 5.2 billion tonnes in 1970 and has increased to about 1.7 times, or 7.4 billion tonnes, in 50 years. This allowed us to clarify the geographical dynamics with high resolution over a long period of time, where and how the material stocks are changing. By overlaying the fundamental information of material stocks with natural- or socioeconomic factors such as disaster risk and population, it contributes to climate change adaptation and mitigation measures.

19.2.4 Natural Disaster Risk and Adaptation Measures in Cities

Climate change has a significant impact on food production, ecosystems, and urban systems. In Japan, the annual number of extreme rainfall events has significantly increased from 1976 to 2023, and there is concern about future flood damage due to

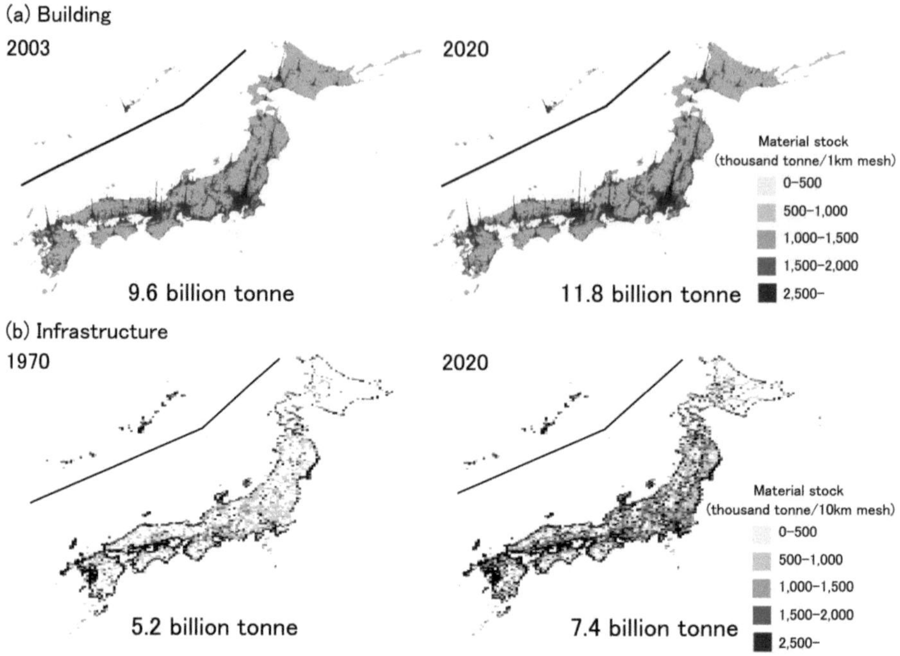

Fig. 19.3 Spatiotemporal distribution of material stock (**a**) Building stock in 2003 and 2020 (1 km mesh), (**b**) Infrastructure stock (roads, railways, airports, ports, fishing ports, dams) in 1970 and 2020 (10 km mesh). (Source: Reproduced from Ota et al. (2023) with permission from Center for Environmental Information Science)

extreme phenomena (Japan Meteorological Agency 2023). In addition, recent studies have pointed out that the population and assets in flood hazard areas are increasing worldwide, and this brings another concern of expanding flood damage and increasing recovery costs in the future. Strategic management of climate change and urban stocks are interconnected, and flood control measures should be tailored to regional situations. It is necessary to predict the amount and location of disaster waste and prepare for the expected changes in land use regulations and building structures.

Figure 19.4 shows the amount of building stock by flood risk, overlaying Japan's building stock distribution with the flood-assumed area data. It has been revealed that about 42% of the total building stock nationwide exists within flood-assumed areas in 2020. The building stocks in flood-assumed areas have been increasing regardless of the flood level over the past 20 years. These greatly affect the increase in damage costs and disaster waste; thus, it will be important to control the population and material inflow into the hazard areas. In addition, damage can be suppressed by introducing adaptive measures such as strengthening the building structures and repositioning and reinforcing levees. Those measures should consider the flood risk and living conditions of each area. Simultaneously, the management of material circulation should be effectively promoted in line with the adaptive

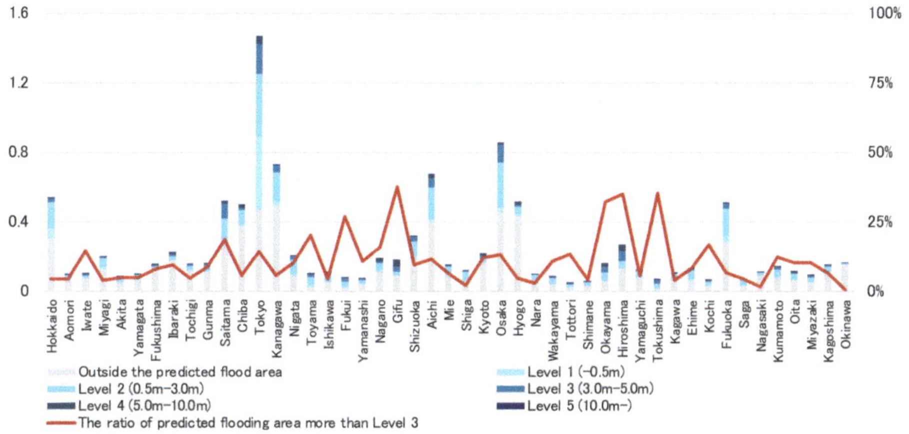

Fig. 19.4 Building stock (billion tonnes) in flood hazard areas in 2020. (Source: Reproduced from Hosokawa et al. (2024))

measures. For example, we can promote the long-term use of existing buildings in areas other than flood-assumed areas while relocating the buildings in the hazard area to other secured areas or reinforcing the structures.

19.3 Urban Metabolism and the Development of Material Stock Indicators

This section explains indicators for understanding the dynamics of material flow and stock. In Japan's Fundamental Plan for Establishing a Sound Material-Cycle Society, the entire material flow from resource extraction to final disposal is evaluated from three aspects: inflow, circulation, and outflow. Resource productivity is the indicator of GDP divided by the amount of natural resource inflow, cyclical use rate is the amount of circular use divided by the total material input or output amount (upstream or downstream of the flow), and the final disposal is the amount of waste that has been finally disposed of directly or after intermediate treatment. These indicators have been annually reported and evaluated over the years. Figure 19.5 shows the transition of each indicator and the numerical target for 2025. Although resource productivity has been somewhat stagnant from 2010 to 2015, it is steadily improving towards the target value of 490,000 yen/tonne for 2025 as a whole. The cyclical use rate was originally an indicator to measure the recycling rate on the inflow side (upstream of resource use), but individual estimates are made to distinguish it from resource recycling on the outflow side (downstream of resource use). The former is the amount of recycled materials accounting for the total material inflow, and the latter is the amount of recycled materials accounting for waste generated. The cyclical use rates on the inflow and outflow sides have been stagnant since 2013 and aim

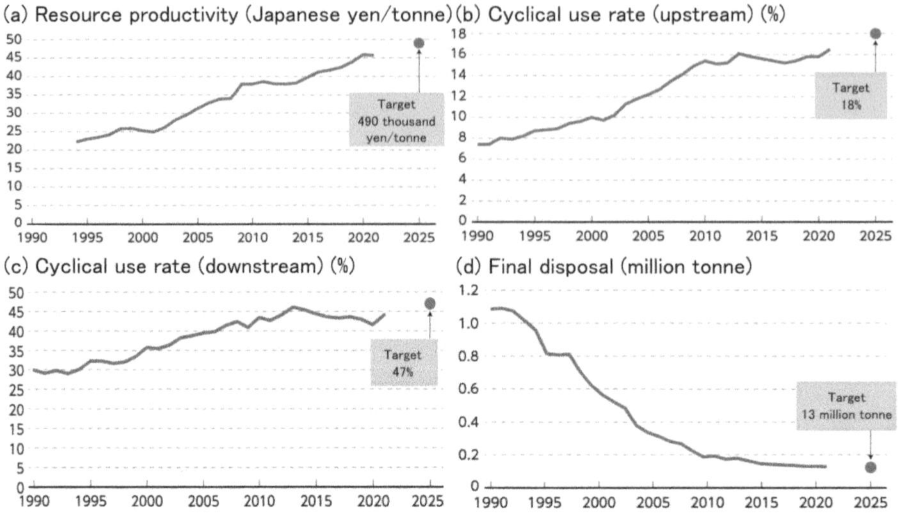

Fig. 19.5 Transition of material flow indicators and numerical targets. (Source: Reproduced from Ministry of the Environment (2024))

to reach 18% and 47% by 2025, respectively. The final disposal amount has decreased by about 80% since 1990, and has already achieved the target value of 13 million tonnes for 2025.

To evaluate the material flow and stocks throughout the lifecycle, it is desirable to use auxiliary indicators that evaluate the efficiency of material stock and its service in addition to the direct effects of resources and energy use by material flow indicators. However, as mentioned earlier, material stock research is still developing domestically and internationally. There is a demand for the development of indicators for organizing the classification of material stock and evaluating its quality and quantity.

19.3.1 Auxiliary Indicators of Material Stock Flow

In an attempt to discuss the efficiency of material stock use, a group of six indicators, including the material use time, which refers to the time the materials remain in society, and the recovery rate of used products, has been developed (Hashimoto and Moriguchi 2004). Material flow and stock are not independent but influence each other throughout the lifecycle. In other words, people's wealth is supported by a combination of material flow and stock; thus, a comprehensive framework that considers the consistency of material flow and stock indicators is necessary.

To evaluate the impact of material stock on society, we introduce the decomposition analysis of material productivity. In the decomposition analysis, material productivity, a material flow indicator, is decomposed into five indicators: primary material use rate, stock retention time, material stock productivity, service utilization, and service productivity. Among those, the three indicators of stock retention time, material stock productivity, and service utilization are proposed as material stock indicators (Yamashita et al. 2021; Tanikawa et al. 2021).

$$\begin{aligned}\text{Material productivity} = {} & \text{Primary material use rate} \times \\ & \text{Stock retention time} \times \\ & \text{Material stock productivity} \times \\ & \text{Service utilization} \times \text{Service productivity}\end{aligned}$$

$$\frac{\text{GDP}}{\text{DMI}} = \frac{\text{DMI} + \text{R}}{\text{DMI}} \times \frac{\text{MS}}{\text{DMI} + \text{R}} \times \frac{\text{Service}(\text{extant})}{\text{MS}} \times \frac{\text{Service}(\text{utilized})}{\text{Service}(\text{extant})} \times \frac{\text{GDP}}{\text{Service}(\text{utilized})}$$

$$\text{DMI} : \text{Direct material input}, \text{R} : \text{Recycled material}, \text{MS} : \text{Material stock}$$

(1) Stock Retention Time

Stock retention time is an indicator for observing the accumulation dynamics of materials by comparing the scale of the material flow in the denominator with that of the material stock in the numerator.

$$\text{Stock retention time} = \frac{\text{Material stock}}{\text{Overall material inflow}}$$

Material stock is measured by weight, and material flow is indicated by weight per unit of time, so the indicator has a unit of time. Stock retention time requires different interpretations depending on the country's physical capital development level. An increase in the stock retention time (a decrease in the ratio of material flow to material stock) means that the demand for material flow for maintenance and replacement of material stock has decreased, and in a mature society, it can be interpreted that the material flow has been suppressed simply by the lifespan extension of the material stock. High-quality material stock intended for long-term use tends to have higher social costs and resource and energy use in the initial stages, and continuous maintenance is required due to physical deterioration. Nevertheless, the inflow and outflow can be less in the long term compared to short-life structures. On the other hand, in emerging countries where the capital is being expanded, the increase in the stock retention time is a concern about an increase in low-quality material stock, i.e., a situation where the material flow is not catching up with the rapid formation of material stock.

(2) Material Stock Productivity

Material stock productivity is an indicator obtained by dividing the designed service volume by the total material stocks, which measures the efficiency of the service provided by the material stock.

$$\text{Material stock productivity} = \frac{\text{Service provision}}{\text{Material stock}}$$

The increase in material stock productivity can be seen as an improvement in the service expected per unit of material stock. For example, technological innovation leads to energy conservation and higher functionality, and it allows to maintain durability and services with less material than before; thus, it can be interpreted that the service capacity has improved. Suppose that the service of a car is a transport function, it can be thought that the service quantity has improved by maintaining the same transport weight while using fewer and lighter materials.

(3) Service Utilization

Service utilization is an indicator that measures to what extent the designed service capacity is being effectively used, indicating the satisfaction of the service provided by the material stock.

$$\text{Service utilization} = \frac{\text{Service provision}\left(\text{utilized}\right)}{\text{Service provision}\left(\text{extent}\right)}$$

Examples of service utilization include the actual traffic volume compared to the designed traffic volume for roads, and the actual number of passengers or load capacity compared to the designed transport capacity for automobiles and railways. Among the material stocks, there are various use stages, ranging from those currently in use (providing services to society), those available but not being effectively used, and those unavailable and unlikely to be used in the future. A decrease in service utilization means an increase in unnecessary stocks for society, and there is concern about the negative impact on material metabolism. The closer the indicator is to 1, the more efficiently the service is being utilized. From the perspective of material circulation, it is desirable to maintain this state. If the service utilization is less than 1, the material stock is excessive and is expected to have many unused stocks, such as vacant houses and abandoned stations and railways in society. On the contrary, if the service utilization is greater than 1, traffic congestion, overloading, and overcrowding of medical facilities may occur. These are all undesirable states for society and may interfere with healthy human life. However, the service utilization of some material stocks, such as disaster prevention facilities, should be interpreted with caution as an exception. These are designed with peak demand times (e.g. in the event of a natural disaster). Therefore, the service utilization during normal times may be extremely low, and if the designed service capacity is reduced to make the indicator appropriate, there is a risk that the stock service will be insufficient when needed, causing confusion in the socioeconomy.

The decomposition formula of material productivity can contribute to environmental policy from two aspects: as a tool for analyzing the current situation that clarifies the impact of changes in material stock on material flow, and for setting medium to long-term goals by backcasting from the future image of a circular economy that Japan aims to achieve. The latter is useful in discussing the long-term sustainability of society, economy, and environment in terms of how and by when we should change each indicator and its components to be effective from the perspective of material metabolism.

19.3.2 The Impact of Changes in Stock Retention Time on New Material Inflow and CO_2 Emissions in the Future

Construction activities are directly linked to resource consumption and GHG emissions, playing a key role in climate change adaptation and mitigation. Several studies have discussed the impact of building lifespans on material metabolism and CO_2 emissions but most of these studies focused on individual buildings and have not reached a long-term quantitative analysis of a whole society. The CO_2 emissions related to construction activities in 2019 are estimated to be 12 GtCO_2-eq (about 32% of total emissions) worldwide. Especially in some industries with high carbon emission intensity, such as cement and steel, it is expected that regulations on GHG emissions will become increasingly strict, requiring urgent measures towards decarbonization. Here, we consider the retention time of materials to be an important key to sustainability. Stock retention time affects future material metabolism and CO_2 emissions throughout the lifecycle; thus, the materials input into society should be effectively utilized on the premise of long-term use.

Figure 19.6 shows the changes in new material inflow and CO_2 emissions by extending the retention time of building stocks in Japan. Scenario I maintains the current situation without changing the lifespan of the building stock in the future, while Scenario II is a lifespan extension scenario that doubles the lifespan of buildings by 2050. For the buildings that have reached their lifespan, we assume (a) a scenario where they are properly demolished and recycled, (b) a scenario where the concrete wastes are crushed and left after demolition, and (c) a scenario where the number of buildings left without any appropriate treatment increases in the future. (C) is the scenario against Japan's background of the current increase in the vacancy rate of buildings. We compare the future new construction material inflow and its accompanying CO_2 emissions by the combinations of those scenarios. Scenario (b) is set by considering the CO_2 absorption of concrete, which has been attracting attention in recent years. Concrete waste is assumed to be crushed with the aim of increasing the surface area of concrete.

By extending the lifespan of building stocks, the amount of new material inflow and CO_2 emissions can both be reduced by about 50% by 2100. The amount of new material inflow decreases slightly more in scenario (a) than in the other scenarios. As for CO_2 emissions, they are about 5% higher in the scenario II that extends the

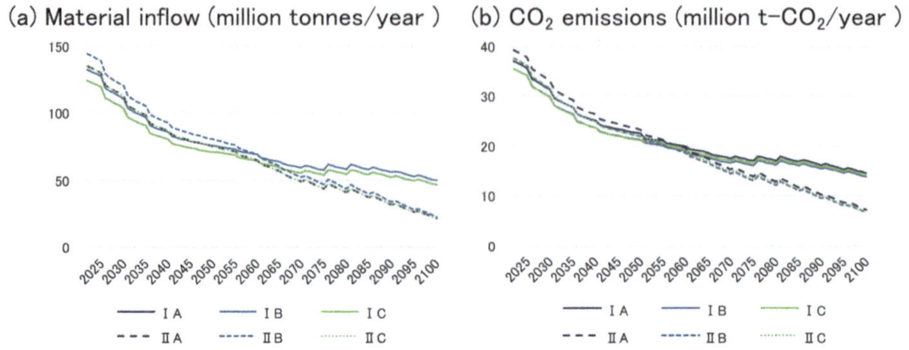

Fig. 19.6 Future material use and CO_2 emissions (**a**) new material inflow (million tonnes/year), (**b**) CO_2 emissions (million t-CO_2/year) in buildings. (Source: Reproduced from Fujikawa et al. (2024) with permission from Center for Environmental Information Science)

building lifespan than in the scenario I that maintains the status quo, for the first few years of future predictions. Then, they are expected to decrease due to the lifespan extension, as they reverse after 2058. There was no significant difference in CO_2 emissions depending on how to deal with buildings that have reached their lifespan, but leaving them demolished and crushed slightly reduces CO_2 emissions compared to actively recycling the demolition waste. This is thought to be because the amount of carbon absorbed temporarily increases as the surface area of concrete exposed to the air increases due to demolition waste crushing. However, leaving demolition waste crushed and abandoned requires more than twice the amount of new materials as other scenarios in terms of material consumption and requires space for storage. It is important to discuss those problems from the multifaceted perspectives of resource recycling and social aspects is necessary, and evaluate the advantages and disadvantages of adaptation and mitigation.

19.4 Conclusion

In a stock-type society, the goal is to shift from short-lived structures that need to be repeatedly rebuilt with each generation to long-lived structures considering material stocks as “assets” in society (Fig. 19.7). Although long-lived structures need to respond to future socioeconomic and natural risks adequately, the accumulation of material stocks beyond generations leads to abundant lives for the next generation through the reduction of resource and energy consumption and environmental load. To use material stocks for a long period of time, it is necessary to adapt structures and facilities for climate change in the future and to design them durable enough for long-term use. This also contributes to climate change mitigation measures through GHG emission reduction and the improvement of urban sustainability.

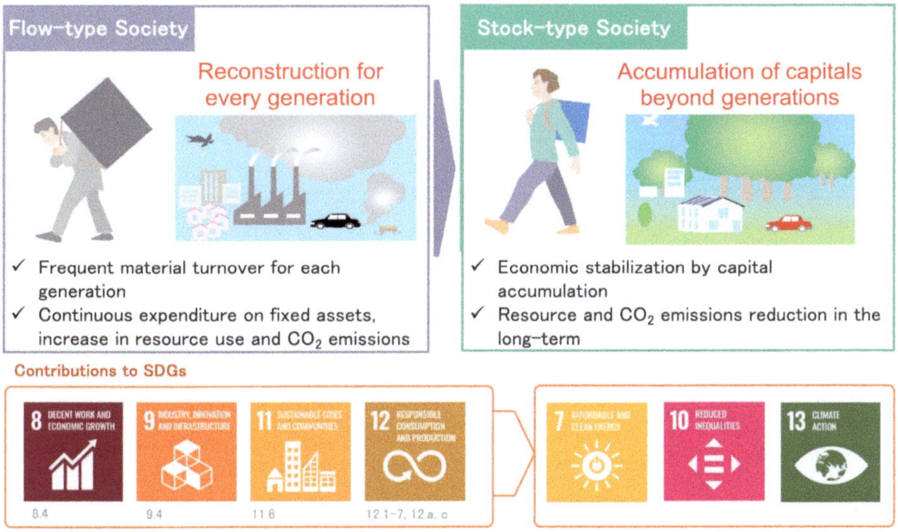

Fig. 19.7 Flow-type society vs stock-type society. (Adapted from Okamoto (2006) with permission form Kajima Institute Publishing Co., Ltd.)

References

Fujikawa N, Yamashita N, Hiruta Y et al (2024) Assessment of the Impact on Future Construction Material Inflow and CO2 Emissions due to Changes in the Retention Time of Buildings. J Environ Inf Sci

Hashimoto S, Moriguchi Y (2004) Proposal of six indicators of material cycles for describing society's metabolism: From the viewpoint of material flow analysis. Resour Conserv Recycl 40:185–200. https://doi.org/10.1016/S0921-3449(03)00070-3

Hosokawa R, Nagata S, Shirakawa H, Tanikawa H (2024) Spatiotemporal analysis of material stock and flow on construction sector in flood hazard areas. J JSCE Ser G (Environ Res). (In press)

Japan Meteorological Agency (2023) Climate Change Monitoring Report 2023

Krausmann F, Wiedenhofer D, Lauk C et al (2017) Global socioeconomic material stocks rise 23-fold over the 20th century and require half of annual resource use. Proc Natl Acad Sci USA 114:1880–1885. https://doi.org/10.1073/pnas.1613773114

Matthews Emily (2000) The weight of nations: material outflows from industrial economies. World Resources Institute

Ministry of the Environment (2024) Annual report on the environment, the sound material-cycle society and biodiversity in Japan 2024

Okamoto H (2006) Transition to a stock-type society: infrastructures in the era of longevity. Kajima Institute Publishing Co., Ltd

Ota Y, Hiruta Y, Yamashita N et al (2023) Material stock and flow estimation by identifying the congruency of urban structures between generations. J Environ Inf Sci 37:195–201. https://doi.org/10.11492/ceispapers.ceis37.0_195

Tanikawa H, Fishman T, Okuoka K, Sugimoto K (2015) The weight of society over time and space: A comprehensive account of the construction material stock of Japan, 1945–2010. J Ind Ecol 19:778–791. https://doi.org/10.1111/jiec.12284

Tanikawa H, Fishman T, Hashimoto S et al (2021) A framework of indicators for associating material stocks and flows to service provisioning: Application for Japan 1990–2015. J Clean Prod 285:125450. https://doi.org/10.1016/j.jclepro.2020.125450
Yamashita N, Guo J, Shirakawa H, Tanikawa H (2021) Empirical research of resource productivity decomposition considering material stock. Case Study of Japan's Residential Buildings, pp II_23–II_31. https://doi.org/10.2208/jscejer.77.6_II_23

Chapter 20
Urban Transport in a Warming World: Adapting to Climate Challenges

Feifan Xu and Hirokazu Kato

Abstract This chapter comprehensively evaluates the risks of urban transport systems on natural hazards—including bus operation bases, railway networks, and road freight transport—and their responses to the increasing threat of climate change. The chapter introduces an integrated risk evaluation methodology that incorporates hazard, exposure, and vulnerability assessments by analyzing climate-induced hazards such as flooding, landslides, and complex multi-hazard scenarios. The findings offer a detailed analysis of how these transport systems are impacted by extreme weather events, highlighting specific vulnerabilities and risks associated with different types of infrastructure. Furthermore, the chapter outlines adaptive strategies to enhance resilience, such as strengthening physical structures, improving flood defenses, and adopting real-time monitoring systems. These strategies provide urban planners and policymakers with actionable insights to mitigate risks and ensure the long-term sustainability of transport networks. We show that a multi-hazard approach and integrated risk management contribute to developing climate-resilient urban transport systems.

Keywords Urban transport resilience · Climate-induced hazards · Flooding and landslides · Risk evaluation · Bus operation bases · Railway networks · Road freight transport

20.1 Introduction

This chapter is part of the research outcomes of the S-18 Project, which aims to provide scientific information to support Japan's climate change adaptation efforts. The S-18 Project was initiated in response to the growing impacts of climate change and the need for Japan to address these challenges, following the enactment of the Climate Change Adaptation Act in 2018. The project supports the development of

F. Xu (✉) · H. Kato
Nagoya University, Nagoya, Japan
e-mail: xu.feifan.r2@f.mail.nagoya-u.ac.jp

N. Mimura, S. Takewaka (eds.), *Climate Change Impacts and Adaptation Strategies in Japan*, https://doi.org/10.1007/978-981-96-2436-2_20

adaptation strategies by local governments and businesses, contributing to both national and international climate goals.

Within this framework, our research under Sub-theme 4–4 focuses on the impacts of climate change on Japan's transportation systems and the evaluation of adaptation strategies. This chapter specifically examines the vulnerabilities of urban transport networks, such as bus operation bases, railway networks, and road freight transport, to climate-induced threats like increased typhoon intensity and frequent heavy rainstorms. The findings are based on comprehensive risk assessments and aim to provide targeted strategies to enhance urban resilience.

The Cabinet Secretariat (2024) outlines Japan's Fundamental Plan for National Resilience, which provides a strategic framework for addressing climate-related risks. This plan has been instrumental in shaping the country's approach to climate adaptation, particularly in sectors like urban transport. Urban transport networks are the backbone of modern cities, enabling the seamless movement of people, goods, and services while supporting economic growth and social development. However, these networks face increasing threats due to climate change, which is driving a rise in the frequency and intensity of extreme weather events such as floods, landslides, and prolonged heavy rainfall. The implications for urban transport systems are profound, as these changes disrupt daily operations, compromise safety, and necessitate costly repairs and maintenance.

In recent years, the impacts of climate change on transport systems have become increasingly evident. Natural disasters, exacerbated by climate change, have caused widespread damage to infrastructure and significant economic losses globally. Japan, for example, has experienced an alarming rise in the frequency and severity of water-induced disasters affecting its transport networks. Between 2008 and 2018, 211 out of 275 railway disasters causing damages exceeding ten million yen were due to water-related incidents, such as floods and landslides. This highlights the urgent need to understand the specific risks posed by climate change to transport infrastructure and develop adaptive strategies to enhance resilience.

Given the critical role of urban transport in ensuring the functionality and sustainability of cities, there is an increasing need to understand the vulnerabilities of different transport systems and develop targeted adaptation measures. This chapter provides a comprehensive analysis of the challenges posed by climate change to urban transport networks, focusing on evaluating risks to specific infrastructures like bus operation bases, railway networks, and road freight transport. By integrating findings from various studies, the chapter aims to provide actionable insights and strategies for enhancing the resilience of urban transport systems in a warming world.

20.2 The Growing Threat of Climate Change

The impacts of climate change are manifesting more frequently, with significant increases in extreme weather events such as heavy rainfall, intense storms, and prolonged droughts. These changes have profound implications for urban transport networks, which are particularly susceptible to disruptions from natural disasters.

20.2.1 Overview of Climate Change Impacts on Urban Transport

Global climate models predict that, without significant mitigation efforts, average global temperatures could rise by 2 °C to 4 °C by the end of the twenty-first century. This increase in temperature will likely lead to more frequent and severe natural disasters, including storms, floods, and landslides. In Japan, the frequency of short-term heavy rainfall events (exceeding 80 mm per hour) has nearly doubled from 1980 to 2020, directly impacting transport infrastructure. The Intergovernmental Panel on Climate Change (IPCC) also warns that urban areas will be disproportionately affected due to their high concentration of people, assets, and economic activities.

Transport networks are especially vulnerable to the effects of floods, landslides, and other climate-induced hazards. The damage to transport infrastructure not only disrupts daily commuting but also affects economic activities, logistics, and emergency responses. Future climate scenarios indicate that the risk of these natural disasters will increase substantially, with more frequent and severe impacts on transport networks expected under a range of potential pathways.

20.2.2 Specific Climate Risks for Urban Transport Infrastructure

1. *Flooding Threats*

   Flooding poses a critical challenge to urban transport systems, particularly in low-lying areas and regions near water bodies. Increased rainfall and rising sea levels due to climate change exacerbate flood risks, threatening both above-ground and underground transport networks. For instance, under the SSP5 scenario, flood risks for railway lines are projected to increase significantly, affecting lines near rivers and other flood-prone areas. Floods can cause severe damage to road surfaces, subways, tunnels, and bridges, leading to prolonged service interruptions and increased maintenance costs.

   The increasing intensity and frequency of rainfall events are expected to overwhelm existing drainage systems, leading to more frequent urban flooding. Flooding of roads and railways not only disrupts daily transportation but also has cascading effects on emergency response and recovery efforts, as damaged infrastructure can delay the delivery of essential supplies and services. Thus, understanding and mitigating flood risks is a priority for urban planners and policymakers.

2. *Landslide Hazards in Mountainous Regions*

   Landslides represent a significant hazard to road and rail networks, especially in mountainous areas where heavy rainfall can trigger soil erosion and slope failures. These landslides can obstruct roads and railways, damage infrastructure, and disrupt transport services for extended periods. Transport networks located in areas with steep terrain and high rainfall are particularly vulnerable. Regions with unstable geological conditions or insufficient vegetation cover are at an increased risk of landslides. Factors such as deforestation, urban expansion, and intense rainfall contribute to the likelihood of slope failures. Consequently, transport routes that traverse mountainous or hilly terrains require focused attention in risk assessments and resilience planning.

3. *Complex Multi-hazard Scenarios*

   Transport networks face complex multi-hazard scenarios, where multiple climate risks, such as floods, landslides, and extreme weather events, occur concurrently or sequentially. For example, intense rainfall following a prolonged dry period can lead to flash floods, landslides, and subsequent infrastructure damage. These overlapping hazards complicate risk assessments and necessitate integrated strategies that address multiple risks simultaneously.

20.3 Methodology for Evaluating Risks to Urban Transport Networks

Risk evaluation is a critical step in understanding how urban transport networks are affected by natural disasters and climate change. By systematically evaluating risks, planners and policymakers can prioritize interventions, allocate resources effectively, and enhance the resilience of critical infrastructure. The methodology for evaluating risks in or research involves a comprehensive analysis of three key indicators: *Hazard*, *Exposure*, and *Vulnerability*. Each indicator provides a distinct perspective on the factors contributing to the overall risk, allowing for a more nuanced and targeted approach to risk management.

20.3.1 Hazard: Understanding Potential Natural Disasters

The term “hazard” refers to the potential occurrence of natural events that could cause harm or damage to urban transport networks. This indicator focuses on identifying and quantifying the types and intensities of natural threats that may affect transport infrastructure, such as floods, landslides, and other extreme weather events. Key Considerations for Hazard Evaluation:

1. *Probability of Occurrence*

   Hazard evaluation begins with understanding the likelihood or probability of a particular natural disaster occurring in a specific location. This involves analyzing historical data on natural events, predictive climate models, and local geographic conditions. For example, if an area has a history of frequent flooding, the probability of future floods is considered high. Advanced climate models provide future scenarios of hazard occurrence under various conditions, such as different greenhouse gas emission pathways.

2. *Intensity and Severity*
   Another critical aspect of hazard evaluation is the intensity or severity of the potential event. For transport networks, the intensity of a hazard determines the extent of damage it could cause. For example, the depth and duration of flooding, the volume and speed of landslides, or the peak temperatures during a heatwave are all factors that influence the severity of the hazard. High-intensity events, such as a flood with significant water depth or a landslide covering a large area, pose greater risks to transport infrastructure.

3. *Spatial and Temporal Distribution*
   Hazards are not uniformly distributed across urban areas; their likelihood and impact vary depending on local geographic and environmental conditions. Hazard evaluation must account for spatial variations, such as areas prone to flooding due to low elevation or poor drainage. Additionally, temporal factors, such as seasonal patterns, are considered. For example, certain areas may experience higher risks during the rainy season or periods of rapid snowmelt.
   By understanding these aspects of hazard, urban planners can better predict where and when natural events may occur and their potential impact on transport infrastructure.

20.3.2 Exposure: Evaluating Impact on Transport Infrastructure

Exposure represents the extent to which transport infrastructure and its users are likely to be affected by a hazard. This indicator considers the presence of people, assets, and activities in areas where natural disasters are likely to occur. The more critical or heavily used a transport route or facility is, the higher its risk exposure. Key Considerations for Exposure Evaluation:

1. *Traffic Volume and Usage Patterns*
   One of the main factors in evaluating exposure is the volume and pattern of traffic on specific transport routes. High-traffic routes, such as major highways, railway lines, and bus operation bases in urban centers, have a higher exposure level due to the greater number of users and economic activities dependent on them. Understanding the flow of people and goods across different routes helps identify which segments are most critical and would be most affected by disruptions.

2. *Criticality of Transport Infrastructure*
   Not all transport infrastructure is equally critical. Some routes or nodes, such as those connecting to major hospitals, airports, or industrial hubs, are considered more critical because their disruption would have broader social and economic

impacts. Evaluating exposure involves identifying these key assets and understanding their role within the wider transport network. For instance, a railway line that serves as the primary commuter route for a large population has higher exposure than a secondary line with less traffic.

3. *Interdependencies and Network Effects*
   Urban transport networks are interconnected systems where the disruption of one segment can have cascading effects throughout the network. Evaluating exposure also involves understanding these interdependencies. For example, if a major road or rail line is disrupted, the impact might extend to alternative routes that become overloaded or to facilities dependent on continuous service. Identifying these network effects is crucial for understanding the full scope of potential exposure.
   By assessing exposure, planners can prioritize investments in more critical infrastructure and develop strategies to minimize the impact on transport services and users.

20.3.3 Vulnerability: Identifying Susceptibility to Damage

Vulnerability refers to how susceptible transport infrastructure is to damage or failure when exposed to hazards. This measure considers various factors related to the infrastructure itself, including its design, condition, level of maintenance, and resilience to specific types of hazards. As highlighted by Suzuki et al. (2020), the evaluation of areas prone to sediment disasters underscores the importance of climate change adaptation in reducing future risks. Key considerations for evaluating vulnerability include:

1. *Physical Condition and Design of Infrastructure*
   The age, materials, and design standards of transport infrastructure significantly influence its vulnerability. Older infrastructure or that built to outdated standards may not withstand modern climate impacts. For example, a railway bridge constructed in the early twentieth century may not be designed to handle the current intensity of floods or storms. The vulnerability evaluation considers these factors to determine how likely the infrastructure is to be damaged under specific hazard scenarios.

2. *Maintenance and Upkeep*
   Regular maintenance is crucial for maintaining the integrity and resilience of transport infrastructure. Neglected or poorly maintained infrastructure is more vulnerable to damage from natural disasters. Evaluating the current state of maintenance, identifying critical weaknesses, and assessing the readiness for repair and reinforcement are key components of vulnerability evaluation.

3. *Adaptive Capacity and Flexibility*
   The ability of transport infrastructure to adapt to changing conditions, such as modifications for flood resistance or the use of heat-resistant materials, reduces its

vulnerability. Assessing how easily infrastructure can be upgraded or adapted is also an important aspect of this evaluation. For instance, can roads be easily elevated or retrofitted with better drainage systems? Can railway lines be relocated or reinforced? The answers to these questions help determine the level of investment needed to improve resilience.

4. *Operational and Functional Dependence*
   Vulnerability also considers how dependent the functionality of the transport network is on certain key segments. If a particular road or railway is critical for the network's overall operation, its vulnerability is higher because its failure would cause widespread disruption. Conversely, having alternative routes or backup systems reduces the vulnerability of the transport network.

By evaluating vulnerability, transport planners and policymakers can identify the most susceptible elements within the network and prioritize them for reinforcement or upgrades.

20.3.4 Integrated Risk Evaluation Approach

The overall risk to urban transport networks is determined by the combined evaluation of hazard, exposure, and vulnerability. This integrated approach allows for a comprehensive understanding of potential threats, their impacts, and the weaknesses within the transport system that need to be addressed.

The risk for each transport segment or facility is evaluated using a combined score derived from the evaluation of hazard (H), exposure (E), and vulnerability (V):

$$\text{Risk}(R) = H \times E \times V$$

This formula provides a quantifiable measure of risk, which can then be used to prioritize areas for intervention, allocate resources, and develop targeted adaptation strategies.

By applying this risk evaluation methodology, urban planners can systematically identify the transport infrastructure most at risk from climate-induced hazards and develop effective resilience-building strategies.

20.4 Risk Assessments of Urban Transports

This section applies the methodology discussed in Sect. 20.3 to evaluate risks for specific transport infrastructures in Japan, focusing on three main categories: *Bus Operation Bases*, *Railway Networks*, and *Road Freight Transport*. The evaluation involves calculating the three indicators—*Hazard*, *Exposure*, and *Vulnerability*—for each type of infrastructure, determining their overall risk levels.

20.4.1 Bus Operation Bases

The research focuses on 2375 bus operation bases nationwide, compiled by the Transport Bureaus of the Ministry of Land, Infrastructure, Transport, and Tourism (MLIT 2020a, b). These bases were confirmed to be in operation on regular and fixed routes as of December 2020, based on data from business websites and local government websites.

The study examines three types of disasters: floods, tsunamis, and landslides, all expected to worsen due to future climate change. Data on "anticipated flood inundation areas (planned scale)," "anticipated tsunami inundation areas," and "predicted landslide disaster areas" were obtained from the National Land Information Division (MLIT 2020a, b) and municipal websites to assess the risk of damage to bus operation bases within these areas (Fig. 20.1).

Approximately 42% of bus operation bases (992 in total) are expected to be affected by various types of disasters, especially in major cities such as Tokyo, Osaka, and Nagoya. Figure 20.2 illustrates the number of bus operation bases predicted to be affected by different types of disasters, categorized by the number of vehicles associated with each base. This visualization highlights the distribution of

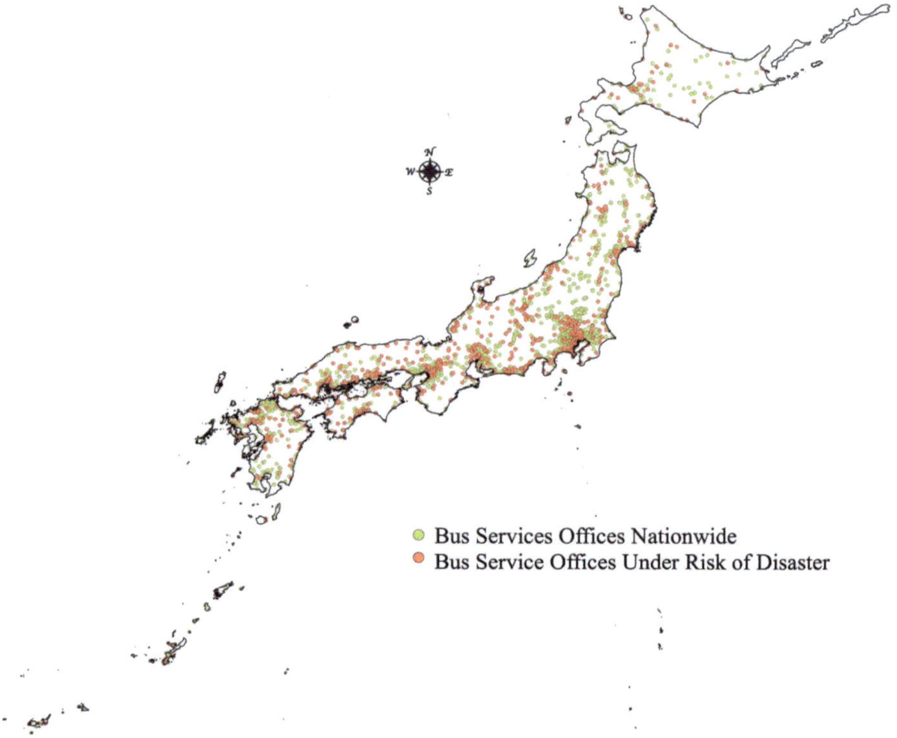

Fig. 20.1 Distribution of bus operation bases expected to be affected by disasters

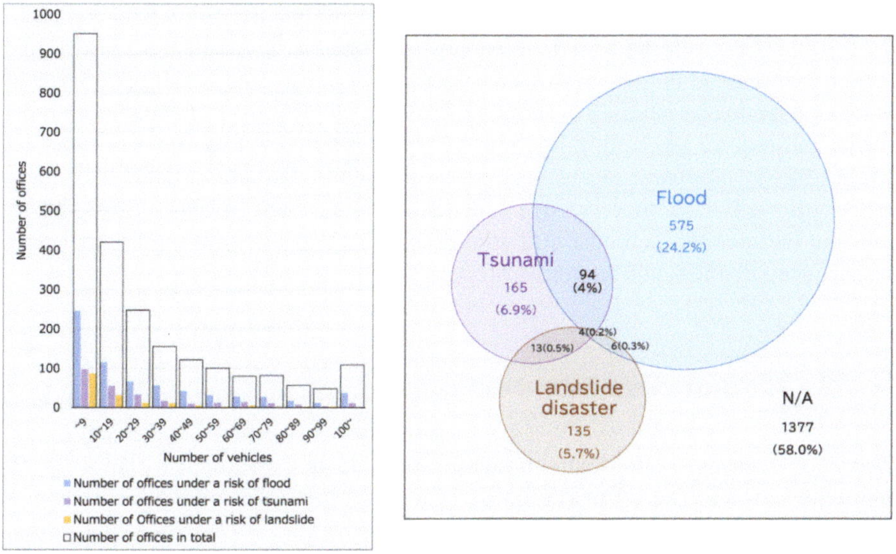

a. by the number of vehicles *b. by the disaster type*

Fig. 20.2 The number of bus operation bases expected to be affected by disasters. (Reproduced from Xu et al. (2023) with permission from Academic Star Publishing Company)

Table 20.1 Bus operation bases with highest disaster risk under each evaluation pattern

	Disaster risk index											
Rank	Pattern1		Pattern2		Pattern3		Pattern4		Pattern5		Pattern6	
1	A	0.558	A	0.530	A	0.454	A	0.544	A	0.564	A	0.675
2	B	0.363	C	0.381	B	0.375	C	0.348	B	0.322	C	0.391
3	C	0.356	B	0.378	C	0.315	B	0.340	C	0.301	D	0.342
4	D	0.287	D	0.320	D	0.261	E	0.294	G	0.278	F	0.325
5	E	0.285	F	0.315	F	0.257	G	0.285	E	0.276	B	0.310

Reproduced from Xu et al. (2023) with permission from Academic Star Publishing Company

risk across different bases, especially those with a higher number of vehicles, which may indicate greater operational importance or vulnerability.

The findings in Table 20.1 are derived from a combination of vulnerability and importance indices to determine the disaster risk levels for each bus operation base. The vulnerability index was calculated based on the probability of occurrence for various disasters—such as floods, tsunamis, and landslides—using data from digital national land information and other relevant sources. The importance index was established by considering multiple factors, including the scale of bus operation bases, the number of passengers, and the population served by each bus stop. Different weighting patterns were applied to emphasize varying priorities (e.g., size, highway bus importance, and the impact on surrounding areas). Table 20.1 provides a ranked list of the top five bus operation bases with the highest disaster risk under different evaluation patterns. Due to the utilization of data that contained confidential information from the railway company, the specific name of the railway line cannot be disclosed in this paper. Consequently, for clarity, they shall be referred to using alphabetic characters. This ranking assists in identifying which bases are most at risk and require targeted mitigation strategies, such as relocation, decentralization, or the development of business continuity plans. The combined use of these indices allows for a comprehensive understanding of which bases are most vulnerable and the specific factors contributing to their risk levels, thus guiding effective disaster preparedness and response efforts.

20.4.2 Railway Networks

The study focuses on "ordinary railways" that are not classified as subways, as defined by MLIT. These railway lines are used as the target routes for flood risk evaluation (Fig. 20.3).

Flood risk evaluation utilizes prediction data from Yanagihara et al. (2022), which includes current and future flood damage projections under different climate models. The study considers five global climate models, two socioeconomic pathways (SSP1 and SSP5), and two land-use scenarios to estimate future flood risks. For this evaluation, the term "standard climate" refers to a baseline climate scenario representing the historical climate conditions before the projected changes due to global warming. This baseline is used as a reference point to assess the recurrence period of flood damage under different future scenarios.

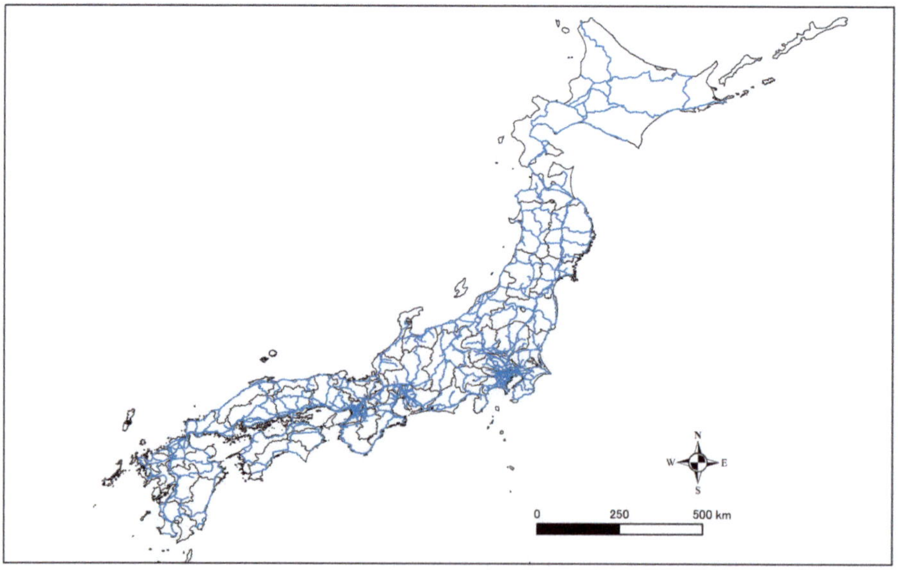

Fig. 20.3 Railway lines subject to evaluation. (Reproduced from Xu et al. (2024) with permission from Eastern Asia Society for Transportation Studies)

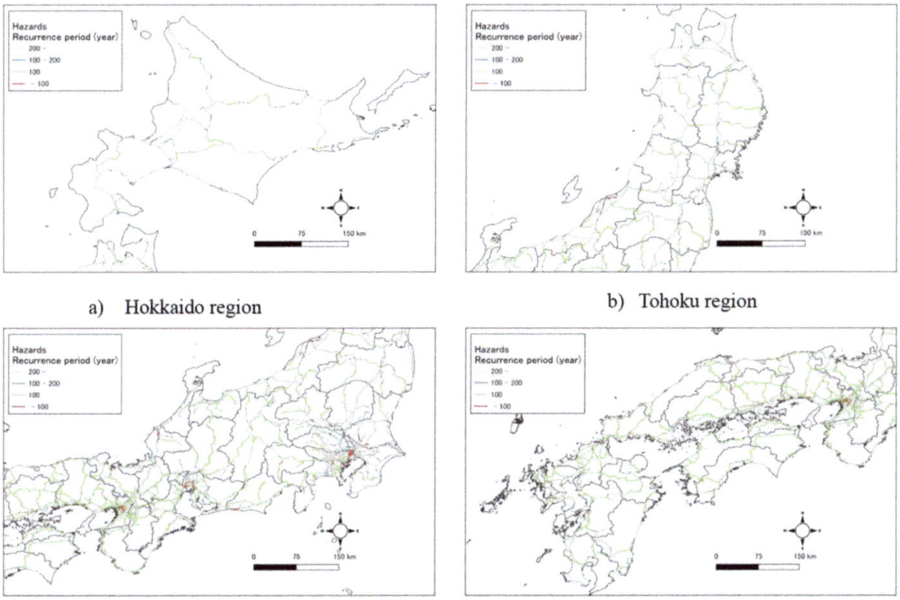

a) Hokkaido region

b) Tohoku region

c) Kanto・Chubu・Kansai region

d) Chugoku・Shikoku・Kyushu region

Fig. 20.4 Recurrence period of flood damage equivalent to the threshold in the standard climate (by region). (Reproduced from Xu et al. (2024) with permission from Eastern Asia Society for Transportation Studies)

Figure 20.4 shows the "Recurrence period of flood damage equivalent to the threshold in the standard climate (by region)." The figure is based on the comparison of flood risks under the standard climate scenario against various future scenarios, incorporating different global climate models, socioeconomic pathways, and land-use conditions. By comparing the set thresholds with corresponding recurrence periods, the study calculates "hazardous sources" for various railway segments under these conditions.

The evaluation indicates that flood risks for railway networks will significantly increase under future climate scenarios, especially along routes near major metropolitan areas (Fig. 20.5). The findings suggest that flood prevention investments should be prioritized in high-risk segments. Table 20.2 lists the top three Railway Routes with Largest Increase in Hazardous Sources.

20.4.3 Road Freight Transport

The study focuses on key logistics routes (ILRs) within the Chukyo metropolitan area in Japan, including major freight transport roads such as highways and essential logistics routes. The risk evaluation draws on data from the National Freight Net Flow Survey (Logistics Census) and the Freight Regional Flow Survey (MLIT 2021).

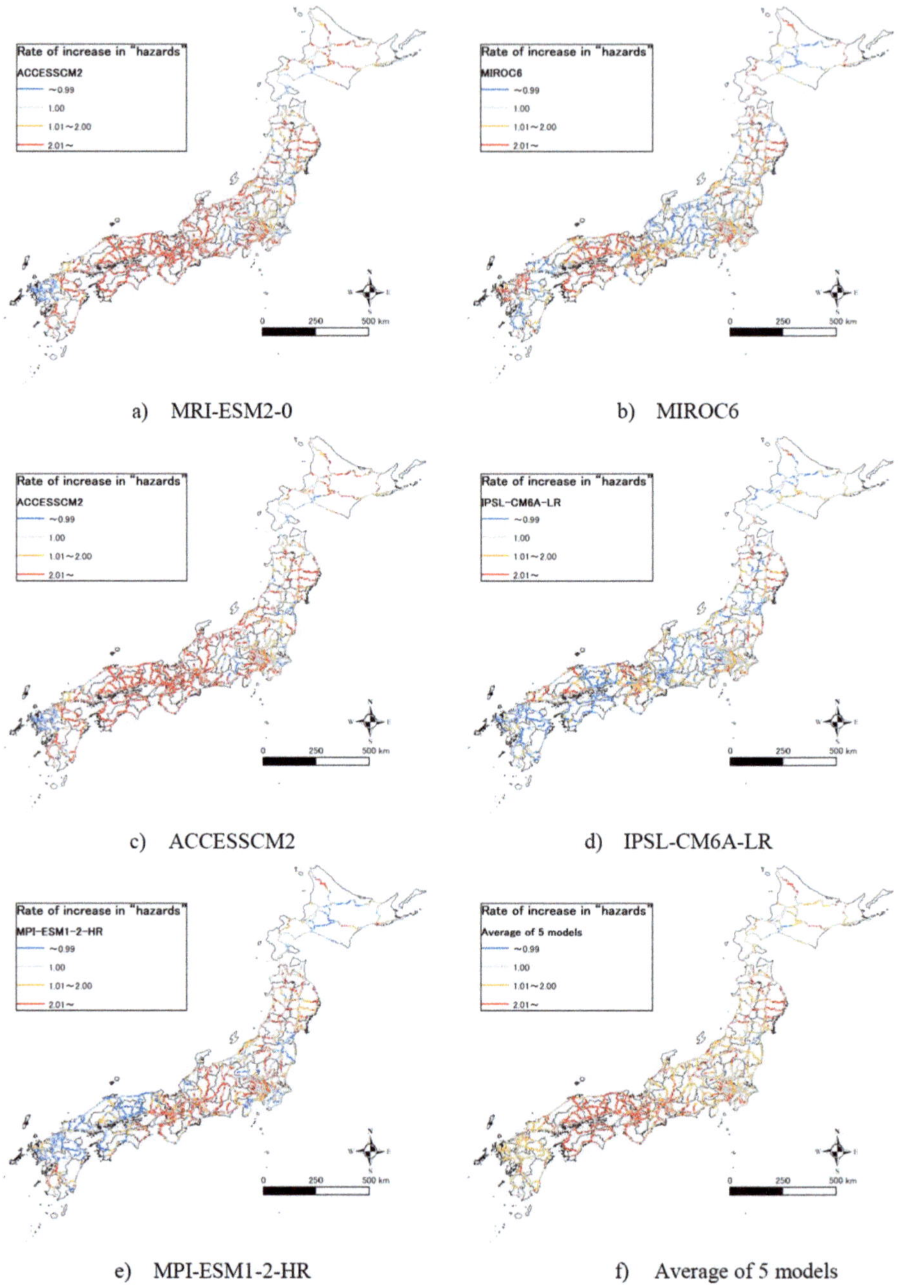

Fig. 20.5 Rate of increase in "hazards" under land use fixed scenario, SSP1 in the near-future climate (by global climate model). (Reproduced from Xu et al. (2024) with permission from Eastern Asia Society for Transportation Studies)

Table 20.2 Top 3 railways with the largest increase in the "hazardous sources" compared to the standard climate in each scenario

(a) Land-use Fixed Scenario					
	No.	SSP1	Increase rate (%)	SSP5	Increase rate (%)
2031 ~ 2050 (Near future climate)	1	TWML	11.84	TYL	10.88
	2	KSL	10.54	FEL	7.74
	3	TYML	10.26	TWML	7.19
2081 ~ 2100 (End of twenty-first century climate)	1	YDL	14.24	TWML	14.55
	2	KGRML	9.84	KSL	13.14
	3	TWML	8.05	KGRML	11.53
(b) Land-use changed scenario					
	No.	SSP1	Increase rate (%)	SSP5	Increase rate (%)
2031 ~ 2050 (Near future climate)	1	TWML	11.38	TYL	10.16
	2	KSL	10.31	FEL	7.60
	3	TYML	9.91	TWML	6.87
2081 ~ 2100 (End of twenty-first century climate)	1	YDL	13.94	TWML	12.80
	2	KGRML	9.86	KSL	12.16
	3	TYML	7.39	KGRML	11.91

Reproduced from Xu et al. (2024) with permission from Eastern Asia Society for Transportation Studies

The evaluation employs future precipitation forecasts from sources like Yanagihara et al. (2022) and slope failure risk evaluation to calculate potential risks for major freight transport roads (Figs. 20.6, 20.7). The analysis synthesizes data to estimate future flood depths and landslide probabilities, evaluating the impact of road closures on freight logistics under future climate conditions.

The study shows that, under future climate change scenarios, approximately 225.2 km of roads are at risk of flooding, accounting for 9% of all evaluated ILRs. Additionally, 173.2 km of roads may face elevated landslide risks. These high-risk areas should be prioritized for preventive measures.

20.5 Adaptive Strategies for Urban Transport Networks

To address the growing risks posed by climate change, it is essential to implement a range of adaptive strategies tailored to different types of transport infrastructure. This section discusses adaptive strategies for bus operation bases, railway networks, and road freight transport.

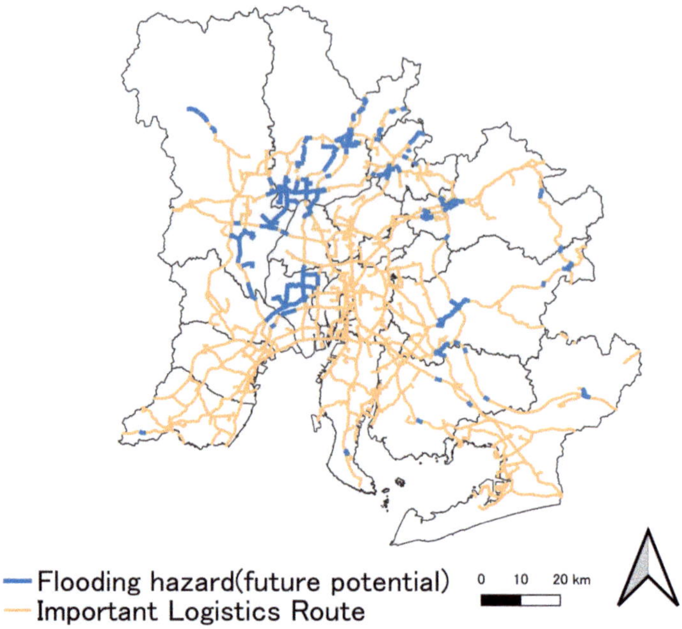

Fig. 20.6 Results of flooding hazard potential for road freight transport

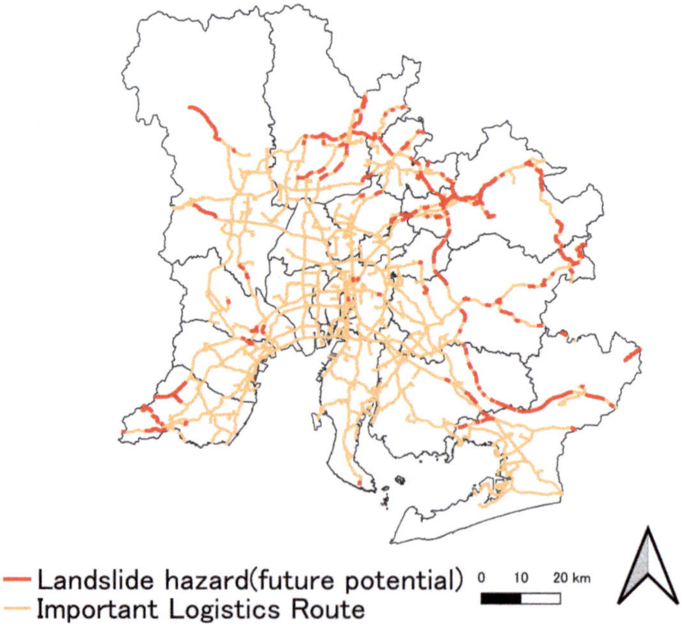

Fig. 20.7 Results of future landslide hazard potential for road freight transport

20.5.1 Bus Operation Bases

Bus operation bases play a crucial role during disasters, serving as hubs for emergency transport and relief efforts. To enhance their resilience, the following strategies are recommended:

Relocation and Elevation: High-risk bases should be relocated to safer areas or elevated above expected flood levels to reduce their vulnerability to flooding.

Improved Infrastructure: Strengthening the physical infrastructure of bus operation bases, such as reinforcing buildings and installing flood barriers, can help protect against natural disasters.

Early Warning Systems: Implementing early warning systems for floods and landslides can provide bus operators with sufficient time to evacuate vehicles and personnel, minimizing potential losses.

20.5.2 Railway Networks

Railway networks require targeted adaptive measures to mitigate the increasing risks of flooding and other natural disasters. Key strategies include:

Strengthening Flood Defenses: Constructing or upgrading flood barriers along vulnerable railway segments can reduce the risk of flooding.

Improving Drainage Systems: Enhancing drainage infrastructure can help manage increased rainfall and reduce the likelihood of flooding.

Integrating Climate-Resilient Design Standards: Future railway construction projects should incorporate design standards that account for the anticipated impacts of climate change.

Network Redundancy: Building parallel routes where feasible can provide alternative paths for trains during network disruptions, enhancing overall resilience.

Real-Time Monitoring: Utilizing real-time monitoring and predictive analytics can help railway operators make timely decisions during extreme weather events, reducing service interruptions.

20.5.3 Road Freight Transport

Road freight transport is highly vulnerable to climate-induced disruptions. The following strategies can help enhance its resilience:

Reinforcing Infrastructure: Strengthening road surfaces, embankments, and bridges, particularly in areas prone to flooding and landslides, is critical for maintaining road functionality.

Developing Alternative Routes: Establishing alternative routes for critical freight corridors can help ensure the continuity of logistics operations during disruptions.

Implementing Advanced Technology: Investing in real-time monitoring systems and weather forecasting tools can provide early warnings and facilitate rapid response measures.

Regular Maintenance: Ensuring timely repairs and maintenance of road infrastructure can prevent deterioration due to weathering effects, reducing vulnerability to natural disasters.

20.6 Conclusion

In conclusion, climate change presents substantial and growing risks to urban transport networks, including bus operation bases, railway lines, and road freight corridors. The increasing frequency and intensity of extreme weather events, such as floods, landslides, and prolonged periods of heavy rainfall, underscore the urgency for proactive risk evaluation and the implementation of adaptive strategies. As these climate-induced hazards become more frequent and severe, the impacts on transport infrastructure will likely become more disruptive, affecting not only daily commuting and logistics but also emergency response capabilities, economic stability, and overall urban resilience.

This chapter has highlighted the specific vulnerabilities of different types of transport infrastructures, showing that certain segments, particularly those in low-lying or mountainous areas, are at a significantly higher risk of damage from natural disasters. The comprehensive risk evaluation framework presented here, which integrates hazard, exposure, and vulnerability assessments, provides a robust basis for understanding these risks in detail. By applying this methodology, urban planners and policymakers can identify the most vulnerable points within the transport network, prioritize high-risk areas for intervention, and allocate resources more effectively to enhance resilience.

Additionally, the targeted strategies outlined in this chapter — including infrastructure strengthening, network redundancy, early warning systems, and climate-resilient design standards — offer a range of solutions tailored to different types of transport infrastructure. These strategies are crucial for not only mitigating the immediate impacts of climate-induced hazards but also ensuring the long-term sustainability and functionality of urban transport networks in a warming world. For instance, relocating highly vulnerable bus operation bases or reinforcing flood defenses along critical railway lines can significantly reduce the risk of catastrophic disruptions.

Moreover, the chapter emphasizes the importance of a holistic, integrated approach to risk management that considers the interplay between various hazards and transport network components. This includes understanding the compound effects of multiple hazards, such as how flooding can exacerbate landslide risks or how heatwaves may weaken infrastructure over time, thereby increasing vulnerability to other events. Urban planners must adopt a multi-hazard perspective to develop adaptive strategies that are flexible and responsive to a range of possible future scenarios.

Ultimately, enhancing the resilience of urban transport networks is not just about protecting infrastructure; it is about safeguarding the essential lifelines that support economic activity, social connectivity, and the overall quality of life in cities. As climate change continues to pose unpredictable challenges, the ability to adapt and respond effectively will become increasingly critical. This chapter calls for a concerted effort from all stakeholders — including governments, private sector entities, and civil society — to invest in robust, forward-looking adaptation measures that can secure the future of urban transport systems in the face of evolving climate threats.

By building adaptive capacity and resilience, cities can not only withstand the immediate impacts of climate change but also thrive in a world where such changes are becoming the new norm. Thus, ensuring the resilience of urban transport networks is a vital component of broader urban sustainability and climate adaptation strategies, aimed at creating cities that are both livable and sustainable for current and future generations.

References

Cabinet Secretariat (2024) Fundamental plan for National Resilience. Cabinet Office, Government of Japan

Ministry of Land, Infrastructure, Transport, and Tourism (MLIT) (2020a) Digital National Land Information. Accessed Dec 2020

Ministry of Land, Infrastructure, Transport, and Tourism (MLIT) (2020b) National Land Information Division: Anticipated Flood Inundation Areas (Planned Scale), Anticipated Tsunami Inundation Areas, and Predicted Landslide Disaster Areas. Accessed Dec 2020

Ministry of Land, Infrastructure, Transport, and Tourism (MLIT) (2021) Freight Regional Flow Survey and National Freight Net Flow Survey. Logistics Census

Suzuki H, Saito Y, Hamada T, Kawagoe S (2020) Development of risk information in sediment disaster-prone areas due to climate change adaptation. J Jpn Soc Civil Eng 76(5):I_211–I_220

Xu F, Tajima H, Kato H (2023) Assessing natural disaster risks for bus operation bases: focus on vulnerability and importance. Modern Environ Sci Eng 9(7–9):99–107

Xu F, Tajima H, Kato H, Khaleghi M (2024) A method for evaluating the future flood risk for railway networks under climate change. J East Asia Soc Transp Stud 15:398–417

Yanagihara H, Kazama S, Tada T, Yamamoto T, Touge Y (2022) Changes in flood damage throughout Japan due to climate change and land-use change using shared socioeconomic pathways. J Jpn Soc Civil Eng 78:I_387–I_395

Chapter 21
Risk Assessment and Adaptation Policies for Dengue Fever

Hiroshi Nishiura and Katsuma Hayashi

Abstract In Japan, while there has not yet been a confirmed increase in dengue fever incidence, the vector mosquito, *Aedes albopictus*, is widely distributed, indicating the potential for more frequent outbreaks in the future. In this study, we develop a mathematical model by constructing the reproduction number from a bottom-up approach, allowing us to evaluate the long-term risk of dengue. Although the model is relatively simple, relying on temperature as the predictive variable, it has proven effective in estimating the likelihood of transmission and epidemics. Even in relatively optimistic scenarios, the inter-epidemic period—during which there is no risk of dengue fever outbreaks—was reduced by approximately 20 days in Tokyo. Geographically, the regions at risk for dengue are expected to shift northward, and in more pessimistic scenarios, the risk could extend as far as Hokkaido. The mosquito biting rate was found to influence the reproduction number in a quadratic manner, mathematically reinforcing the importance of focusing efforts on mosquito bite prevention. Considering future regional adaptation strategies is essential.

Keywords Epidemic · Dengue fever · Basic reproduction number · Extinction probability · Inter-epidemic period · Outbreak · *Aedes albopictus*

21.1 Introduction

The World Health Organization (WHO) has listed several serious concerns regarding the epidemiological impacts of climate change on health. These concerns include the increase in heatstroke as a result of heat stress due to rising temperatures, the expansion of vector-borne diseases influenced by the high temperature sensitivity of vectors, food shortages and child malnutrition caused by increased droughts, and the rise in respiratory diseases due to worsening environmental

H. Nishiura (✉) · K. Hayashi
Kyoto University, Kyoto, Japan
e-mail: nishiura.hiroshi.5r@kyoto-u.ac.jp

N. Mimura, S. Takewaka (eds.), *Climate Change Impacts and Adaptation Strategies in Japan*, https://doi.org/10.1007/978-981-96-2436-2_21

pollution. In Japan, there has not yet been a confirmed increase in the incidence of mosquito-borne diseases, such as dengue fever or Zika fever. However, the vector mosquito, *Aedes albopictus*, is widely distributed from the north to the south of the country, and in southern France, which shares similar environmental conditions, dengue outbreaks have become commonplace every summer. This suggests the possibility that a future where outbreaks within Japan, triggered by imported cases, may soon become a constant concern. In this chapter, we focus on dengue fever, reviewing past research findings. Furthermore, through the introduction of our latest research that we carried out as part of the strategic research program S-18 on the adaptation to climate change, we hope to explore adaptation policies for dengue fever in Japan together with the readers.

21.2 Dengue and Japan

21.2.1 Ecological Dynamics of Dengue Fever

Many viruses classified as flaviviruses, such as dengue fever and Zika fever, are transmitted through the bites of vector mosquitoes, specifically *Aedes aegypti* or *Aedes albopictus*. Approximately 30 species of flaviviruses are found in the Asia region (Mackenzie 2009). Classically, dengue fever has been a mosquito-borne viral infection prevalent in tropical and subtropical areas. According to WHO estimates, hundreds of millions of people are infected annually, and tens of thousands lose their lives, indicating a significant disease burden (WHO 2024; Sutherst 2004). Recent research and systematic reviews have scientifically demonstrated that climate factors—such as rising temperatures, changes in rainfall amounts and patterns, and humidity fluctuations—are gradually increasing the risk of dengue epidemics (Bhatia 2022; Abdullah 2022). These climatological factors not only alter the geographic distribution of vector mosquitoes but also extend the activity period of blood-sucking female mosquitoes and increase their activity, as measured by biting frequency. Additionally, it is known that higher temperatures accelerate the replication rate of the virus within the gut of infected mosquitoes, further contributing to the increased risk of dengue epidemics in various ways.

Risk assessments of dengue fever are often carried out through epidemiological studies that identify the risk factors permitting dengue transmission and epidemics, as well as predictions based on these studies. Predictions for dengue fever are generally classified into (i) statistical forecasts using time-series models, and (ii) mechanistic model forecasts that describe the mechanisms of dengue epidemics as phenomena (Morin 2013). Many studies have been reported from endemic regions where dengue fever is present year-round, such as Southeast Asia and Central and South America. In these regions, where climate change is progressing, not only are long-term time-series data on dengue cases available, but also mosquito surveillance data, making it technically easier to statistically discuss the relationship between climate change and dengue outbreaks (Rochlin 2016). On the other hand, in non-endemic regions, such time-series data do not yet exist, requiring predictions through mechanistic models, the validation of which is principally difficult (Medlock 2015). Although temperature has a significant impact on the predictors of dengue epidemics (Butterworth 2017), it has been suggested that factors such as the region's poverty level and population density are also closely related to the incidence of dengue fever, beyond just climate data (Naish 2014).

21.2.2 Dengue Fever in Japan

Japan, traditionally a temperate region, was known for successfully eradicating dengue fever by the early twentieth century. However, due to rising temperatures and significant fluctuations in dengue fever epidemics in Southeast Asia, the situation surrounding dengue fever in temperate Japan is rapidly changing. Notably, the habitat range of dengue-transmitting mosquitoes, *Aedes aegypti* and *Aedes albopictus*, has been expanding northward (Kobayashi 2002). In 2014, dengue fever transmission was confirmed domestically for the first time in 70 years. This was not merely a case of imported infections being diagnosed; sustained domestic transmission occurred, centering around Yoyogi Park in Tokyo, resulting in an expanding epidemic within that year. In 2019, two Japanese individuals without travel history were reported to have contracted dengue. These events clearly indicate that dengue fever is becoming a tangible threat in Japan. The possibility of domestic outbreaks expanding due to climate change is no longer a mere theoretical concern but a situation that must be recognized as a realistic risk.

Additionally, the number of travelers from Southeast Asia to Japan is increasing year by year, and 200–300 cases of imported dengue infections are reported annually. Given the often mild symptoms of dengue, it is estimated that more than 20 times the number of confirmed cases may have occurred among travelers visiting Southeast Asia, with the virus being brought into Japan (Yuan 2018).

In infectious diseases, the population can be classified into susceptible hosts who can become infected upon exposure, infectious hosts who can spread the disease, and immune hosts who have acquired immunity through previous infection. In the case of dengue fever, there are four distinct serotypes. Once infected with one serotype, a person is known to acquire lifelong immunity to that particular serotype. Therefore, in countries where dengue fever is endemic, most adults have already experienced three or four times of dengue fever in childhood, and the primary threat of dengue fever lies with small children. However, the situation in Japan is quite different. Nearly 100% of Japan's population is susceptible to dengue fever. This means that in Japan, the risk of dengue fever extends equally across all age groups, not just children. Due to this unique situation, it may not be appropriate to directly apply transmission parameters from studies conducted in dengue-endemic countries to Japan. Specific research and countermeasures tailored to Japan's circumstances are necessary.

21.3 Research Methods Incorporating Temperature

Against this backdrop, we conducted a projection study aimed at quantitatively assessing the impact of climate change on the risk of dengue fever outbreaks in Japan (Hayashi 2022). A key feature of this study is the use of a model that describes the transmission mechanism of dengue fever from the ground up in mathematical form, allowing us to assess risk without heavily relying on observational data from countries with different immune conditions. At the core of the analysis is an epidemiological indicator called the basic reproduction number (R_0), which represents the average number of secondary human cases caused by one infected human (through mosquitoes). The reproduction number for a mosquito-borne disease like dengue can be described mathematically, modeling the interaction between mosquitoes (vectors) and humans (hosts), as shown in Eq. (21.1) (Wang 2021):

$$R(T) = \frac{ma^2bc}{r\mu} e^{-\mu EIP} \tag{21.1}$$

Here, the reproduction number $R(T)$ on the left-hand side is shown as a function of temperature T. The variables on the right-hand side include the mosquito biting rate a, the transmission probability from mosquitoes to humans b, the transmission probability from humans to mosquitoes c, the mosquito mortality rate μ, the recovery rate r from the infectious period in humans, and the extrinsic incubation period (EIP), which is the time from when a mosquito feeds on an infected person until it can transmit the virus to another person. Although the temperature dependency for T is omitted from the right-hand side, all the parameters have been quantified as functions of temperature based on laboratory observations.

Furthermore, to assess the future risk of dengue outbreaks, it is useful to introduce two indicators: extinction probability and the interepidemic period (IEP). Extinction probability refers to the likelihood that an infection chain will naturally die out without intervention after one infected person is introduced. For example, with a disease that has a reproduction number of 2, theoretically, each generation produces twice as many new infections (two in the first generation, four in the second, and so on), leading to geometric growth in the number of cases by generation. However, in reality, the number of secondary cases produced by one infected person varies. Some individuals may infect several others, while others may not transmit the infection at all. Thus, even with a reproduction number of 2, this is just an average value, and there is a chance that the infection chain will not grow exponentially and may die out. In contrast, "non-extinction" means that when one infected person is introduced into a population susceptible to dengue, the infection chain continues, leading to a major epidemic and, eventually, a large-scale epidemic. "Extinction," on the other hand, means that the chain of infection is interrupted, either because no secondary infections occur or because the outbreak fades naturally. The distribution of secondary cases can be described by a negative binomial distribution with the mean value as

$$q = \left(1 + \frac{R}{k}(1-q)\right)^{-k} \tag{21.2}$$

For the sake of mathematical simplicity, we set $k = 1$, which corresponds to the assumption that the distribution of secondary transmissions follows a geometric distribution. Other results can be found in the supplement to reference (Hayashi 2022). The IEP (Interepidemic Period) is defined as the number of days in a year during which the extinction probability is 100%, meaning there is no risk of dengue transmission. Subtracting the IEP from 365 allows us to estimate the number of days in a year during which there is a risk of dengue transmission.

For predicting temperature changes due to climate change, we used the climate model "MIROC6" (Model for Interdisciplinary Research on Climate version 6), developed by Japan's climate modeling community (Ishizaki 2022). This model, based on the CMIP6 (Climate Model Intercomparison Project Phase 6), provides bias-corrected climate scenarios at a 1 km grid resolution across Japan. We examined three future scenarios based on classifications of CO_2 emissions and socioeconomic activities: RCP2.6 (optimistic), RCP4.5 (intermediate), and RCP8.5 (pessimistic) climate change scenarios. Regarding the uncertainty of the climate model, we treated the projected data as observational data, using a state-space model to analyze the mean temperature as a latent variable, along with its 95% confidence interval.

21.4 Results

Projections using the MIROC6 model showed a warming trend in daily average temperatures in Tokyo from 1990 to 2100 across all climate change scenarios. This temperature increase directly impacts the risk of dengue fever epidemics. In future projections of the reproduction number in Tokyo, an upward trend in the reproduction number was observed with rising temperatures. Figure 21.1a shows the average and 95% confidence interval of the reproduction number in Tokyo for July from 1990 to 2100. In all three RCP models, the reproduction number increased over time, rising from approximately 1.4 in 1990 to around 3.2 under the RCP8.5 scenario and 2.6 under the RCP2.6 scenario by 2100. Figure 21.1b shows the projections for October. The reproduction number rises from approximately 0.4 in 1990 to about 2.3 in the RCP8.5 scenario by 2100, while it remains around 0.7 in the RCP2.6 scenario. Since a reproduction number greater than 1 is required for a large-scale epidemic, the October projections differ significantly between the three RCP scenarios. If the RCP2.6 scenario plays out, the risk of dengue fever in Tokyo in October will likely remain low even by 2100. The more pessimistic the scenario, the higher the probability of dengue epidemics, and in addition, the risk of dengue transmission may become sufficiently high during periods like early summer and early autumn, which are not currently considered dengue transmission seasons.

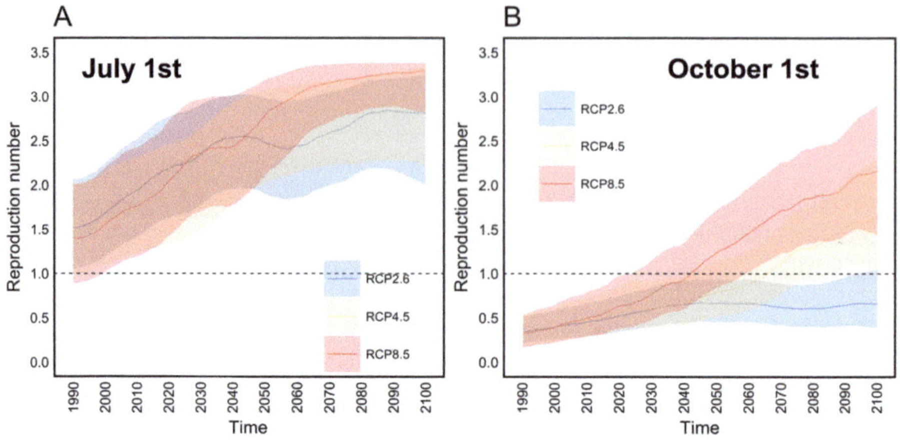

Fig. 21.1 Trends in the reproduction number of dengue fever. (Created by the authors from the original data in Hayashi et al. 2022)

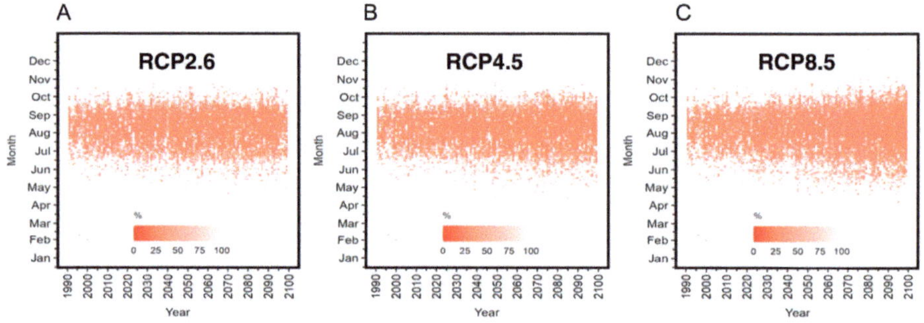

Fig. 21.2 Trends in the extinction probability of dengue fever. (Created by the authors from the original data in Hayashi et al. 2022)

Figure 21.2 shows the extinction probability expressed on a daily basis throughout the year. The vertical axis represents the days of the year, with 365 grids, and the horizontal axis shows the years from 1990 to 2100. The red areas indicate regions with a low extinction probability. When the effective reproduction number falls below 1, the extinction probability is 100%, and these areas are depicted in white. As explained in Figs. 21.1 and 21.2 also shows that the risk period for dengue fever lengthens year by year. Under the RCP8.5 scenario, the risk of dengue fever outbreaks becomes notably high from mid-May to the end of October by 2100. On the other hand, under the RCP2.6 scenario, there appears to be little change in the length of the high-risk period throughout the year by 2100.

Figure 21.3 shows the annual trend of the interepidemic period (IEP), which quantifies the non-risk period during which dengue outbreaks are unlikely. In 1990,

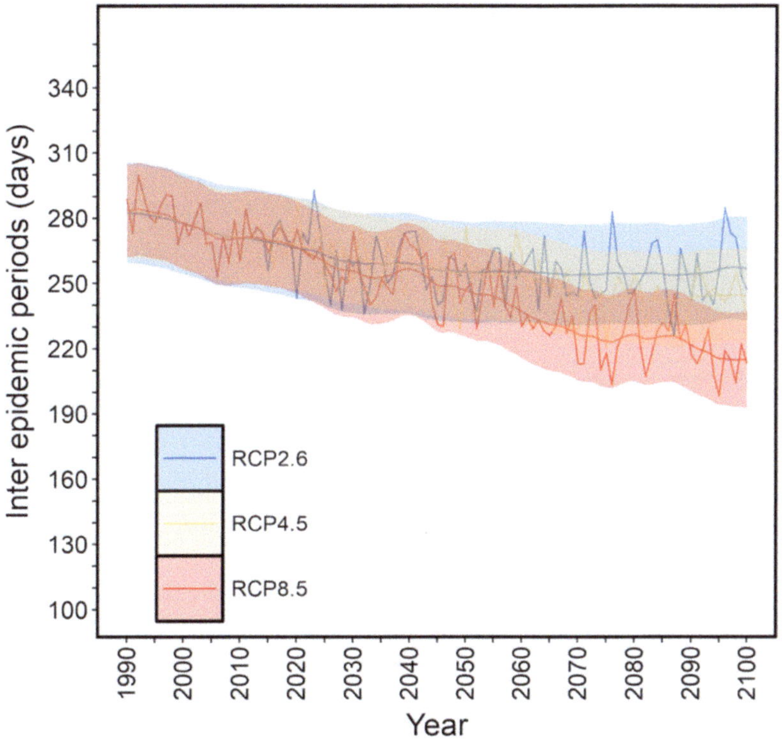

Fig. 21.3 Trends in interepidemic periods. (Created by the authors from the original data in Hayashi et al. 2022)

it was estimated that there were around 280 non-risk days per year, but by 2100, the IEP shortens to about 220 days under the RCP8.5 scenario and 260 days under the RCP2.6 scenario. Even under the RCP2.6 scenario, the non-risk period is expected to shorten by about 20 days, with significant changes occurring by around 2050, followed by a more gradual change. In contrast, the RCP8.5 scenario shows a steeper decline from around 2050 onwards.

From a nationwide perspective, the IEP projections indicate a shortening trend across the country due to rising temperatures. Figure 21.4 shows the nationwide distribution of IEP under the RCP8.5 scenario. The closer the color is to white, the fewer days there are where the risk of dengue transmission exists. Figure 21.4a shows the distribution in 2030, Fig. 21.4b for 2050, and Fig. 21.4c for 2100. In 2030, the IEP in northern regions of Japan, such as the Tohoku region and Hokkaido, and in cooler inland areas, is nearly 300 days or more, but over time, the IEP steadily shortens, and the geographic range of dengue transmission risk is expected to expand. In Japan's southernmost regions, it is suggested that they could face dengue outbreak risks for more than 300 days per year.

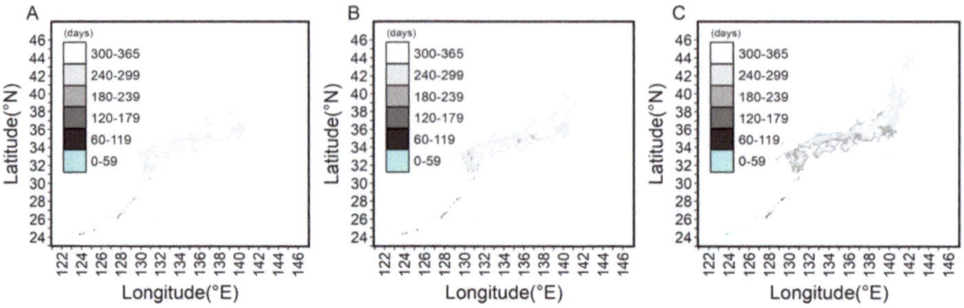

Fig. 21.4 Distribution of interepidemic periods. (Created by the authors from the original data in Hayashi et al. 2022)

21.5 Discussion and Conclusion

Through our study (Hayashi 2022), we constructed a mathematical model by describing the reproduction number and extinction probability in a bottom-up manner, enabling us to evaluate the future risk of dengue fever over a long time series. Although the model is simple, using temperature as the predictive variable, it has proven useful in estimating whether transmission and epidemics will occur. Significant differences in dengue fever risk were observed across different RCP scenarios in Japan. Even under relatively optimistic scenarios, the IEP period, during which there is no risk of dengue fever epidemics in Tokyo, was reduced by about 20 days, while under pessimistic scenarios, the epidemic period extended by over 60 days. Geographically, the regions at risk of dengue fever epidemics are expected to shift northward, and under pessimistic scenarios, the risk may extend as far as Hokkaido. In Japan's southernmost regions, public health measures against dengue fever may be required year-round, except for the two coldest months of winter. This study utilized a relatively simple model that did not account for factors like precipitation or the suitability of habitats for mosquitoes in different regions. However, it successfully demonstrated that the risk of dengue fever in Japan will increase as climate change scenarios become more pessimistic. While this study focused on Japan, the model can be applied to many temperate regions in Europe and North America, which are not currently endemic for dengue fever.

Based on this study, we can consider future countermeasures for dengue fever in Japan. Measures against mosquito-borne diseases like dengue fever can be understood by analyzing Eq. (21.1), which reveals "whether large-scale epidemics will occur." The mosquito biting rate contributes to the reproduction number in a quadratic manner, while the total number of mosquitoes (m) influences the reproduction number linearly. This mathematically supports the principle that it is more efficient to focus efforts on preventing mosquito bites rather than directly exterminating mosquitoes. Consequently, the decision to implement measures comes down to comparing the cost of dengue fever prevention with the number of cases prevented by such measures. As of 2024, the period when the risk of dengue fever epidemics begins, marked by an average reproduction number exceeding 1, is typically around late June. However, it is possible to determine when this period may shift earlier, for example to late May, depending on the region. Each municipality can estimate when the dengue risk will increase, allowing them to time their public campaigns for mosquito bite prevention accordingly.

One major research theme for the future would be the development of predictive models for the time series changes in the total number of mosquitoes. The ecological model for mosquitoes in a laboratory setting is becoming theoretically understood (Falcón-Lezama 2017). However, in reality, mosquito ecology is closely linked to human activity. *Aedes albopictus*, for example, is prevalent in urban areas, preferring to lay eggs in puddles, abandoned tires, and drainage ditches (Suresh 2023). In the long-term, given the expected population decline in Japan, urban structures will inevitably change. The ease with which dengue fever could emerge will likely depend heavily on land use, and quantifying such outcomes could be useful for future urban planning.

Assessing the risk of dengue fever in Japan, where no large-scale outbreak has occurred for over 80 years, is a challenging task. However, it is a crucial research topic for considering future regional adaptation strategies, and further research is needed to evaluate specific policy measures.

References

Abdullah NAMH, Dom NC, Salleh SA et al (2022) The association between dengue case and climate: a systematic review and meta-analysis. One Health 15:100452

Bhatia S, Bansal D, Patil S et al (2022) A retrospective study of climate change affecting dengue: evidences, challenges and future directions. Front Public Health 10:884645

Butterworth MK, Morin CW, Comrie AC (2017) An analysis of the potential impact of climate change on dengue transmission in the southeastern United States. Environ Health Perspect 125:579–585

Falcón-Lezama JA, Santos-Luna R, Román-Pérez S et al (2017) Analysis of spatial mobility in subjects from a dengue endemic urban locality in Morelos State, Mexico. PLoS One 12:e0172313

Hayashi K, Fujimoto M, Nishiura H (2022) Quantifying the future risk of dengue under climate change in Japan. Front Public Health 10:959312

Ishizaki NN, Shiogama H, Hanasaki N, Takahashi K (2022) Development of CMIP6-based climate scenarios for Japan using statistical method and their applicability to heat-related impact studies. Earth Space Sci 9:e2022EA002451

Kobayashi M, Nihei N, Kurihara T (2002) Analysis of northern distribution of Aedes albopictus (Diptera: Culicidae) in Japan by geographical information system. J Med Entomol 39:4–11

Mackenzie JS, Williams DT (2009) The zoonotic flaviviruses of southern, south-eastern and eastern Asia, and Australasia: the potential for emergent viruses. Zoonoses Public Health 56:338–356

Medlock JM, Leach SA (2015) Effect of climate change on vector-borne disease risk in the UK. Lancet Infect Dis 15:721–730

Morin CW, Comrie AC, Ernst K (2013) Climate and dengue transmission: evidence and implications. Environ Health Perspect 121:1264–1272

Naish S, Dale P, Mackenzie JS et al (2014) Climate change and dengue: a critical and systematic review of quantitative modelling approaches. BMC Infect Dis 14:167

Rochlin I, Faraji A, Ninivaggi DV et al (2016) Anthropogenic impacts on mosquito populations in North America over the past century. Nat Commun 7:13604

Suresh S, Meraj G, Kumar P et al (2023) Interactions of urbanisation, climate variability, and infectious disease dynamics: insights from the Coimbatore district of Tamil Nadu. Environ Monit Assess 195:1226

Sutherst RW (2004) Global change and human vulnerability to vector-borne diseases. Clin Microbiol Rev 17:136–1734

Wang X, Nishiura H (2021) The epidemic risk of dengue fever in Japan: climate change and seasonality. Can J Infect Dis Med Microbiol 2021:6699788

WHO Dengue and severe dengue (2024). https://www.who.int/news-room/fact-sheets/detail/dengue-and-severe-dengue

Yuan B, Nishiura H (2018) Estimating the actual importation risk of dengue virus infection among Japanese travelers. PLoS One 13:e0198734

Part VI
Economic and Policy Analysis of Climate Change Impacts and Adaptation

Chapter 22
Determinants of Farmers' Strategies for Adapting to Climate Change

Katsuhito Nohara, Akira Hibiki, Shinsuke Uchida, and Jun Yoshida

Abstract The increasing frequency and severity of disasters caused by climate change have become a significant management risk for farmers, potentially affecting their future operations. On the other hand, adaptation measures offer a way to mitigate these risks. This study uses structural equation modeling to explore how factors such as past disaster experiences, years of farming, and search for successors influence farmers' decisions to adopt adaptation measures. The findings show that the challenge of finding a successor and dealing with heat stress increases management risks for rice farmers. As a result, these farmers are more likely to register as certified farmers or adopt adaptation strategies, such as switching to heat-tolerant crops. Conversely, heavy rain and typhoons increase vegetable farmers' management risks. In response, these farmers prefer implementing adaptation measures like adjusting water volume and temperature, introducing heat-tolerant vegetable varieties, or switching to other heat-tolerant crops.

Keywords Structural equation modeling · Management risk of farmers · Decision-making · Past disaster experiences · Successor · Heat-tolerant crops

K. Nohara (✉)
Rikkyo University, Niiza, Saitama, Japan
e-mail: noharak@rikkyo.ac.jp

A. Hibiki
Tohoku University, Sendai, Japan

S. Uchida
Nagoya City University, Nagoya, Japan

J. Yoshida
Tohoku Gakuin University, Sendai, Miyagi, Japan

N. Mimura, S. Takewaka (eds.), *Climate Change Impacts and Adaptation Strategies in Japan*, https://doi.org/10.1007/978-981-96-2436-2_22

22.1 Introduction

Climate change is expected to induce many serious problems globally. Its impact on agriculture is one of the major concerns among others, since it directly affects our daily lives. In 2023, the world experienced extremely high temperatures. In Japan, the average percentage of first-class rice was about 60%, the lowest on record (Ministry of Agriculture, Forestry and Fisheries, 2023a). Among fruit trees, apple production decreased by 18% compared to 2022 due to sunburn caused by high temperatures and low rainfall in summer (Ministry of Agriculture, Forestry and Fisheries, 2024).

In Hokkaido, the northernmost of Japan's four main islands and located in the cold belt, farmers have traditionally adapted to low temperatures by cultivating crops that are more resistant to cold than heat. Hokkaido accounts for about 25% of Japan's arable land. However, even in this typically cooler region, the extreme heat of 2023 has raised concerns about Japan's future food security. In Nanporo Town, central Hokkaido, farmers abandoned broccoli cultivation due to heat and disease, and are now considering breeding crops more commonly grown in warmer regions, such as Kyushu in southwestern Japan. Similarly, in Ishikari City, located in central Hokkaido on the Sea of Japan, apple and cherry yields were severely affected by softening flesh, sunburn, poor coloration, and increased pest and disease pressure, leaving some farmers struggling. These adverse weather effects are expected to intensify in the future, leading to more severe weather-related disasters. As a result, farmers will need to adopt adaptation measures to reduce their vulnerability to these events (Di Falco et al. 2011).

It is important to note that farmer characteristics, such as age, are likely to influence the adoption of adaptation measures. In Japan, only about 20% of farmers rely on agriculture as their primary source of income. Additionally, according to the Ministry of Agriculture, Forestry and Fisheries (2023b), the number of farmers engaged in self-employed agriculture as their main occupation in 2023 had fallen to half of what it was in 2000, with approximately 70% of these farmers aged 65 or older. The small number of full-time farmers, the aging farming population, and the lack of successors are likely to negatively impact the implementation of climate change adaptation measures.

It is important to understand farm decision-making, specifically which types of farms adopt particular adaptation measures. Several previous studies have analyzed such decision-making (e.g., Bryan et al. 2009; Obayelu et al. 2014; Baba et al. 2015; Azadi et al. 2019; Pakmehr et al. 2020). However, there have been limited quantitative studies that examine the structure of farm decisions regarding how management risks—various external risks that affect the continuity of farming—are related to climate change, and how these risks are linked to adaptation measures. Japanese farms are expected to face increasing management risks due to natural and socioeconomic factors in the future. Therefore, clarifying the structure of farm decision-making is essential for deriving effective policy implications.

Management risks influence farm behavior, such as the purchase of crop damage insurance and the adoption of adaptation measures. We apply Structural Equation Modeling (SEM) to comprehensively analyze how the experience of crop damage caused by weather-related disasters impacts farmers' perception of risk management and drives their adaptive behavior. SEM is an analytical method that models relationships between multiple variables as a linear combination, making it useful for testing the validity of hypotheses. Through this analysis, we aim to identify the most effective adaptation measures based on farmers' characteristics.

22.2 Overview of the Survey

To understand the current situation and adaptive actions of farmers, we conducted an online survey of 1479 farmers, including 1227 crop farmers, between February 20 and 27, 2023, using the Job Panel provided by Rakuten Insight Inc. The agricultural workers include livestock industry (7.6%), agricultural service (2.1%), and horticultural service (2.7%), but we included them in the analysis if they produce rice or vegetables.

Table 22.1 summarizes the types of farmers, registration as a certified farmer, types of insurance purchased by farmers, whether they have a successor, their education, and age. The numbers in parentheses in the table indicate the percentage of the 1178 individual farmers. Table 22.2 summarizes the number of years of farming for each farmer and the types of crops, vegetables or fruits that each farmer mainly produced. About 50% of the respondents had been farming for more than 20 years and had extensive experience. Most farmers, 47.6%, produced mainly vegetables, followed by rice farmers (43.9%). Table 22.3 summarizes the types of past extreme weather events and natural disasters that caused significant damage to crop/vegetable/fruit production. Typhoons were the most common, followed by heat stress and heavy rains. Individual farmers are expected to be more vulnerable to extreme weather events and disasters because their financial budget to cope with them is smaller than that of corporate farmers.

Following Torres et al. (2020), we selected four adaptation measures and asked respondents to indicate whether their production was relevant to each adaptation measure. Respondents were also asked whether they were currently implementing each adaptation measure or not. If they had implemented it, they were asked whether they would continue to implement it or not. If they had not implemented it, they were asked whether they had already decided to implement it in the future, were considering it, or were not considering it. Four adaptation measures are control of water volume and temperature (Adaptation 1), change in planting time (Adaptation 2), introduction of high temperature tolerant varieties (Adaptation 3), and conversion to other high temperature tolerant crops (Adaptation 4). It should be noted that few respondents indicated that they are implementing adaptation measures but will terminate them in the future (0.2%, 0.4%, 0.9%, and 0.6% for Adaptation 1 ~ 4 respectively). Their main reason is that they will quit farming in their own generation.

Table 22.1 Data description

Type of agricultural management	Non-corporative farmer	79.6%
	Agricultural corporations and others	20.4%
Certified farmer	Yes	33.7% (31.7%)
Insurance type	Mutual aid	53.2% (56.5%)
	Income	12.8% (12.4%)
	Private	18.4% (18.0%)
	I don't know	13.7% (9.1%)
	Uninsured	22.5% (24.9%)
Successor	Yes	18.6% (16.1%)
	No	41.6% (47.2%)
	Others	39.8% (36.7%)
Education	Junior high school graduate	2.2% (2.4%)
	High school graduate	34.9% (35.8%)
	University graduate	40.6% (40.2%)
	Graduate school graduate	3.0% (2.3%)
	Others	19.3% (19.3%)
Age	20 s	2.4% (1.4%)
	30 s	11.3% (9.3%)
	40 s	21.2% (19.8%)
	50 s	22.4% (22.6%)
	60 s	26.8% (29.5%)
	70 s	15.1% (16.4%)
	80 s	0.8% (1.0%)

Table 22.2 Overview of data on agricultural management

Years of farming	Less than 1 year	2.2% (1.9%)
	1 to less than 5 years	8.6% (8.2%)
	5 to less than 10 years	15.8% (15.6%)
	10 to less than 20 years	23.7% (23.1%)
	20 to less than 30 years	26.8% (26.7%)
	More than 30 years	23.1% (24.5%)
Main crops grown	Rice	43.9%
	Wheat	6.7%
	Beans	8.3%
	Vegetable	47.2%
	Fruits vegetable	12.3%
	Fruits	15.1%
	Flower	7.9%
	Livestock	8.0%
	Other	5.9%

Table 22.3 Past experiences of extreme weather events and natural disasters that caused substantial damages to crops/vegetables/fruits production

Type of extreme weather events and disasters experienced	Heat stress (Heat)	47.5% (47.0%)
	Cold damage (Cold&Sun)	28.1% (29.2%)
	Heavy rain (Rain&Typh)	33.3% (34.9%)
	Pest infestation and disease damage (Pest)	27.5% (29.5%)
	Typhoon (Rain&Typh)	53.3% (55.4%)
	Water shortage (Water)	24.5% (25.5%)
	Lack of sunlight (Cold&Sun)	25.8% (27.5%)
	Not experienced	14.0% (12.5%)
	Other	6.6% (7.0%)

22.3 Analysis by Structural Equation Modeling

Potential management risks related to climate change are directly affected by the type of farmer, such as the experience of having suffered actual damage from weather disasters, the presence or absence of successors, and the number of years of farming experience. For example, farmers who have experienced weather disasters in the past may have a greater potential management risk than those who have not, because farmers with extensive agricultural experience know how to avoid business risks. On the other hand, farmers can reduce potential management risks by contracting various agricultural insurance, registering as certified farmers,[1] and introducing adaptation measures. Therefore, the attributes of farmers affect their adaptation behavior through potential management risks.

In this chapter, we use the MIMIC model (multiple indicator multiple cause model) of SEM (SEM-MIMIC) to analyze which type of farmer has a high potential management risk and how the farmer's behavior contributes to reducing that risk. However, to focus on management risks related to climate change, we highlight three actions: the number of insurance enrollments, whether or not they are registered as certified farmers, and whether or not they have introduced adaptation measures. In the following, we will analyze the following steps. As shown in Table 22.2,

[1] This is a system where farmers with plans to improve their management with ingenuity in line with the basic concepts of the municipality are certified. This certification makes it easier for them to receive subsidies, financing, and tax benefits.

various types of farmers have responded to the survey. The adaptation measures that farmers can introduce depend on their products (e.g., rice, vegetables, fruits, or livestock). Therefore, we first asked about the adaptation measures related to their farming practices and then classified the farmers into two categories based on their main income source. Specifically, for the analysis of Adaptation 1, we targeted farmers who responded in the survey that “Adaptation 1 is related to my agricultural production,” and analyzed separately for farmers who responded that the income ratio of rice is the highest and for farmers who responded that the income ratio of vegetables is the highest. The analyses of Adaptations 2–4 are conducted in the same way. In these analyses, we created a dummy variable that equals 1 for those who have introduced or decided to introduce each adaptation (Adaptations 1–4) in the future, and 0 otherwise. For the estimation of various variables, we used the robust Satorra-Bentler adjustment to address data non-normality, and the parameters were standardized. The hypotheses we established are as follows:

Hypothesis 1: The situation where farmers are looking for successors affects management risk

Hypothesis 2: Farmers’ past disaster experiences with actual damage affect management risk

Hypothesis 3: The number of years of farming (a proxy variable for years of agricultural experience) affects management risk

Hypothesis 4: Farmers with high management risk will contract insurance, register as certified farmers, and introduce adaptation measures

Hypothesis 5: Farmers’ past disaster experiences with actual damage affect management risk and increase the introduction of adaptation measures

22.3.1 Farmers with the Highest Income Proportion from Rice

The estimated results are shown in Fig. 22.1.[2] The rectangles on the left and right side represent observed variables, and the numbers within them (the numbers in the upper part of the left-side rectangle) represent the constants when standardized. The observed variables on the left are factors that affect the latent management risks (ManagRisk). If each estimated value on the arrow is positive, it means that observed variables increase management risk, and if it is negative, it means reducing that risk, with the size of the number indicating the degree of impact. These variables, in order from the top left, represent a dummy that takes 1 if a farmer looks for a successor and 0 otherwise (Look_for) and other dummies of disaster experience created in the same way (water shortage (Water), heat stress (Heat), heavy rain & typhoon (Rain&Typh), pest and disease damage (Pest), cold damage & lack of sunlight (Cold&Sun), and the number of years of farming (Farm_year). On the other

[2] For detailed estimation results, refer to the DP of the Policy Design Research Center, Tohoku University

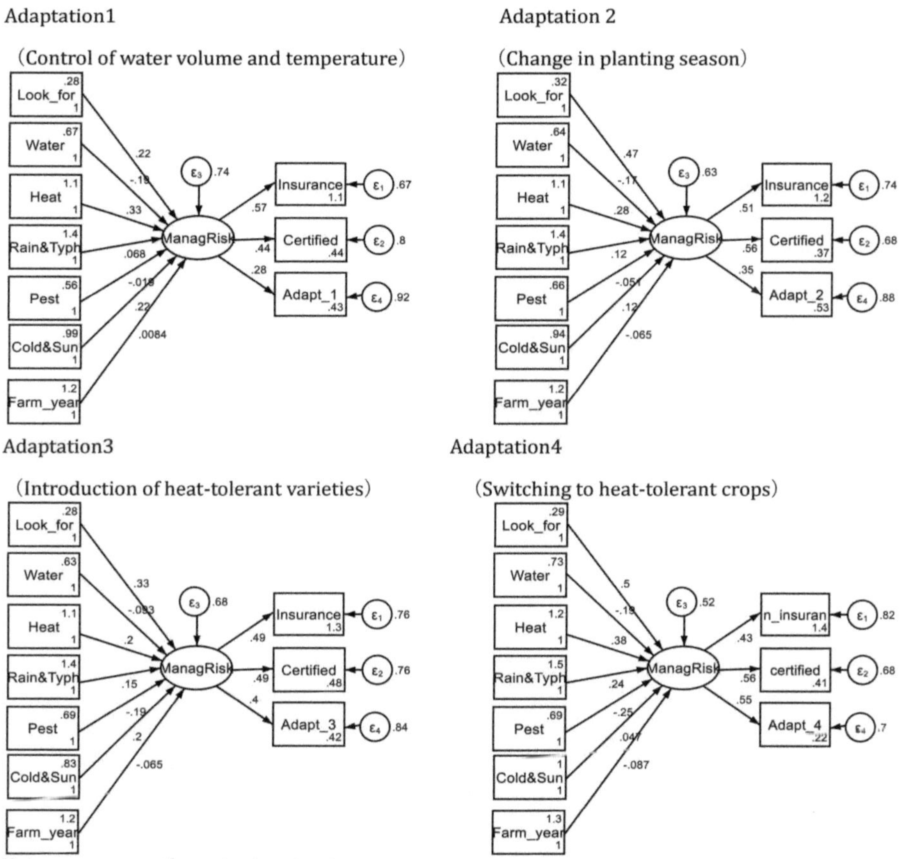

Note: ε 1, ε 2, ε 3, and ε 4 are error terms.

Fig. 22.1 SEM results for farmers with the highest proportion of income from rice

hand, the right-side observed variables represent the actions of farmers (the number of various agricultural insurance subscriptions (Insurance), registration to certified farmers (Certified), and the introduction of adaptation measures 1–4 (Adapt_1~4) (including decisions for future introduction)) and show the extent to which farmers taking each action are affected by management risk.

What is common to all adaptation measures is that the situation of looking for a successor has a positive and significant impact on management risk. This means farmers looking for a successor have a higher management risk. In addition, management risk has a positive and significant impact on all adaptation measures 1–4, and it was found that high management risk affects the introduction of adaptation measures. On the other hand, farmers' past disaster experiences were not necessarily factors that increased management risk. However, for farmers with a high proportion of income from rice, heat stress had a significantly positive impact on management risk in Adaptations 1, 2, and 4. In particular, for Adaptation 1, the impact of experience of heat stress on management risk was the largest, and it was found that certified farmers and those enrolled in insurance are more affected by management risk compared to farmers who implement Adaptation 1. In terms of the introduction of Adaptation 4, there was almost the same strong impact as certified farmers, and it was found that the impact of management risk was large.

From the above, for farmers who primarily grow rice, a lack of successors and experiences of heat stress increase management risk. As a result, it can be said that they tend to register as certified farmers or introduce adaptation measures such as switching to other crops that are heat-tolerant. In Adaptations 1 and 3, the experience of lack of sunlight and cold damage had a significantly positive impact on management risk, and it was found that they have a strong impact on that risk. Two reasons can be considered regarding this point. First, particularly in the case of cold damage, adaptation is possible through the regulation of water volume and temperature. Second, in the Tohoku region, which produces a large amount of rice, there is a history of cold damage caused by cold winds (In Japan, this cold, moist wind is called "Yamase.") and cold water, which reduced rice yields. For example, to address this issue, the cold-tolerant rice variety "Fujisaka No. 5" was developed by Mr. Minoru Tanaka. Farmers who have experienced cold damage in the past and have introduced cold-tolerant rice may find it relatively easy to decide to introduce heat-tolerant rice varieties. In fact, among the 63 farmers who have either introduced or decided to introduce Adaptation 3 and who reported having the highest proportion of income from rice, 38 were located in the Tohoku region (60.3%). For such farmers, it would be effective to promote the adoption of already developed heat-tolerant rice varieties in the event of a reduction in rice production due to further temperature rise in the future.

In Adaptations 3 and 4, the experience of pests and diseases had a significantly negative impact on management risk. Damage to rice by pests such as stink bugs and diseases like rice blasts have long been issues, and farmers have accumulated know-how to deal with them. Furthermore, with the availability of pesticides and fungicides that can easily control these problems today, experiences with pests and diseases might not be perceived as a significant management risk. In cases other than Adaptation 1, management risk has the strongest impact on certified farmers, and it has become clear that there is a tendency to become a certified farmer to hedge management risk.

In all cases of Adaptations 1–4, the results indicate that all of the fitting model indicators were appropriate, suggesting an adequate model fit for the empirical data (i.e., the Satorra-Bentler scaled test (a statistical test method when the data is not normally distributed), RMSEA (Root Mean Square Error of Approximation), CFI (Comparative Fit Index), TLI (Tucker-Lewis Index), and SRMR (Standardized Root Mean Square Residual)). These indicators generally verify how well the data fits the model from various angles. The cutoff criteria follow conventional rules of thumb: RMSEA ≤ 0.08, SRMR ≤ 0.08, CFI ≥ 0.90, and TLI ≥ 0.90.

22.3.2 *Farmers with the Highest Income Proportion from Vegetable*

Next, the results for farmers with a high proportion of income from vegetables are shown in Fig. 22.2. The situation of looking for a successor had a significantly positive impact on management risk only in Adaptation 2. Unlike farmers with a high proportion of income from rice, farmers with a high proportion of income from vegetables generally don't recognize a high management risk in the situation of looking for a successor, and those farmers seldom introduce adaptation measures. It

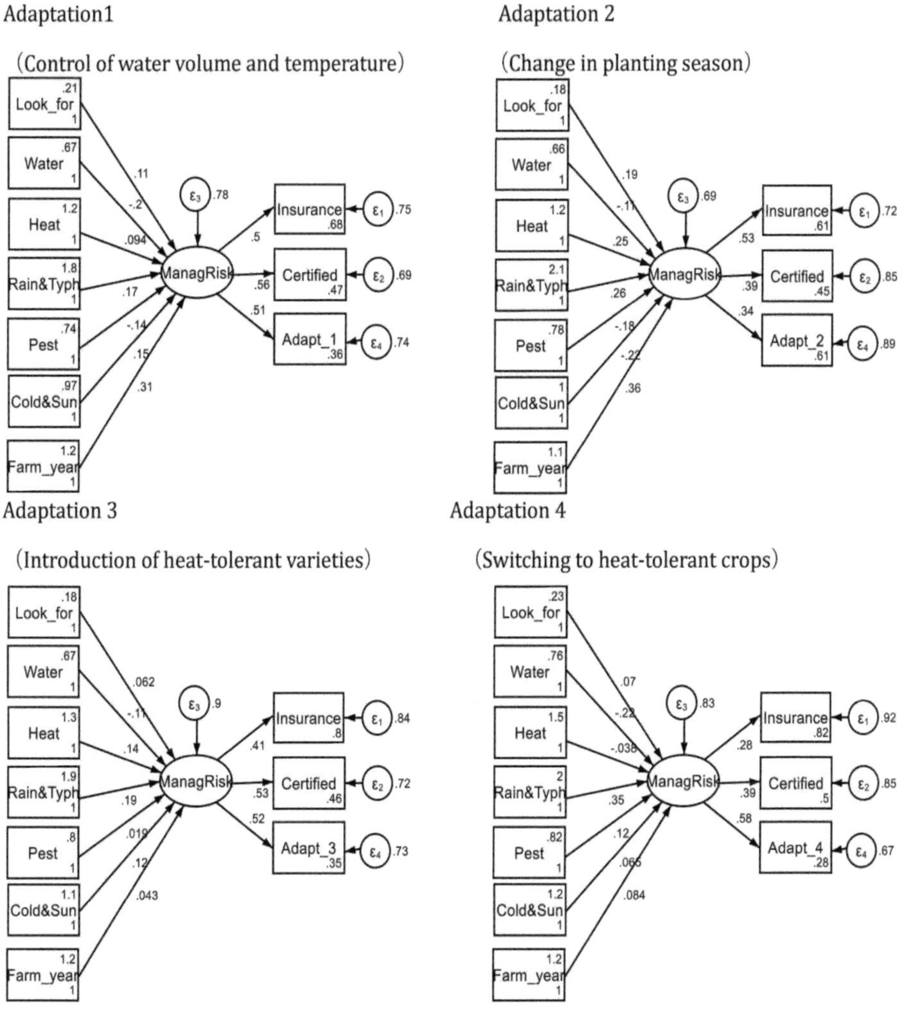

Note: ε_1, ε_2, ε_3, and ε_4 are error terms.

Fig. 22.2 SEM results for farmers with the highest proportion of income from vegetables was found that the experience of heavy rain and typhoons has a significantly positive effect on management risk in all Adaptations 1 to 4. In particular, for farmers related to Adaptation 4, past experiences of heavy rain and typhoons have the strongest impact on management risk, and that risk has the greatest impact on adaptation measures (conversion to other heat-tolerant crops).

In Adaptations 1 and 2, the number of years of farming has a significantly positive effect on management risk, suggesting that farmers with longer farming experience may recognize management risk. In Adaptation 1, although almost the same strength of influence from management risk was seen in all of the insurance enrollment, registration as a certified farmer, and introduction of adaptation measures, insurance was most affected in Adaptation 2. From these results, it can be seen that farmers with longer farming experience who primarily grow vegetables tend to hedge management risk by increasing the number of insurance enrollments rather than by changing the planting seasons. Similarly, for Adaptation 2, past experiences of heat stress also have a strong impact on management risk; however, it can be inferred that that risk is covered by insurance rather than changing the planting seasons.

From the above, for farmers who primarily grow vegetables, the experience of heavy rain and typhoons increases management risk, and as a result, there is a tendency to introduce adaptation measures such as adjusting water volume and temperature, introducing heat-tolerant vegetable varieties, and converting to other heat-tolerant crops. In particular, farmers related to Adaptations 1 and 2 are more likely to have management risk as the number of years of farming increases, and tend to introduce water volume and temperature adjustments rather than changing the planting seasons.

In the case of Adaptation 2, it has met all goodness-of-fit indices of the Satorra-Bentler scaled test, RMSEA, CFI, TLI, and SRMR. Almost all of the models have met goodness-of-fit indices, suggesting a good fit for the empirical data. Therefore, it can be inferred that our results generally indicated an acceptable model fit. According to Diamantopoulos and Siguaw 2000, the RMSEA has been regarded as one of the most informative fit indices because it is sensitive to the number of estimated parameters in the model. In all eight models in this chapter, the RMSEA was 0.08 or lower.

22.4 Conclusion

In this chapter, we used SEM-MIMIC to analyze how the management risk of farmers affects their decisions to implement four types of adaptation measures. Farmers with a high proportion of income from rice tend to hedge their management risk with crop conversion adaptation measures, based on their past experiences of actual damage caused by heat stress. Similarly, farmers with a high proportion of income from vegetables tend to hedge their management risk with crop conversion adaptation measures, based on their past experiences of actual damage caused by heavy rain and typhoons. Since many of these farmers are relatively young certified farmers operating in Hokkaido and the Tohoku region, it can be said that the government needs to design policies that facilitate their transition to heat-tolerant crops and create an environment, including financial support, that encourage the adoption of adaptation measures. To maintain vegetable production, it will be necessary to promote the introduction and spread of heat-tolerant vegetable varieties such as Okina and Saiho cabbages or heavy rain-tolerant Chinese cabbage such as Sunny Yellow 85 (Tomaru (2020)), targeting farmers who have suffered from heavy rain and typhoon damage in the past.

Furthermore, it was found that farmers who look for successors or have been farming for many years face higher management risks, and their past disaster experiences also influence these risks. On the other hand, both rice and vegetable farmers tend to cover these risks by enrolling in multiple insurance plans rather than adopting adaptation measures such as changing planting seasons. In the future, it will be important to provide more weather information to make it easier for farmers to decide to change planting seasons because concerns about weather instability due to climate change issues such as rising temperatures and changes in rainfall increase.

In the future, in order to encourage farmers to introduce more adaptation measures, it will be important to continue to understand in detail the factors that influence farmers' decisions, the attributes of farmers, and regional characteristics, and to continue efficient and effective outreach.

References

Azadi Y, Yazdanpanah M, Mahmoudi H (2019) Understanding smallholder farmers' adaptation behaviors through climate change beliefs, risk perception, trust, and psychological distance: evidence from wheat growers in Iran. J Environ Manag 250:109456

Baba K, Kawai Y, Kobayashi M, Tanaka M (2015) Factors to determine risk perception of climate change, and attitude toward adaptation policy of the citizens in rural areas and farmers. J Soc Civ Eng G (Environ) 71(5):143–151. (in Japanese)

Bryan E, Deressa T, Gbetibouo G, Ringler C (2009) Adaptation to climate change in. Ethiopia and South Africa: options and constraints. Environ Sci Pol 12:413–426

Di Falco S, Veronesi M, Yesuf M (2011) Does adaptation to climate change provide food security? A micro-perspective from Ethiopia. Am J Agric Econ 93(3):829–846

Diamantopoulos A, Siguaw JA (2000) Introducing LISREL. Sage Publications, London

Ministry of Agriculture, Forestry and Fisheries (2023a) Results of the Agricultural Product Inspection for Rice, etc. https://www.maff.go.jp/j/seisan/syoryu/kensa/kome/. Accessed 17 July 2024

Ministry of Agriculture, Forestry and Fisheries (2023b) Statistics of Agriculture, Forestry and Fisheries (Results of the 2023 Survey on the Dynamics of Agricultural Structure). https://www.maff.go.jp/j/tokei/kekka_gaiyou/noukou/r5/index.html. Accessed 6 Oct 2023

Ministry of Agriculture, Forestry and Fisheries (2024) Statistics of Agriculture, Forestry and Fisheries. https://www.maff.go.jp/j/tokei/kekka_gaiyou/sakumotu/sakkyou_kajyu/ringo/r5/index.html. Accessed 16 July 2024

Obayelu OA, Adepoju AO, Idowu T (2014) Factors influencing farmers' choices of adaptation to climate change in Ekiti State. Nigeria J Agric Environ Int Dev 108(1):3–16

Pakmehr S, Yazdanpanah M, Baradaran M (2020) How collective efficacy makes a difference in response to water shortage due to climate change in Southwest Iran. Land Use Policy 99:104798
Tomaru Y (2020) Measures for heat and humidity control in cabbage and Chinese cabbage. https://www.takii.co.jp/tsk/saizensen_web/cultivation/variety_selection-2/. Accessed 10 Oct 2023
Torres MAO, Kallas Z, Herrera SIO (2020) Farmers' environmental perceptions and preferences regarding climate change adaptation and mitigation actions; towards a sustainable agricultural system in México. Land Use Policy 99:105031

Chapter 23
Direct and Indirect Economic Impacts of Sea Level Rise

Ken Itakura, Akira Hibiki, Jun Yoshida, Makoto Tamura, and Hiromune Yokoki

Abstract This chapter evaluates the economic impacts of sea level rise on Japan's 47 prefectures using a computable general equilibrium (CGE) model, focusing on the direct and indirect effects of land loss. By analyzing two future scenarios—SSP1-2.6, which assumes a sustainable socioeconomic pathway, and SSP5-8.5, which represents high greenhouse gas concentrations—the study examines how the loss of agricultural and developed land affects regional economies. The findings indicate that sea level rise could reduce Japan's real GDP by 3.6% to 4.1% by 2100, with Saga, Aichi, and Tokyo among the most severely impacted regions. While the direct effects of land loss are significant, the analysis also emphasizes the importance of indirect effects, such as disruptions to supply chains and diminished interregional trade, which play a crucial role in shaping regional outcomes. Interestingly, some prefectures without direct land loss may experience economic gains as production demands shift from affected areas. The study highlights the need to consider both the negative and positive indirect effects across regions.

Keywords Sea level rise · Economic impact · Computable general equilibrium · Land loss · Regional economy

K. Itakura (✉)
Nagoya City University, Nagoya, Japan
e-mail: itakura@econ.nagoya-cu.ac.jp

A. Hibiki
Tohoku University, Sendai, Japan

J. Yoshida
Tohoku Gakuin University, Miyagi, Japan

M. Tamura
Ibaraki University, Mito, Japan

H. Yokoki
Ibaraki University, Hitachi, Japan

N. Mimura, S. Takewaka (eds.), *Climate Change Impacts and Adaptation Strategies in Japan*, https://doi.org/10.1007/978-981-96-2436-2_23

23.1 Introduction

According to the sixth Assessment Report of the IPCC (IPCC 2021), relative to the period 1995–2014, sea levels are projected to rise by 0.28 to 1.01 m (under SSP1-2.6 to SSP5-8.5 scenario) by the end of the twenty-first century. Mimura et al. (1994) estimated that a 30 cm rise in sea level would result in the loss of half of Japan's sandy beaches, while more than 90% is lost if sea levels rise by 1 m. Kodama et al. (2022) found that the potential inundated area could reach 2261–2598 km^2 by 2100, affecting approximately 3.76–4.92 million people. Tamura et al. (2024) projected that by 2100, the largest potential inundated area would be in Fukuoka and Saga, covering 481–487 km^2, followed by Gifu, Aichi, and Mie with 452–469 km2 (under SSP5–8.5 scenario).

The loss of land due to inundation caused by sea level rise will halt production in the affected areas, leading to economic losses. It is important to note that such economic impacts will also affect production in regions that are economically linked to the inundated areas through supply chain relationships. For example, companies outside of the affected areas will no longer be able to purchase inputs produced in those regions, forcing them to reduce their own production. To mitigate this negative impact, they may seek substitutes from other companies located outside the inundated areas, leading to increased production for those suppliers. Additionally, companies that previously sold products to businesses in the lost areas will no longer be able to do so after the sea level rise. In this way, sea level rise will affect regional economies not only through the direct impacts of land loss but also through changes in the supply chain network, with both negative and positive effects. Therefore, to understand the overall economic impact of sea level rise on regional economies, it is necessary to analyze both the direct impact of land loss and the indirect impacts arising from changes in supply chain networks.

In this chapter, we project the economic impacts (total, direct and indirect) of land loss caused by future sea level rise for the 47 prefectures in Japan. We used a computable general equilibrium (CGE) model as our analytical tool. The CGE model is particularly useful because it incorporates the interdependence of economic activities not only among multiple industries but also among multiple regions (prefectures in this study), using the Input-Output table of each region. The indirect impacts through the supply chain can be captured by the interdependencies among the 47 prefectures. For our analysis, we compared two future scenarios: the sea-level rise scenario, which causes land loss (agricultural land and built-up area[1]), and the no-sea-level rise scenario, which includes no land loss. The latter scenario serves as the baseline for comparison.

For our simulation, we calculated the loss of agricultural land (rice fields and others), as well as that of built-up area (commercial, industrial, and residential area), from 2030 to 2100 using the findings of Kodama et al. (2022) and future land use

[1] Built-up area is where commercial, industrial, and residential buildings are densely located in urban or rural area.

information from the Japanese version of the 3rd Mesh Land Use Scenario for each Shared Socioeconomic Pathway (SSP), released by the National Institute for Environmental Studies (NIES), Japan, in September 2021. Kodama et al. (2022) provided information on the lost areas at 1 km mesh resolution, which was calculated using: (1) elevation data (National Land Numerical Data (Elevation/Slope 3rd Mesh)), (2) administrative area data (National Land Information), (3) tidal data (Egbert and Erofeeva 2002), (4) sea-level rise data (MIROC-ESM-CHEM, RCP2.6 and RCP8.5), and (5) a list of mesh codes for each city (Statistics Bureau of the Ministry of Internal Affairs and Communications). We overlaid the future land loss areas with future land use scenarios to identify the lost areas for each land use category. Finally, we aggregated the results at the prefecture level.

It should be noted that the lost area is calculated under the assumption that all current protective measures, such as sea walls, do not exist and that no additional protective measures will be implemented in the future. Therefore, the economic loss captured by the difference in regional value added between the sea-level rise scenario and the no- sea-level rise scenario should be interpreted as the potential economic value of such protective measures.

Figure 23.1 reports the aggregated land loss, based on the potential inundated area in Japan from 2030 to2100, under two sea-level rise scenarios. These scenarios are defined by the combination of future socioeconomic pathways (SSP: Shared Socioeconomic Pathway) and representative concentration pathways for greenhouse gases (RCP: Representative Concentration Pathway). Although there are multiple scenarios in both SSP and RCP, we used the combination of SSP1 and

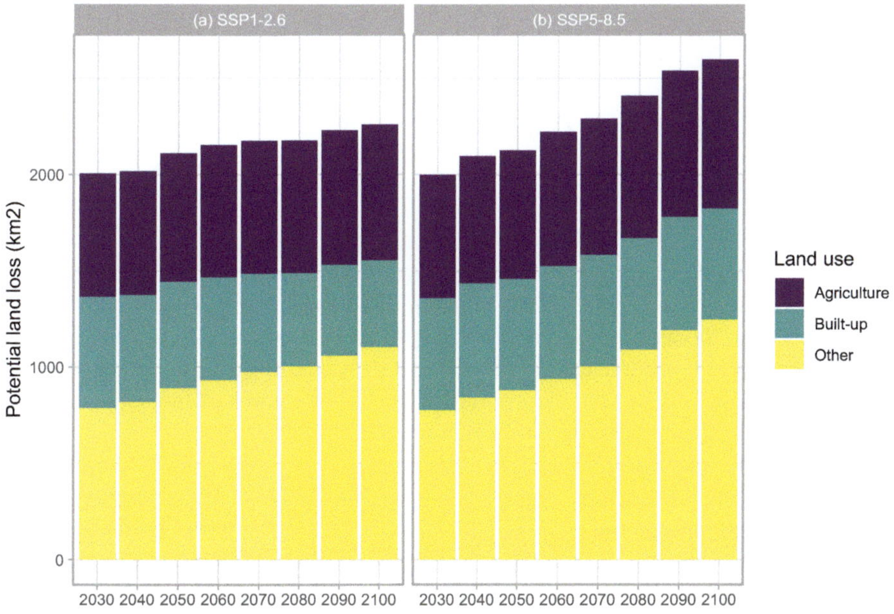

Fig. 23.1 Potential Land Loss (km^2). (The data used is from Kodama et al. (2022))

RCP2.6 (SSP1-2.6) and SSP5 and RCP8.5 (SSP5-8.5) for our analysis. SSP1 represents a scenario of a sustainable socioeconomic society, while SSP5 reflects a society dependent on fossil fuels. RCP2.6 assumes low concentrations of greenhouse gases in 2100, whereas RCP8.5 represents a high concentration scenario.

Figure 23.1 shows that, in total, the potential land loss in SSP5-8.5 is larger than in SSP1-2.6, with a notable increase after 2060. These trends appear uneven across prefectures. The figure also shows the land loss by land use: agricultural land, built-up area of commercial, industrial, and residential buildings, and others (such as forests, wastelands, roads, railways, other lands, rivers and lakes, beaches, sea areas, and golf courses). In SSP1-2.6, the proportion of built-up area is lower than in SSP5-8.5. It should be noted that we did not incorporate the loss of other land use into our simulation, therefore, our results may underestimate the economic impacts because of this omission.

Figure 23.2 shows the projected proportions of the land loss of agricultural land and built-up area in 2100, relative to the total land area for each land use under the no-sea-level-rise scenario. In the figures, uncolored prefectures indicate no lost area for agricultural land and built-up area. The prefecture with the highest land loss ratio is Saga prefecture, with 25.4% in SSP1-2.6 and 26.3% in SSP5-8.5, followed by Aichi (13.3% and 13.8%) and Tokyo (8.7% and 10%). It is remarkable that the ratio of lost agricultural land is highest in Saga, and that Tokyo does not suffer from losing agricultural land. Since we impose the change in ratio of land loss for our simulation analysis, these results will play an important role in the economic impact. In the CGE simulations, the economic impact is computed by comparing the scenario without land loss to the scenario with land loss due to sea level rise in SSP1-2.6 and SSP5-8.5.

23.2 CGE Model and Database

23.2.1 Overview of CGE Model

The GTAP Model (Corong et al. 2017; Hertel 1997; McDougall 2003) and the GTAP Database (Aguiar et al. 2019) form a global computable general equilibrium (CGE) model of the world economy, linking production, trade, and consumption for 65 industries in 141 countries/regions. The CGE model used in this chapter is an extension of the GTAP model, incorporating the 47 prefectures of Japan (see Itakura and Iwamoto (2024) for detailed information).

The theoretical framework of the CGE model is derived from the optimized behavior of economic agents. Households maximize utility, and producers minimize costs under constant returns to scale technology, given market prices. Production activities use intermediate goods from both domestic and overseas industries and primary factors of production such as labor, capital, energy, land, and natural resources. Figure 23.3 illustrates the structure of value added (VA), where

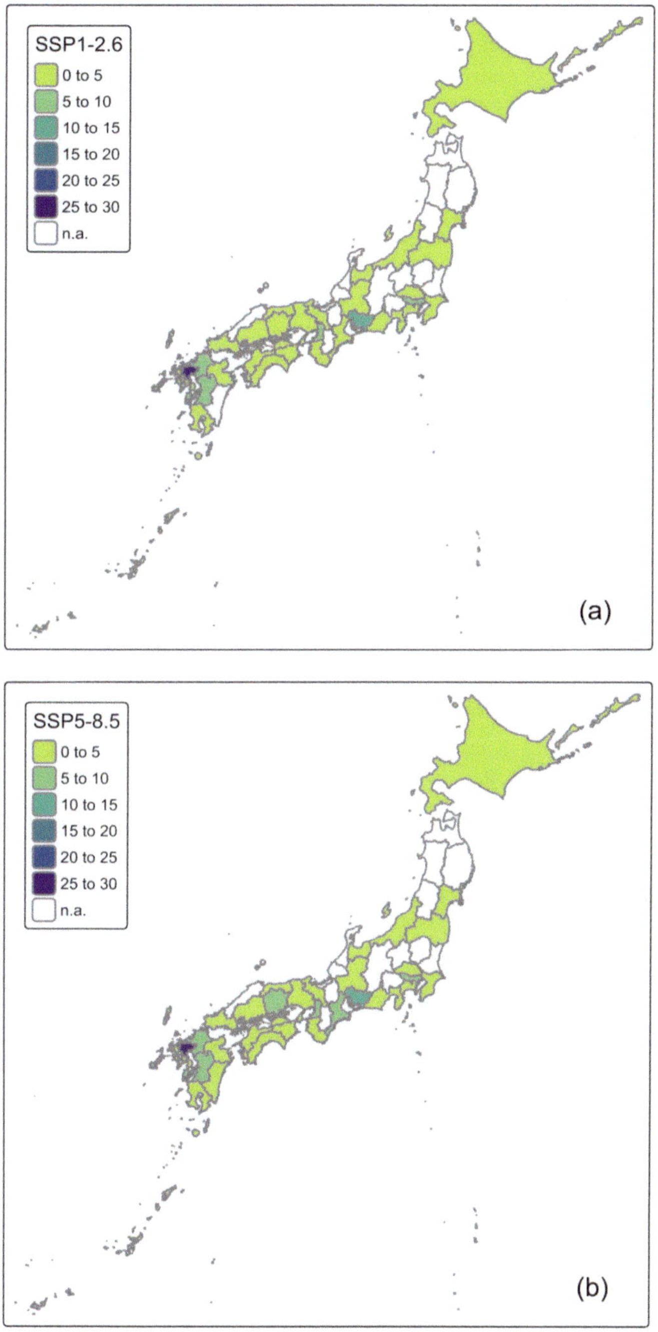

Fig. 23.2 Ratio of Land Loss in Land Use in 2100 (%). (The data used is from Kodama et al. (2022) and NIES (2021))

primary factors are combined. Starting from the bottom of the figure, capital and energy are substitutes, meaning the input amount is adjusted by changes in the relative price of capital and energy. The capital-energy (KE) subproduct is with skilled labor (SL), unskilled labor (UL), land, and natural resources (NR) to form the value added (VA). This VA composite is used in production activities alongside intermediate inputs such as raw materials and parts from other industries.

Goods and services produced in each prefecture are supplied to itself as well as to other prefectures as outflows. They are also supplied to foreign countries as exports. Inflows of goods and services from other prefectures and imports from abroad are for the final demand such as consumption and investment in each prefecture, and for intermediate demand for raw materials and parts. By assuming that the products of each industry are differentiated by the place of production, we can accommodate the intra-industry trades observed in data (Armington 1969).

Our aim in this chapter is to estimate the economic impacts of land loss in agricultural land and built-up area. Using Fig. 23.3 to explain how the impacts disseminate, damage to agricultural land can be represented as a decrease in land, while damage to built-up area can be represented as a decrease in capital. Damage to land and capital results in a decrease in value added in production activities, leading to a reduction in output. This decrease in supply affects not only the prefecture where the land loss occurred but also other prefectures through supply chains of the outflows and inflows of goods and services, as well as foreign countries through international trade. Prefectures where no land loss occurs may still be indirectly affected by changes in outflows, inflows, and trade. However, the degree and direction of these impacts are likely to vary depending on their economic interdependence with other prefectures.

23.2.2 Overview of the Database

By reviewing the database used for the analysis, we can grasp the economic linkages between prefectures. Figure 23.4 shows a simplified version of the inter-prefectural Input-Output table. The figure includes the 47 prefectures from Aichi to Yamanashi, and aggregates foreign regions into a single entry ("Foreign"). For

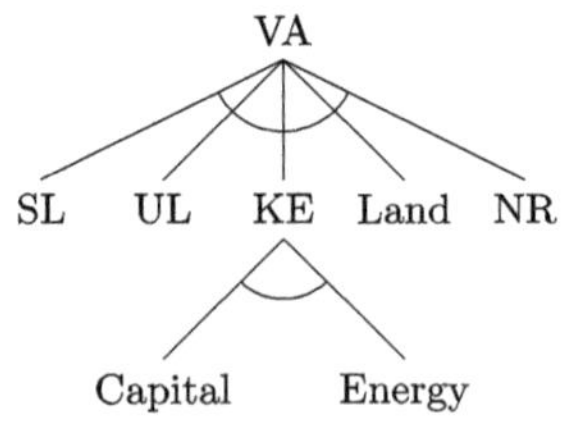

Fig. 23.3 Structure of value added in CGE model. Note: *A* Agriculture, *M* Manufacturing, *S* Services, *VA* Value added, *O* Output, *FD* Final Demand, D^M Domestic Intermediate Transaction, D^{FD} Domestic Final Demand

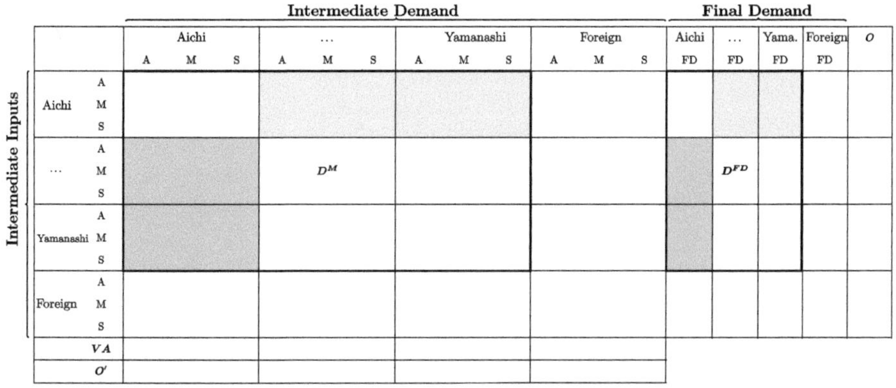

Fig. 23.4 Multiregional input-output table

simplicity, it illustrates only three industries: agriculture (A), industry (M), and services (S). The blocks enclosed by bold lines indicate domestic intermediate transactions (D^{M}) and domestic final demand (D^{FD}) for goods and services produced in Japan. Taking Aichi prefecture as an example, the outflows to other prefectures are represented by the light gray cells, while the inflows from other prefectures are represented by the dark gray cells.

When agricultural land and built-up area are lost due to sea level rise in Aichi prefecture, the supply of output (O) decreases because the value-added input to production activity becomes lower. This reduction in supply leads to a decrease in outflows for intermediate and final demands in other prefectures. As production activities in Aichi decrease, the inflow of intermediate goods from other regions also declines. Thus, the land loss in Aichi spreads its impact to other prefectures through these outflows and inflows. The decrease in production also reduces the demand for primary factors of production, such as labor and capital, leading to a decline in corresponding factor prices, including wage rates and capital rental rates. This reduction results in lower income of household, which supplies these production factors, thereby reducing consumption. Additionally, the decrease in the capital rental rate lowers the rate of returns to investment, leading to a decrease in investment through fixed capital formation. As a result, these decreases in consumption and investment will negatively affect Gross Regional Product (GRP). By aggregating the GRP of each prefecture, we can estimate the change in Japan's Gross Domestic Product (GDP).

In this chapter, we extend the GTAP Database (Aguiar et al. 2019) by incorporating Japan's 47 prefectures, and we aggregate the 141 countries/regions and 65 industries of the GTAP database into 49 regions and 21 industries. In addition to Japan's 47 prefectures, we have two aggregated foreign regions: RCEP[2] and the rest of the world (ROW). To maintain consistency in industry classification, we

[2] RCEP (Regional Comprehensive Economic Partnership) is a free trade zone consisting of the 10 ASEAN countries, China, South Korea, Australia, New Zealand, and Japan.

aggregated the Input-Output table of each prefecture and the GTAP Database. For our analysis, we used the GEMPACK economic modelling software (Horridge et al. 2018).

23.3 Baseline and Sea-Level Rise Scenarios

Future scenarios are the settings for simulations that combine assumptions about expected future changes in socioeconomic conditions and land loss due to sea level rise over an extended period, from the benchmark year of 2011 to the final year of the analysis, 2100. These future scenarios are constructed for both the baseline scenario, which serves as a basis of comparison, and the sea-level rise scenarios for SSP1-2.6 and SSP5-8.5.

23.3.1 Baseline Scenario

In the baseline scenario, future socioeconomic prospects are described in terms of total population, working-age population, and real GDP. The actual and projected values of total population and working-age population of countries worldwide from 1950 to 2100 were sourced from the U.N. World Population Prospects (2022), using the medium variant. The actual and projected values of population and working-age population by prefecture were obtained from the National Institute of Population and Social Security Research (2018) until 2045, and extended to 2100 by applying the growth rate from the U.N. World Population Prospects (2022). Real GDP projections were extended to 2060 using the Prefectural Economic Accounts (2023) from the Economic and Social Research Institute (ESRI) and the OECD Real GDP Long Term Forecast (2018), with the growth rate of the final year applied through to 2100. Capital stock is determined endogenously by investment and depreciation, with total savings equal to total investment globally. The allocation of investment is determined by the expected rate of return on investment in each region.

The baseline scenario serves as a basis for comparison with the subsequent sea-level rise scenarios, and thus assume no future land loss. It should be noted that the baseline scenario is not intended to be an accurate long-term prediction, but rather a counterfactual future scenario for comparative purposes.

23.3.2 Sea-Level Rise Scenario

In the sea-level rise scenarios, the loss of agricultural land and built-up area is estimated under the climate change scenarios, SSP1-2.6 and SSP5-8.5. As shown in Fig. 23.2, these scenarios specify the proportion of land loss that is imposed exogenously from 2030 to 2100. The proportion of the lost land is calculated as the ratio of potential land loss to the land use that does not anticipate sea level rise for each of SSP1-2.6 and SSP5-8.5 (Kodama et al. 2022; NIES 2021).

Compared to the baseline scenario which assumes no-sea-level rise, agricultural land and built-up area are subject to decline under the sea-level rise scenarios, leading to a decrease in production activities, except in certain prefectures. In prefectures where land loss is anticipated, it is expected that the real GRP, an indicator of economic activity, will be lower than in the baseline scenario. There are 15 prefectures that either do not face the sea or experience no land loss. These prefectures are not directly affected by land loss due to sea-level rise; however they are expected to face indirect impacts via supply chains of outflows and inflows.

23.4 Results

We implement simulations using the CGE model for both the baseline and the two sea-level rise scenarios, and estimate the economic impacts by comparing the results. Figure 23.5 shows the impact on Japan's real GDP. Land loss due to sea level rise reduces real GDP in both SSP1-2.6 and SSP5-8.5 scenarios. As land loss expands over time, real GDP continues to decline relative to the baseline until 2100. Compared to the decrease in the SSP1-2.6 scenario, which assumes sustainable economy, the decline in SSP5-8.5 scenario, which assumes an economy dependent on fossil fuels, is more pronounced after 2050. The impact on Japan's real GDP in 2100 reaches −3.6% in SSP1-2.6 and −4.1% in SSP5-8.5.

The impact on Japan's real GDP can be broken down into the real GRP (Gross Regional Product) of the 47 prefectures. Figure 23.6 shows the impact on the real GRP of each prefecture in 2100. The greatest damage from land loss is observed in Saga prefecture, where real GRP decreases by −32.5% in SSP1-2.6 and −34.7% in SSP5-8.5. This is due to Saga having the highest ratio of land loss in agricultural land and built-up area at 26%, as shown in Fig. 23.2, which leads to a significant decrease in real GRP. The fall in real GRP is also notable in Aichi and Tokyo, which have the second-highest land loss ratios. Aichi's real GRP decreases by −10.9%, −11.3% and Tokyo at −7.6%, −8.4% in SSP1-2.6 and SSP5-8.5, respectively. Since the areas projected to lose land due to sea level rise are primarily located in western Japan, the negative impact on real GRP tends to be greater in that region.

Some prefectures experience a slight increase in real GRP, as indicated by the yellow color in Fig. 23.6. For the prefectures without expected land loss, demands are shifting from the affected prefectures for their goods and services, thereby raising their real GRP.

An interesting result is observed in prefectures where real GRP increases despite experiencing land loss. For example, Hokkaido shows an increase in real GRP of 0.4% in SSP1-2.6 and 0.5% in SSP5-8.5. Similar results were confirmed in five other prefectures. In these regions, the predicted land loss of agricultural land and built-up area is relatively small and limited. As a result, the positive effect of demand diversions from other prefectures outweighs the negative effect of land loss.

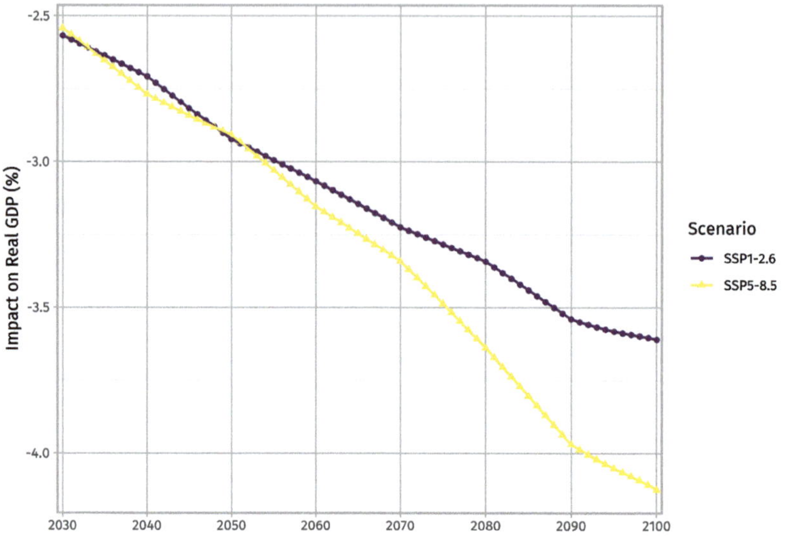

Fig. 23.5 Impact on real GDP in Japan (%, relative to the baseline)

The impact on real GRP is decomposed into direct impact due to land loss and indirect impact arising from changes in outflows and inflows. In Fig. 23.7, the sum of the direct and indirect impacts corresponds to the total impact on real GRP shown in Fig. 23.6. It is evident that the total impact varies greatly by prefecture. In prefectures where the decrease in real GRP is significant, it is not surprising that the direct impact is greater than the indirect impact. In prefectures where the total impact is small, the indirect impact can be both negative and positive. Some prefectures show zero direct impact because no land loss is expected. However, since all prefectures are economically interdependent, no prefecture has zero indirect impact.

To analyze the strength of economic linkages between prefectures, additional simulations are conducted assuming land loss due to sea level rise that occurred only in a specific prefecture. We take Aichi prefecture and Tokyo as examples for this counterfactual experiment, using the SSP5-8.5 scenario. The direct impact on real GRP will only appear in Aichi or Tokyo, so other prefectures will only observe indirect impacts. This allows us to analyze the pattern and strength of economic linkages between a specific region and other regions. Figure 23.8 shows the indirect impact on real GRP originating from land loss in Aichi or Tokyo. The direct impact on Aichi's real GRP is −11.3%. It is confirmed that the indirect impact originating from Aichi spreads nationwide. These indirect impacts are divided into prefectures with negative effects (blue) and prefectures with positive effects (green). The indirect impact originating from Aichi is less than 0.5 in absolute value, which contrasts with Tokyo. The direct impact on Tokyo's real GRP is −8.3%, but the indirect impact on other prefectures is larger than Aichi's case because of the large economic size of Tokyo. The positive indirect impact is pronounced in the Tohoku region, while the negative indirect impact is more significant in western Japan. These results indicate that each prefecture exhibits heterogeneity in the structure of

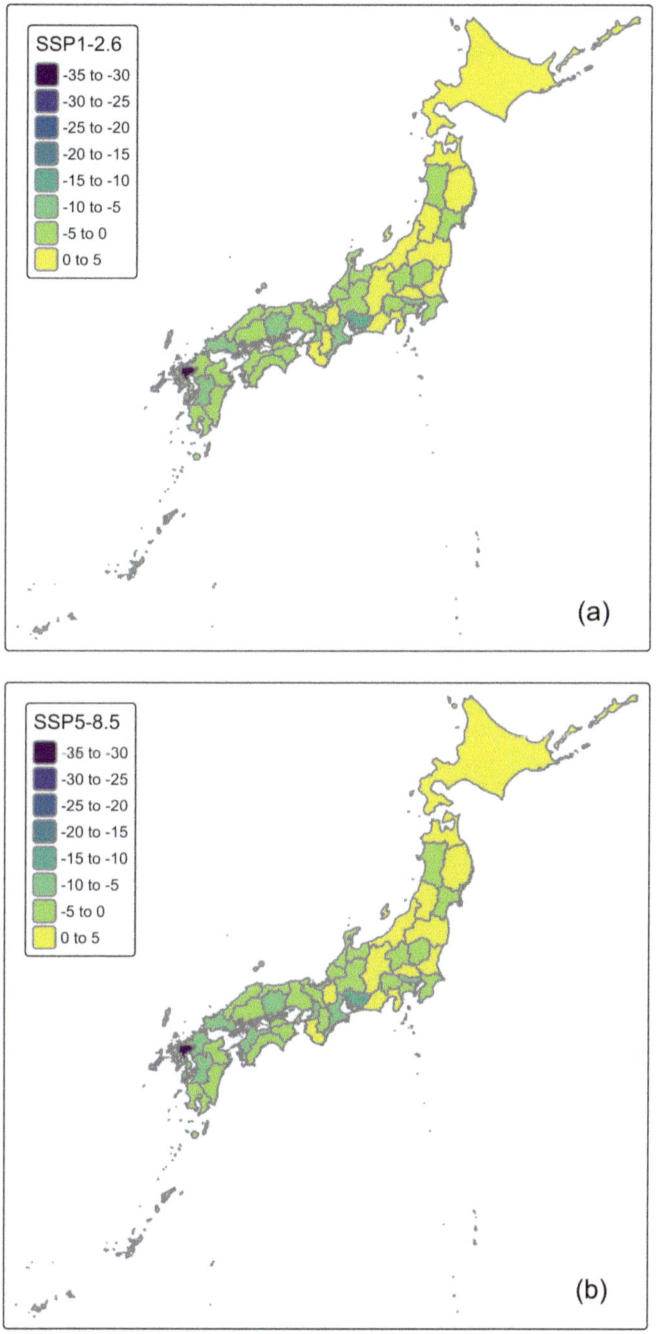

Fig. 23.6 Impact on real GRP by prefectures in 2100 (%, relative to the baseline)

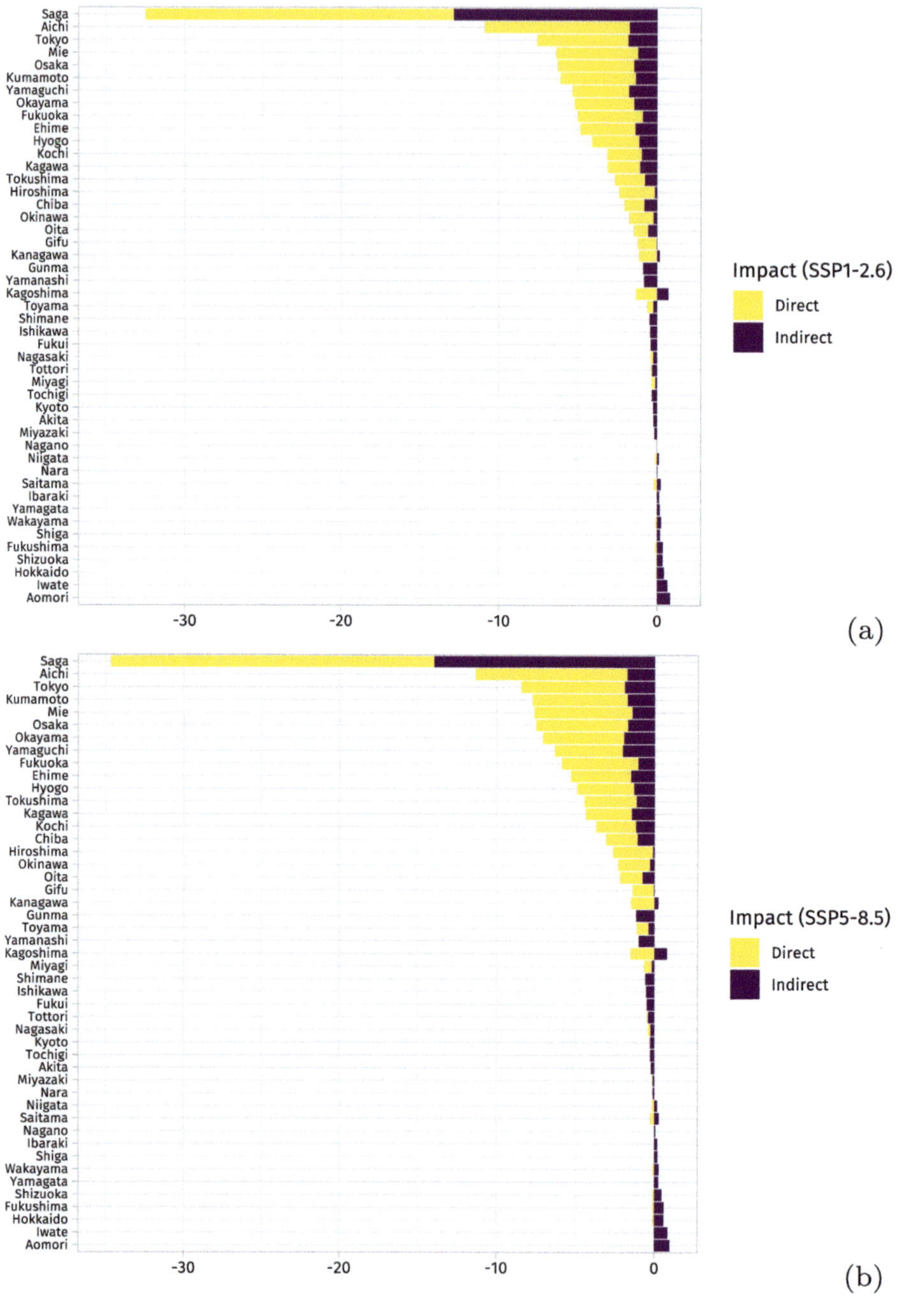

Fig. 23.7 Direct and indirect impacts on real GRP by prefecture in 2100 (%, relative to the baseline)

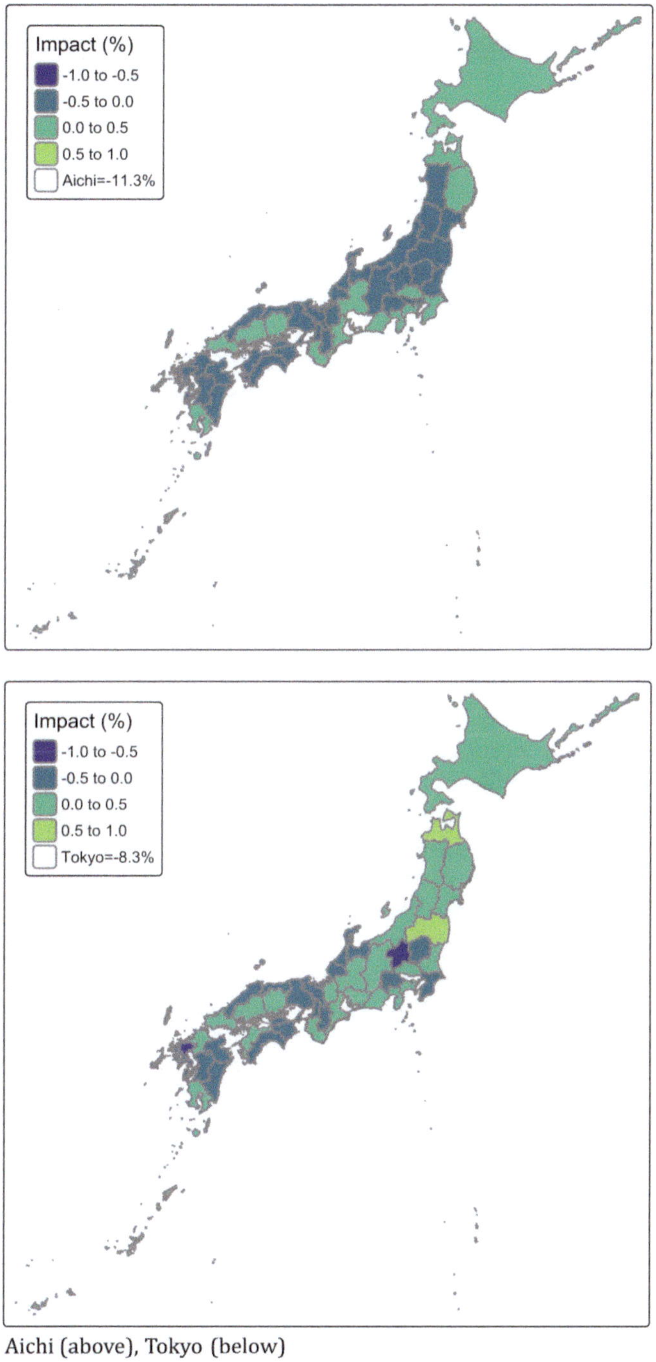

Fig. 23.8 Indirect impact on real GRP originating from Aichi or Tokyo, 2100 (%, SSP5-8.5)

economic linkages with other regions. Therefore, it is crucial to consider the indirect impact between prefectures when estimating the overall economic impact of sea level rise.

23.5 Conclusion

In this chapter, we estimated the economic impact of future land loss caused by sea level rise on agricultural land and built-up area of commercial, industrial, and residential buildings, by conducting a set of simulations using a global CGE model, incorporating all 47 prefectures of Japan. We considered the future land loss scenarios of SSP1-2.6 and SSP5-8.5 and imposed the projected land loss for each. As a result of the potential future expansion of the land loss, Japan's real GDP is projected to decrease by −3.6% in SSP1-2.6 and −4.1% in SSP5-8.5 by 2100. When this decrease in real GDP is decomposed into the real GRP of each prefecture, the negative impact of land loss was most significant in Saga, Aichi, and Tokyo, with a clear downward trend observed in western Japan. The impact on real GRP was divided into direct impacts in prefectures where land loss is expected and indirect impacts reflecting economic linkages between prefectures. In regions where the decrease in real GRP was significant, the direct impact was greater than the indirect impact. Conversely, in regions where the impact on real GRP was small, the indirect impact determined whether real GRP increased or decreased. Our analysis of the spread of indirect impact originating from Aichi and Tokyo confirmed heterogeneity among prefectures, reflecting the structure of economic linkages. These results indicate that when estimating the economic impact of future land loss due to sea level rise, it is important to consider the indirect impact in addition to the direct impact on prefectures.

The CGE model and database used in this chapter are suitable for ex ante analysis of potential economic impacts, but further refinement of the analytical method is necessary. Specifically, it would be beneficial to expand the coverage of land use beyond agricultural land and built-up area of commercial, industrial, and residential buildings. Damage to infrastructure is likely to have a considerable economic impact and should be included in future analyses. Continuous updates and improvements to the CGE model and database are also important.

References

Aguiar A, Chepeliev M, Corong E et al (2019) The GTAP data base: version 10. J Global Econ Anal 4(1):1–27

Armington P (1969) A theory of demand for products distinguished by place of production. IMF Staff Pap 16(1):159–178

Corong E, Hertel T, McDougall R et al (2017) The standard GTAP model version 7. J Global Econ Anal 2(1):1–119

Egbert G, Erofeeva S (2002) Efficient inverse modeling of barotropic ocean tides. J Atmos Ocean Technol 19(2):183–204
Hertel T (1997) Global trade analysis: modeling and applications. Cambridge University Press, New York
Horridge M, Jerie M, Mustakinov D et al (2018) GEMPACK manual. GEMPACK Software, Melbourne
Intergovernmental Panel on Climate Change (2021) Climate change 2021: the physical science basis. Contribution of Working Group I to the Sixth Assessment Report. Cambridge University Press, Cambridge
Itakura K, Iwamoto T (2024) Regional computable general equilibrium model of Japan and the global economy. TUPD Discussion Paper 2024-16:1–20
Kodama K, Yokoki H, Tamura M (2022) Assessing socioeconomic impacts of sea level rise on Japanese coasts via scenarios of population and land use. J Jpn Soc Civil Eng 78(5):I_349–I_357. (in Japanese)
McDougall R (2003) A new regional household demand system for GTAP. GTAP Tech Paper 20:1–57
Mimura N, Inoue K, Kiyohashi M, Nobuoka H (1994) Assessment of the impact of sea-level rise on sandy beaches (2): verification of the validity of the prediction model and nationwide evaluation. Proc Coast Eng 41:1161–1165. (in Japanese)
Tamura M, Imamura K, Kumano N, Yokoki H (2024) Assessing the effectiveness of adaptation against sea level rise in Japanese coastal areas: protection or relocation? Environ Dev Sustain 26:23561–23577

Data Source

National Institute of Population and Social Security Research (2018) Regional Population Projections for Japan: 2015–2045. https://www.ipss.go.jp/index-e.asp. Accessed 20 Sept 2022
NIES: Japan-specific land use scenarios—Comprehensive Environmental Research Promotion Fund 2-1805 results, 2021. Accessed 8 July 2024
OECD (2018) GDP long-term forecast. OECD, Paris
United Nations (2022) World population prospects: the 2022 revision. United Nations, New York

Chapter 24
Climate Security Policy in Japan: Toward Climate-Based Policymaking

Seiichiro Hasui

Abstract This chapter discusses the relationship between climate security and human security, focusing on how Japan can effectively integrate climate change measures into its broader security policies. Climate security, a concept increasingly recognized in international discourse, highlights the risks climate change poses to sociopolitical stability, such as exacerbating conflicts and causing socioeconomic disruptions. These issues are linked to human security, which emphasizes protecting people's fundamental livelihoods, rights, and freedoms. This chapter argues that climate security and human security are interdependent: the impacts of climate change undermine human security by threatening livelihoods, health, and safety. Constrained by its pacifist constitution, Japan has not fully embraced climate security in its national policy despite the increasing international focus on the topic. The chapter proposes that Japan adopt a "human security" lens to address climate challenges, aligning its foreign and domestic policies with global climate security efforts. This approach emphasizes nonmilitary means of addressing climate threats, fostering international cooperation, and enhancing social resilience.

Keywords Climate security · Human security · Geopolitical risks of climate change · Critical security studies · Cascading risks · Environmental security · Comprehensive security

---

S. Hasui (✉)
Ibaraki University, Mito, Japan
e-mail: seiichiro.hasui.irps@vc.ibaraki.ac.jp

N. Mimura, S. Takewaka (eds.), *Climate Change Impacts and Adaptation Strategies in Japan*, https://doi.org/10.1007/978-981-96-2436-2_24

24.1 Introduction

How can Japan's environmental policy effectively utilize the various climate change impact measures investigated in each theme of the S-18 Project? This study aims to demonstrate the effectiveness of framing this issue within the context of "human security," which has long been adopted in Japan's diplomatic strategy, and "climate security," which is increasingly recognized in international politics.

In international political science, the concept of "security" has been repeatedly redefined according to the changing circumstances of international society, even in the post-World War II era, making it a highly debated and diverse concept. The emergence of the concept of climate security stems from the growing international recognition that climate change impacts various sectors, including politics, economics, and society, ultimately leading to violent conflicts. However, the term "climate security" is inherently multifaceted, with many politicians, scholars, and bureaucrats using the term in different contexts. Several studies in Japan have also attempted to classify the term to analyze a wide range of discourses (Hasui 2011; Kameyama and Ono 2021; Kanie 2007; Sekiyama 2023).

The potential for climate change to cause severe impacts around the world has been repeatedly discussed in the assessment reports (ARs) published by the Intergovernmental Panel on Climate Change (IPCC) since 1990. In Chap. 12 of AR5, the relationship between climate change and security was seriously discussed for the first time. The lead authors, who had previously addressed environmental security by recognizing that environmental issues pose security threats, expanded the discussion under the chapter "Human Security"(Adgar 2014). As the chapter deepened their discussion at the 2007 AR4, they analyzed the factors that increase conflict as climate change-sensitive while noting that climate change has little direct impact on the increase in violent conflict. They also highlighted that indigenous peoples, local communities, and traditional knowledge systems are essential resources for climate change adaptation, and certain forms of migration may undermine human security. In conclusion, they argued that conditions for security and national security policies would inevitably need to be transformed.

The "human security" concept first appeared in the "Human Development Report 1994" by the United Nations Development Programme (UNDP 1994). Later, the "Human Security Now" report by the Ogata-Sen Commission on Human Security broadly defined human security as "to protect the vital core of all human lives in ways that enhance human freedoms and human fulfillment" (CHR 2003). By utilizing this broad concept, Chap. 12 of AR5 aimed to define climate change's wide-ranging and diverse impacts on people as broader security threats.

On the other hand, by the time AR5 was published in 2014, "climate security" had already become a significant issue in policy making and research in Europe and the United States. A key point that emerged from these discussions was that cooperation from nonmilitary organizations is essential to effectively address the threats of climate change (Hasui and Komatsu 2021). As shown in Fig. 24.1, climate change exerts various impacts, and for these to lead to traditional security threats such as

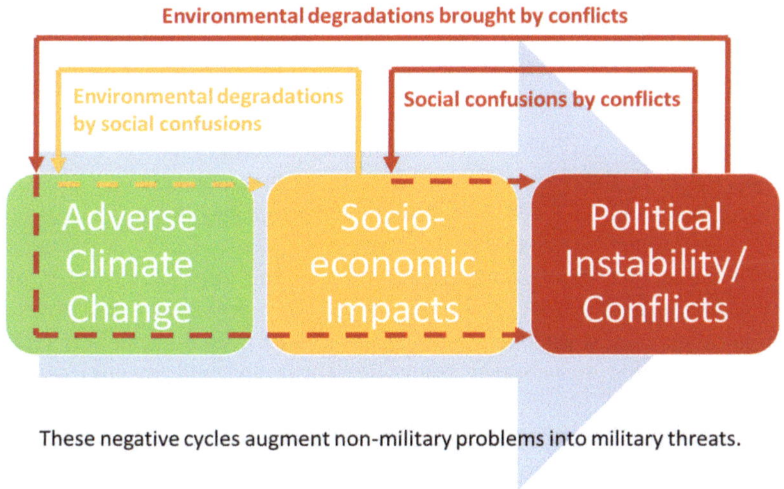

Fig. 24.1 The process by which climate change leads to conflict and the negative cycle. (Source: Reproduced from Homer-Dixon (1999) with permission from Prinston University Press)

political instability and conflict, the deterioration of socioeconomic conditions is a necessary intermediate step. This process creates a cycle where the socioeconomic impacts, political instability, and conflicts generate negative feedback, further worsening the situation. In this context, the role of military organizations in addressing these issues is very limited.

The AR6 Synthesis Report highlighted the increasing complexity of risks and the cascading effects of climate change as follows.

> B.2.3 With further warming, climate change risks will become increasingly complex and more difficult to manage. Multiple climatic and non-climatic risk drivers will interact, resulting in compounding overall risk and risks cascading across sectors and regions. Climate-driven food insecurity and supply instability, for example, are projected to increase with increasing global warming, interacting with non-climatic risk drivers such as competition for land between urban expansion and food production, pandemics and conflict. (high confidence) (IPCC 2023: 15)

Based on this international consensus, how can climate security be effectively incorporated into Japanese policy? As divisions grow between Western countries and within domestic contexts, can climate change measures serve as a common language of security that facilitates cooperation among opposing nations? Why has the climate security policy in Japan not been established?

Although climate security, which emphasizes nonmilitary means, is a subject of active research internationally, the concept of climate security, while beneficial for Japan—where the use of military force is constitutionally restricted—in participating in international security activities, has not yet become a comprehensive area of interest (Kameyama and Ono 2021). Researchers from the National Institute for Environmental Studies and the National Institute for Defense Studies have noted this point. The delay in discussions in Japan is shared among Japanese researchers in this field. In 2007, the Ministry of the Environment invited experts to compose a report titled "Report on Climate Security" (Sub-Committee on International Climate Change Strategy Global Environment Committee 2007), but as pointed out by Kameyama and Ono, as well as Sekiyama (2023: 12–17), further discussions have not progressed. It cannot currently be considered a comprehensive matter of concern. What might be the reason for this? This is the first question.

On the other hand, in recent years, climate change has been discussed almost monthly at the United Nations Security Council, and the UN Secretary-General has repeatedly made speeches linking climate change and security. In Japan, climate security also frequently appears in the speeches of politicians and government reports. In this way, discussions are politically driven both domestically and internationally. Since this is a concept for policy formation and promotion, political leadership is essential. However, in AR6, the relationship between climate change and violent conflict is rated as having "medium confidence." There are concerns about policies advancing without robust scientific evidence. Why has this situation of political precedence occurred? This is the second question.

Considering the above situation, the third question is how to effectively position various measures against the impacts of climate change as part of Japan's climate security policy.

24.2 Climate Change and Security

24.2.1 The First Question: Why Hasn't Climate Security Become a Comprehensive Concern in Japan?

Research has confirmed a consistent quantitative increase in academic discussions on climate security globally from the 1990s to the 2020s (Sekiyama 2023: 11), indicating that research on climate change, conflict, and security has become established in the global academic community. However, contributions from Japan to the international discourse remain limited. As of 2021, only a few English-language papers authored by Japanese scholars could be found, and the author himself only added one paper in 2021 (Hasui and Komatsu 2021). There are structural reasons for this limitation.

In Japan's academic community, there has been little contribution to discussions on "Critical Security Studies," which emphasize nonmilitary aspects. In Japan, the image of traditional military state security remains deeply ingrained when discussing security, and Critical Security Studies have not been widely debated in the academic field. The publication in which the author participated in 2022 became one of the earliest specialized textbooks on the subject (Minamiyama and Maeda 2022).

The primary reason is that many researchers hold the preconceived notion that in a mid-latitude developed country like Japan, it is difficult to imagine climate change causing political instability or violent conflict. In many discussions at the United

Nations, it is also believed that African countries are the most severely affected by climate change and that it can easily escalate into a security threat.

Furthermore, according to a public opinion survey conducted by the Cabinet Office, interest in climate change tends to be lower among younger Japanese generations. In the 2023 survey, 48.0% of all respondents were "interested" in climate change, while 41.4% said they were "somewhat interested." However, among those aged 18 to 29, only 31.0% were "interested," and 39.7% were "somewhat interested." This trend of lower interest among younger generations has remained unchanged since 2005, the earliest year for which generational data is available (Cabinet Office Public Opinion Survey Website).

Thus, the lack of academic contributions and the stagnation of interest in climate change are structural factors contributing to the low interest in climate security within Japanese society.

24.2.2 The Second Question: Why Do Political Discussions Progress Ahead of Robust Scientific Evidence?

Japan's postwar security policy, constrained by Article 9 of the Constitution, has continually sought nonmilitary means. Historically, during the Ohira administration in the 1980s, discussions on "comprehensive security" positioned food, energy, and large-scale earthquakes as security threats (Cabinet Office, 1980). This effort laid the foundation for developing Japan's Critical Security Studies. Furthermore, the 1989 "Report on Diplomacy and Comprehensive Security" mentioned global warming (House of Councilors Research Committee on Diplomacy and Comprehensive Security 1989: 68–69). Additionally, the Obuchi administration (1998–2000) made "human security," a concept proposed by the United Nations Development Programme in 1994, one of the important policies of Japanese diplomacy. This formed the undercurrent of Japan's Critical Security Studies up to the present day. However, during the 2000s, there was a return to the traditional concept of military state security under administrations like Koizumi and Abe, which gained national support. The so-called Japan's Legislation for Peace and Security in 2015 was a major outcome of this shift.

However, there were new developments during the Suga administration, which succeeded the Abe administration in 2020. In April 2021, shortly after the inauguration of the Biden administration, Defense Minister Kishi (at the time) participated in a session on climate security during the Climate Summit and stated, "Climate change is a 'cascading risk.' Climate change goes beyond just being an environmental issue; it threatens the peace and stability of individual countries and the world at large" (Ministry of Defense 2021), thus expressing the Japanese government's recognition of the issue. This address marked an important step for Japan toward climate security policy. Under the subsequent Kishida administration from 2021, the Climate Change Adaptation Plan was formulated, recognizing the link between climate change and security, including human security (Prime Minister Office of Japan 2021: 11–12). However, the report did not outline specific climate security policies. Meanwhile, in 2022, the same administration, citing Russia's invasion of Ukraine as a factor, made a cabinet decision to double defense spending over the next five years starting in 2023 (Saito et al. 2023). Additionally, under the recognition that the scope of security is expanding, laws were enacted in the same year to secure supply chains, respond to cyberattacks, and prevent hegemonic competition over advanced technologies to ensure Japan's military security under the framework of "economic security." The same year's National Security Strategy and National Defense Strategy also positioned climate change as a security threat. Still, the primary focus remained on military threats from foreign powers.

In this way, within Japanese politics, the concept of security has continued to be influenced by traditional military and geopolitical risks, even while maintaining the undercurrent of Critical Security Studies in this century. This trend, summarized through a literature review of various policy documents, is illustrated in Fig. 24.2.

Viewing climate change as a national security threat is not unique to Japan. For example, the United States has long regarded climate change as a national security threat, framing it as a "threat multiplier" that exacerbates existing social issues and leads to political conflict (Busby 2007; The CNA Corporation 2007). However, the primary focus has been on U.S. military adaptation to climate change and on foreign countries, particularly those in the Global South.

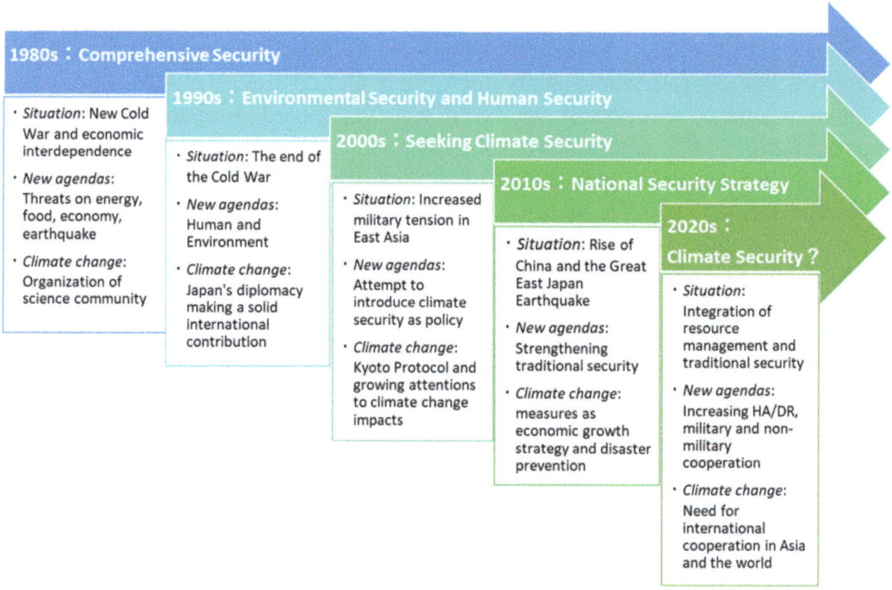

Fig. 24.2 Transition of Japan's climate and security policies. (Source: Hasui and Komatsu (2021))

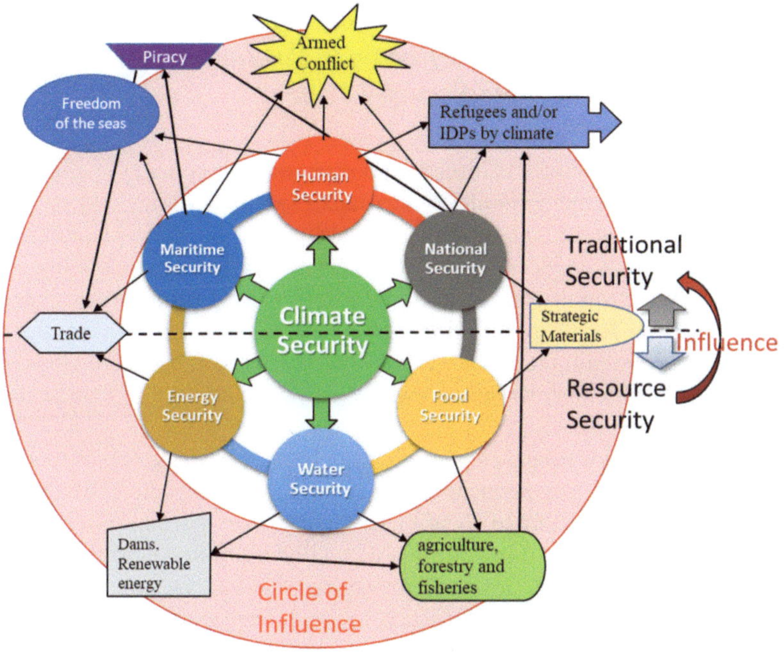

Fig. 24.3 Relationship among security concepts. (Source: Modified from Hasui and Komatsu (2020))

On the other hand, in examining various research outcomes related to climate security, many studies have shown that climate change impacts include not only direct effects such as changes in temperature, precipitation, and the frequency of typhoons, but also affect nonmilitary security sectors such as food and energy production and consumption, land use, fishery resources, migration and refugees, and trade. These studies have linked climate change to these areas.

In summary, the findings suggest climate security, as shown in Fig. 24.3, lies at the origin of various interrelated security concepts. Climate change has a wide-reaching and powerful influence on Earth's ecosystems. Its fluctuations affect almost all aspects upon which human survival depends. In this sense, Fig. 24.3 can be seen as extracting the elements of security that are recognized as arising from the complex relationships surrounding climate change. The worsening of these security issues spreads into various problems, interconnected with each other. These problems within this "circle of influence" constitute the urgent global issues that current international politics must address. The frequent and interconnected nature of these issues is why discussions on climate security are advancing as an urgent political response without waiting for further academic accumulation.

24.2.3 The Third Question: How Can Measures Against Climate Change Be Effectively Positioned as Part of Japan's Climate Security Policy?

However, there are challenges in directly incorporating the concept of climate security from the international community into Japan's climate policy and securitizing it. First, it is not easy to see the benefits of securitization in policy planning, implementation, and evaluation. Second, policymakers tend to focus on geopolitical risks, and it is believed that significant geopolitical risks resulting from climate change are unlikely to occur in Japan soon. Third, as a result, there is a possibility that some aspects of the current mitigation and adaptation measures, which have already been formulated into policy, may be lost or overlooked when reorganized under the concept of climate security.

Therefore, the author proposes incorporating human security, one of the broader approaches within Critical Security Studies, into climate policy, similar to the approach taken in AR5. Human security is already an important pillar of Japan's foreign policy, primarily used in relations with Global South countries through initiatives such as Official Development Assistance (ODA). On the other hand, by using this concept to reorganize domestic climate policy, it may be possible to formulate a climate-based security policy that consistently aligns both foreign and domestic policy.

In the 2003 Ogata-Sen "Report on Human Security," "Protection," and "Empowerment" were proposed as the central concepts. In the 2022 Special Report, "Solidarity" was added as a core concept for the next generation of human security. This new concept considers the Anthropocene context, which includes various issues such as threats from digital technology, violent conflict, intergroup inequality, and health threats.

In human security theory, enhancing people's "Agency" is emphasized. The Special Report defines "Agency" as "the ability to hold values and make commitments, regardless of whether they advance one's wellbeing, and to act accordingly in making one's own choices or in participating in collective decision-making" (UNDP 2022: 6). Through this concept, the next-generation human security framework has evolved to focus on increasing individual agency, expanding proactive participation based on mutual trust, and engaging in decision-making. Figure 24.4 summarizes the Special Report's understanding, illustrating the interrelationship between "Protection," "Empowerment," and "Solidarity," with "Trust" as the underlying foundation.

The incorporation of the next-generation concept of human security into climate change measures is outlined in Table 24.1. While the main actors in "Protection" are national and local governments, the opportunities for citizens to take the lead increase in the field of "Empowerment" to "Solidarity." Through the implementation of these policies, the "Agency" of each actor is strengthened, which in turn enhances the capacity for climate policy formulation. This will likely serve as a critical driving force in promoting climate-based policy development in the future.

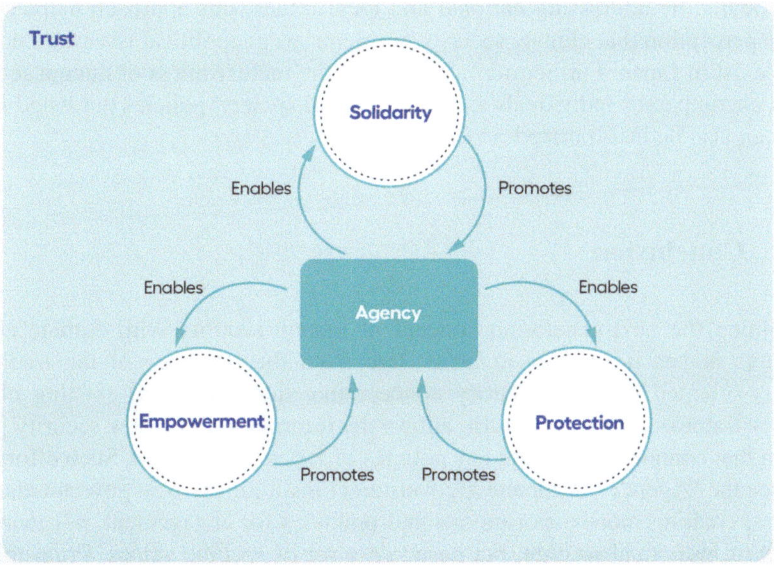

Fig. 24.4 Enriching human security for the Anthropocene context: Adding solidarity to protection and empowerment. (Source: UNDP 2022: 30)

Table 24.1 Matrix of climate security based on human security and climate change countermeasures

		Human security		
		Protection	Empowerment	Solidarity
Climate change countermeasures	Mitigation (Common climate security policies)	Renewable energy policies	Promotion of green economy	Implementation of the Paris agreement
		CO_2 emission regulations	Environmental value education programs	International environmental protection movements and agreements
		Strengthening environmental laws	Support for green technology dissemination	Global confidence building in climate diplomacy
				Promoting SDGs
	Adaptation (National climate security policies)	Development of disaster prevention infrastructure	Climate adaptation techniques in agriculture	Climate refugee support programs, institutions
		Implementation of disaster prediction institutions	Strengthening resilience of local governments and communities	Inter-regional climate cooperation networks
		Early warning systems	Community-based adaptation programs	Coordination of international adaptation financial systems
		Land use change		

Source: Created by the author

Additionally, by addressing national and local issues, this approach helps correct the misperception that climate security problems, as geopolitical issues, do not currently exist in Japan. Furthermore, because of the inclusiveness of human security, which encompasses individuals and international society, policies not listed in this table can also be incorporated.

24.3 Conclusion

Combining the next-generation concept of human security with climate change measures makes it possible to break free from the influence of the traditional military-oriented national security concept that has dominated existing climate security frameworks. This, in turn, allows the reformation of a new security policy system that connects with other key policies of Japan's diplomacy. Such efforts will enhance the "Agency" of Japanese government institutions, local governments, and citizens, creating more opportunities and practices for engagement, not merely in pursuit of individual welfare, but based on a set of specific values. From the perspective of climate security, this leads to a reorganization of adaptation measures into "National Climate Security Policies" tailored to each country's characteristics. On the other hand, mitigation measures are restructured as "Common Climate Security Policies" premised on international cooperation.

At this point, an essential social prerequisite is the "mainstreaming climate security." In other words, as discussed in Fig. 24.3, the global recognition that the impacts of climate change affect all aspects of security and that mitigation and adaptation policies are security policies vital to human survival must be shared and prioritized worldwide. One of the key conclusions of AR6 was Climate Resilient Development; similarly, Climate Resilient Security will become an essential aspect of the practice and making of new security policies.

In future climate security research, it will be important to determine how to incorporate the perspective of next-generation human security and engage the countries of the Global North. At that time, it will be crucial for Japan, which has long adopted the human security framework, to contribute actively through research, policy implementation, and advocacy. This will be essential for Japan's role in shaping international rulemaking.

References

Adger WN et al (2014) Human security. In: Field CB et al (eds) Climate Change 2014: Impacts, Adaptation, and Vulnerability. Part A: Global and Sectoral Aspects. Contribution of Working Group II to the Fifth Assessment Report of the Intergovernmental Panel on Climate Change. Cambridge University Press, Cambridge, New York, pp 755–791

Busby J (2007) Climate Change and National Security: An Agenda for Action, Council on Foreign Relations. https://www.cfr.org/report/climate-change-and-national-security/. Accessed 15 Sep 2024
Cabinet Office, Cabinet Secretariat, Assistant to the Prime Minister's Office (1980) Comprehensive Security Strategy—Report of Prime Minister Ohira's Policy Study Group—5. Ministry of Finance Printing Bureau
Commission on Human Security (CHR) (2003) Human security now. https://digitallibrary.un.org/record/503749?ln=en&v=pdf. Accessed 17 Jun 2024
Hasui S (2011) Climate security and its implications for integrating paradigms of development and security. In: Iai S (ed) Achieving global sustainability: Policy recommendations. United Nations University Press, New York, pp 279–321
Hasui S, Komatsu H (2020) Preparing for the compound risks of climate change. National Institute for Environmental Studies https://www.nies.go.jp/social/news/compound_risks_of_climate_change.pdf. Accessed 25 Jun 2024
Hasui S, Komatsu H (2021) Climate security and policy options in Japan. Politics and Governance 9(4):79. https://doi.org/10.17645/pag.v9i4.4414
Homer-Dixon T (1999) Environment, scarcity, and violence. Princeton University Press, Princeton, NJ
House of Councilors Research Committee on Diplomacy and Comprehensive Security (1989) Research Report on Diplomacy and Comprehensive Security
IPCC (2023) Summary for policymakers. In: Lee H, Romero J (eds) Climate Change 2023: Synthesis Report. Contribution of Working Groups I, II and III to the Sixth Assessment Report of the Intergovernmental Panel on Climate Change. IPCC, Geneva, pp 1–34. https://doi.org/10.59327/IPCC/AR6-9789291691647.001
Kameyama Y, Ono K (2021) The development of climate security discourse in Japan. Sustain Sci 16(1):271–281. https://doi.org/10.1007/s11625-020-00863-1
Kanie N (2007) Towards the formation of international order surrounding climate security: the deep layer of environmental politics becoming high politics. contemporary thought. Seidosha 35(12):210–221
Minamiyama A, Maeda Y (2022) Understanding of critical security studies. Houritsubunkasya, Kyoto
Ministry of Defense (2021) On the attendance of the minister of defense at the 'Climate Summit Climate Security Session'. https://www.iges.or.jp/jp/projects/summit-climate. Accessed 20 Jul 2024
Prime Minister's Office of Japan (2021) Climate change adaptation plan. Cabinet Decision on October 22, 2021
Saito S et al (2023) Special measures law for securing the financial resources necessary for the fundamental strengthening of Japan's defense capabilities. Research Bureau Ronkyu 20:221–241
Sekiyama K (2023) The logic of climate security—geopolitical risks of climate change. Nikkei Newspaper Publishing, Tokyo
Sub-Committee on International Climate Change Strategy Global Environment Committee, Central Environment Council (2007) Report on climate security. Ministry of the environment of Japan. http://www.env.go.jp/en/earth/cc/CS.pdf. Accessed 14 Aug 2024
The CNA Corporation (2007) National security and the threat of climate change, center for naval analysis. https://www.cna.org/reports/2007/national-security-and-the-threat-of-climate-change/. Accessed 15 Sep 2024
United Nations Development Programme (UNDP) (1994) Human Development Report 1994: New Dimensions of Human Security. New York. https://hdr.undp.org/content/human-development-report-1994. Accessed 19 Sep 2024
United Nations Development Programme (UNDP) (2022) New threats to human security in the anthropocene: demanding greater solidarity: 2022 special report. https://digitallibrary.un.org/record/3958751?ln=en%3Fln%3Den&v=pdf. Accessed 14 Sep 2024

Correction to: Impact on Brown Macroalga *Undaria pinnatifida* Farming Under Changing Ocean Climate

Shigeho Kakehi, Goh Onitsuka, and Hideaki Kidokoro

Correction to:
Chapter 7 in: N. Mimura, S. Takewaka (eds.),
***Climate Change Impacts and Adaptation Strategies in Japan*,**
https://doi.org/10.1007/978-981-96-2436-2_7

The abstract and keywords for Chapter 7 were omitted from the printed book (PDF version). Chapter 7 is updated with the abstract and keywords.

The updated version of this chapter can be found at
https://doi.org/10.1007/978-981-96-2436-2_7

N. Mimura, S. Takewaka (eds.), *Climate Change Impacts and Adaptation Strategies in Japan*, https://doi.org/10.1007/978-981-96-2436-2_25

Index

N. Mimura, S. Takewaka (eds.), *Climate Change Impacts and Adaptation Strategies in Japan*, https://doi.org/10.1007/978-981-96-2436-2

MIX
Papier aus verantwortungsvollen Quellen
Paper from responsible sources
FSC® C105338

If you have any concerns about our products,
you can contact us on
ProductSafety@springernature.com

In case Publisher is established outside the EU,
the EU authorized representative is:
Springer Nature Customer Service Center GmbH
Europaplatz 3, 69115 Heidelberg, Germany

Printed by Libri Plureos GmbH
in Hamburg, Germany